I0729482

AMERICAN DENIM

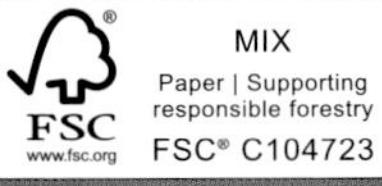

FOR GREENSBORO

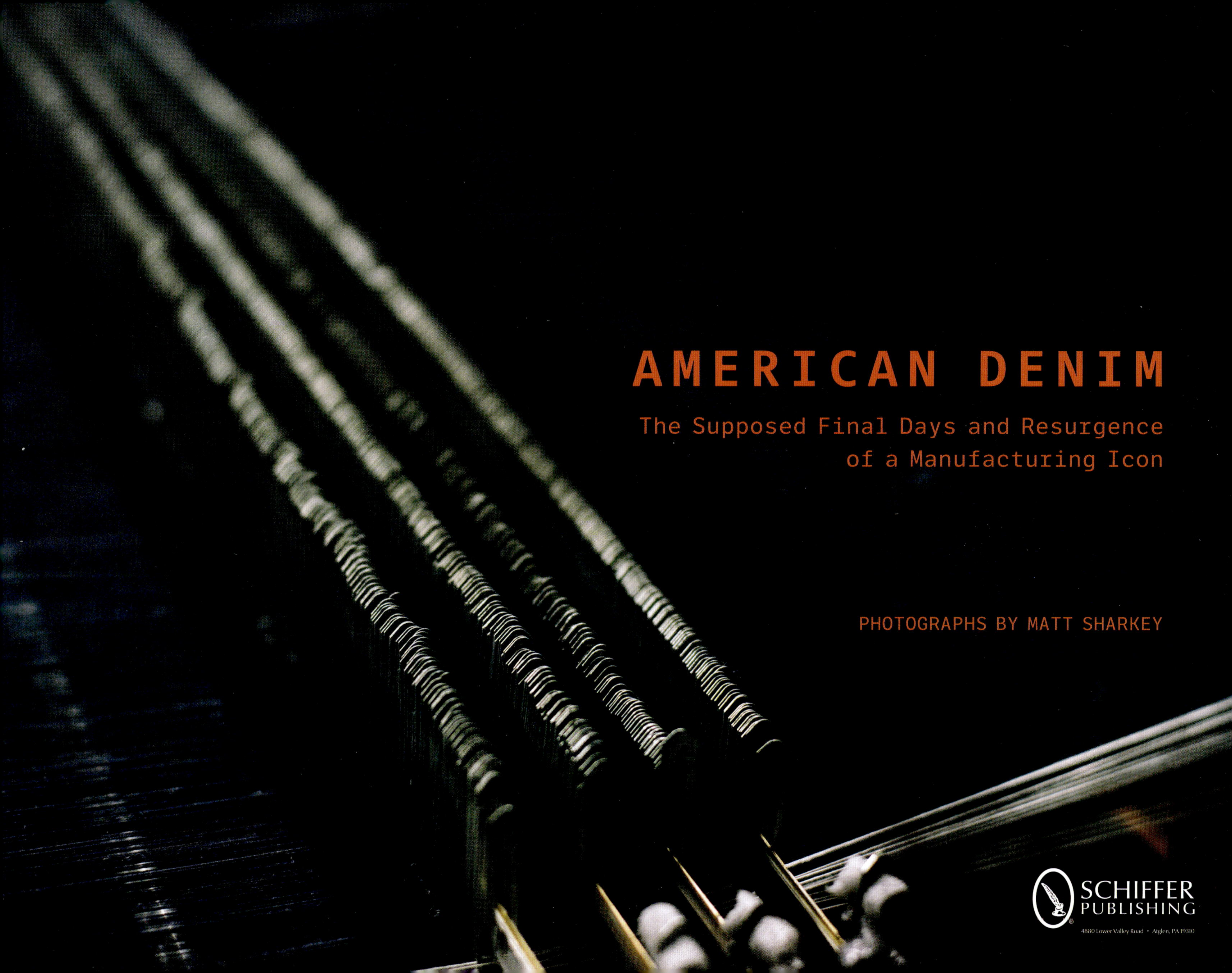
AMERICAN DENIM
The Supposed Final Days and Resurgence
of a Manufacturing Icon
PHOTOGRAPHS BY MATT SHARKEY
SCHIFFER PUBLISHING
4880 Lower Valley Road • Atglen, PA 19310

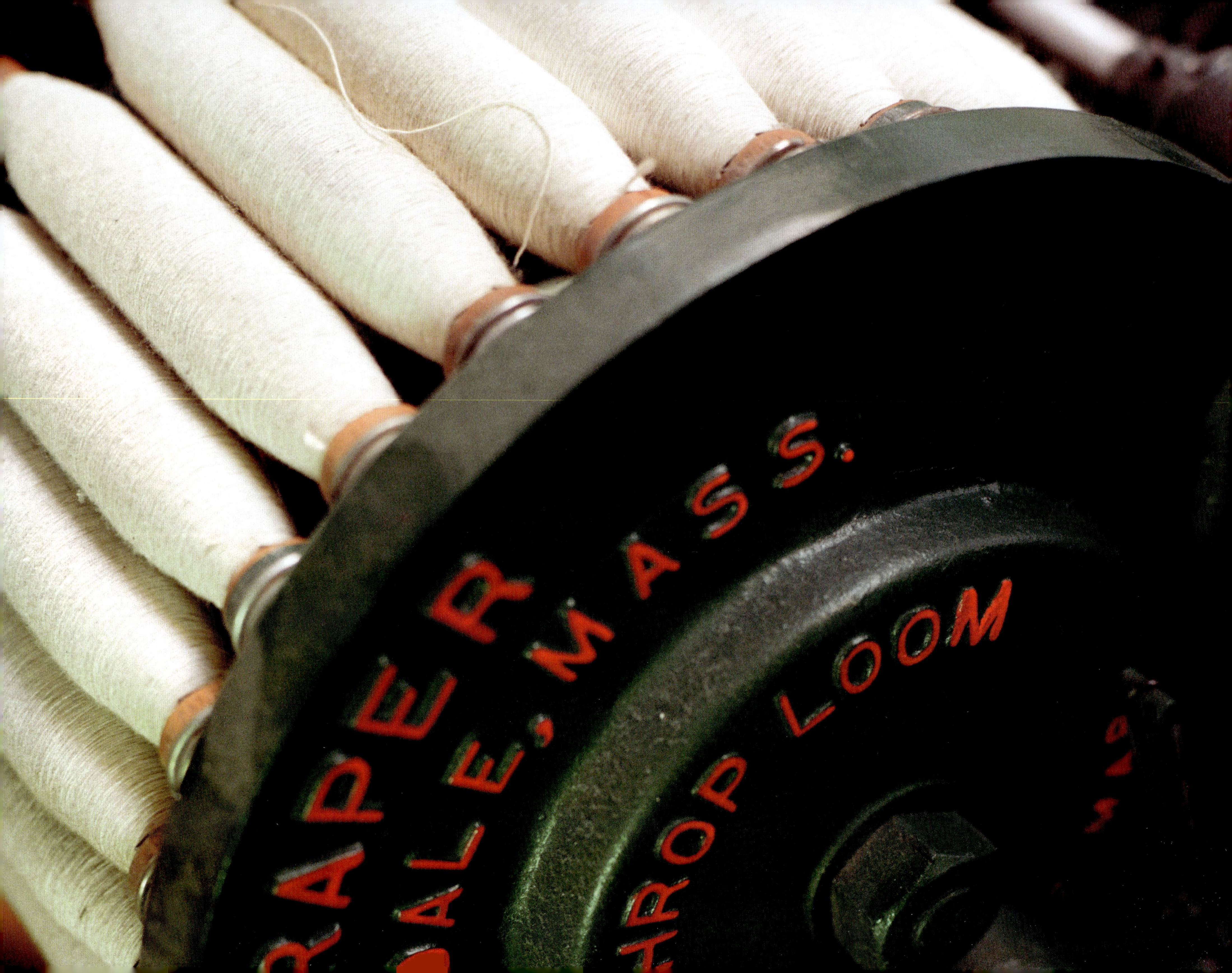
RAPER, MASS.
DALE, MASS.
NORTHROP LOOM

FOREWORD *By Evan Morrison & Ralph B. Tharpe Jr.*

While the sturdy blue cloth known commonly as denim has global origins, it was in New England where the indigo-dyed cotton yarn was first woven together in the late 1790s at a factory toured personally by our founding father, George Washington. One hundred years later, the manufacture of denim cloth had migrated south for cheaper labor, and overall factories were popping up in small towns all across the country, supplying the nearby working class with functional clothing for long wear. It was in Greensboro, North Carolina, that two brothers would begin to realize their vision of building the greatest mills in the world. Moses and Ceasar Cone began erecting their own denim mill here in 1895, after operating as sales agents for other mills. A few years later they would construct White Oak Cotton Mill, slated to be the world's largest denim mill, a title it earned just three years into operation in 1908.

In November 1913, Cone Export & Commission Company, agent for Proximity and White Oak, Greensboro's two denim mills, sold and shipped the first order of denim fabric to Levi Strauss & Company in San Francisco, California, by rail. By 1915, Levi's began to entrust the Cone's White Oak plant to produce and furnish all of their cloth for their prized 501 jean and complementing jacket. By 1917, the Levi Strauss company catalog dropped other competitor mill names from its pages, and in 1922 their jeans went through a modern upgrade and received belt loops, and Cone's White Oak had firmly established its role as Levi Strauss's exclusive supplier of premium denim.

In 1921, White Oak yet again revolutionized the denim industry by developing and commercializing the long-chain compound dye range, a disruptive innovation in dye technology that created a massive improvement in color uniformity of cloth. This meant no more streaky, discolored denim, a common result of dyeing warp yarn in hard-to-regulate indigo vats. This invention, known as the Touchstone dye apparatus, remains the worldwide industry standard today.

The mills in northeast Greensboro were bustling. Cotton came in by the bale, and boxes filled with denim went out by rail daily. Millions of yards of cloth were produced in but a few short blocks every week. If you wore anything made out of cotton denim, flannel, corduroy, chambray, or chamois, or if you dried off from showering with a towel, chances are you touched something that was made with fabric produced right here. The factories were surrounded by neighborhoods constructed for employees and their families, along with

"

schools, churches, YMCAs, shopping centers, and even public transit. At one point in time, about a quarter of the city's population either worked for the mills or were one of their many suppliers.

The Cones, children of Jewish immigrants who had taken up residence in the Deep South, employed all races and genders, encouraged children's education by employing the best teachers in their schools, and offered vacations, bonuses, and summer camps, making a genuine effort to build trust with their staff. While they were making a name for their business worldwide, they were taking care of their people here at home. All of the Cones were known among the mill workers as kind, caring people, who prided themselves in memorizing workers' first names, visiting their mills in person frequently, and making sure their people were well cared for.

White Oak would continue to grow rapidly through the years, modernizing by replacing the Whitin power looms that first stood atop its weave floors with E models in the 1930s and later X models from the 1940s through 1970s from the lauded Draper Corporation. Those iconic green machines would become the common thread that linked its past to present day.

World War II would see an influx of government contracts for the production of sturdy cotton fabrics, and the Cones leased 500+ acres of land adjacent to the mill village for $1 per year to the United States Army for Basic Training Camp #10. The mills would go on to receive the Army-Navy Production "E" Award for their efforts to do their part.

Postwar demand for blue jeans skyrocketed, especially as the decades progressed and blue jeans became not just the working man's clothing choice, but also the choice of youth counterculture and Better Housekeeping, alike. Greasers, hippies, bikers, and protestors all wore denim, and it began to transcend its reputation as a symbol of the blue collar. White Oak plant prided itself in besting production volume year after year.

As fashion took center stage in the 1970s, paralleling new taste was the rise of global manufacturing, with other countries embracing their opportunity to advance and develop a middle class of their own. In an effort to stay relevant, White Oak's management elected to cut out the aged wooden floors and replaced the old, narrow looms with wider, faster models without shuttles, creating an enormous uptick in production volume and decreasing marginal costs.

But Greensboro would see three of its four Cone Mills Corporation–owned plants shutter in the 1980s, and it seemed that the odds were beginning to stack against American denim, as firms looked to lower costs and boost profits by offshoring labor and building mills overseas.

The enactment of NAFTA further eroded the domestic textile industry and effectively created the beginning of the end of the golden era of American apparel manufacturing. With lean-manufacturing-practice principles applied, mills followed brands to locate closer to their cut-and-sew operations, and USA-made denim production began to dwindle.

Seizing the opportunity, White Oak's management looked to embrace the past to find new success. Old, narrow looms that built iconic cloth, from which the likes of Lee, Wrangler, and Levi's jeans were made with for decades, were brought out of storage, refurbished, and then placed back into production. Those old Draper looms clicked and clacked away on a portion of original, triple-thick wooden floors that remained in the southern half of the weave room, and the selvage denim flowed yet again off the machines. For those who aren't familiar with the term, selvage is a contraction that describes the self-edged nature of cloth produced on shuttle looms, whereby the natural-colored yarn projected back and forth, interlacing to become cloth, and turned back on itself each pass, creating a finished, or self edge, without any fray. This self edge is commonly found in the outseam, waistband, coin pocket (on older pairs of jeans), plackets and pockets on jackets, and straps, outseams, pockets, and flys on bib overalls.

The industry legend witnessed resurgence in demand. In the 1990s, Levi's began a subbrand, Levi's Vintage Clothing, to thoughtfully reproduce archived garments from bygone eras in their collection, and small heritage brands began to pop up all across the country, all using this iconic fabric. Small-batch, premium jeans took fashion by storm, and selvage denim became a commonly known term in consumer's vocabulary. Ralph Lauren launched their RRL line, building collections of staple denim jeans and jackets with this storied cloth, and True Religion and Lucky Jeans created five-pocket selvage jeans for the masses, riding a denim wave into the early 2000s. White Oak cloth was spoken for before it was finished being woven, and it seemed as if the mill had found its cruise control for a third consecutive century.

In 2004, a man named Wilbur Ross, who had previously invested in the American steel, coal, and automotive components industries through buyouts and mergers, would outbid Warren Buffett and purchase Cone Mills Corporation and Burlington Industries, merging them into what would become known as International Textile Group. The textile business looked to be safe and secure in the hands of "the man who saved Pennsylvania steel."

But not all was what it seemed. Beneath the surface the industry's gears were churning, and domestic demand dwindled. Selvage denim produced overseas at much-lower costs caused brands to change suppliers, coupled with localized apparel manufacturing and cheap labor serving as the main factors. Swift, ACG, Avondale, West Point, Graniteville, Galey & Lord, and Denimatrix all shuttered within just a few years, along with many of their dedicated suppliers. But White Oak held fast, with its name constantly mentioned in news articles and trade journals, and the cloth was lauded by designers for its legendary quality.

Then, the outcome of the 2016 presidential election produced an unthinkable variable. Wilbur Ross accepted the honorable appointment to serve as secretary of commerce for the United States of America. In order to carry out the duties of this appointment, he sold off his assets that were viewable as a conflict of interest prior to taking his seat. This divesture included his prized International Textile Group, which was practically sold overnight in a fire sale for pennies on the dollar. Its new owner, Platinum Equity Group, would begin to quickly restructure the conglomerate.

Without considering any of its history, White Oak was abruptly closed within a calendar year of its acquisition. Employees were given their notice, and the mill operated on a lean crew during its final few months, in order to finish producing all of the "too little, too late" order influx that occurred at the time of its closure announcement.

By the end of January 2018, White Oak would be no more. It was sold off immediately to lean the business, a key skill of M&A (merger and acquisition) firms that engage in "buy in bulk and piecemeal off" business tactics.

Its new owner, J.W. Demolition, immediately viewed White Oak as an asset once again and embarked on the task of helping chart the course for its new chapter. Within months of its purchase, the mill transformed, all of the textile machinery was liquidated into the secondhand market, and as buildings and structures were demolished, nearly every brick, steel beam, and stick of wood was salvaged for resale, the firm's specialty. But those 46 narrow Draper X-3 looms sat in their same place on the wooden floors, protected from the elements with thick plastic sheeting.

The property underwent numerous renovations and constructions, turning the town-sized plant on 100 acres from a ghost town full of inanimate machinery to a massive logistics hub for turnaround trucking. Departments that once housed yarn now house hundreds of pallets of consumer goods destined for stores all across the region. The spinning, dyeing, slashing, and beaming departments are all but memories, since their structures are no longer, but the giant weave room with its iconic sawtooth roof remains. Those signature triple-thick wooden floors on the south end of the weave floor are still there. And so are a few of those narrow shuttle looms, still clicking and clacking away, producing that same, storied denim cloth.

A group of retired Cone employees and local supporters established a nonprofit organization in 2020, which would come to be known as the White Oak Legacy Foundation, or W.O.L.F., for short. In an effort to preserve local textile history specific to denim and Cone Mills, W.O.L.F. received the last Draper looms at White Oak as a gift from Cone Denim and established a for-profit subsidiary under another storied name, Proximity Manufacturing Company, to begin producing selvage denim once again in Greensboro.

Fast forward two years, and Proximity Manufacturing has woven thousands of yards of selvage denim fabric that has since been cut and sewn into hundreds of articles of clothing by brands still cutting and sewing their products right here in the USA.

Together, W.O.L.F. and Proximity have become the caretakers of this rich history for a third consecutive century. Protecting industry-related knowledge culminated in the reestablishment of the Denim 101 course, a multiday denim educational program that brings design teams in from around the world. W.O.L.F. has produced several history exhibits at Revolution Mill, another one-time Cone-owned mill, showcasing denim and textile history. Through this effort, W.O.L.F. and Proximity will continue to inspire the next generation of denim innovations.

Through photographical documentation, this book tells the story of the rise, demise, and rebirth of White Oak Cotton Mill, the world's proverbial Mecca of Denim.

True grit, visionary servant leadership, and an inextinguishable spirit. Now, that is American.

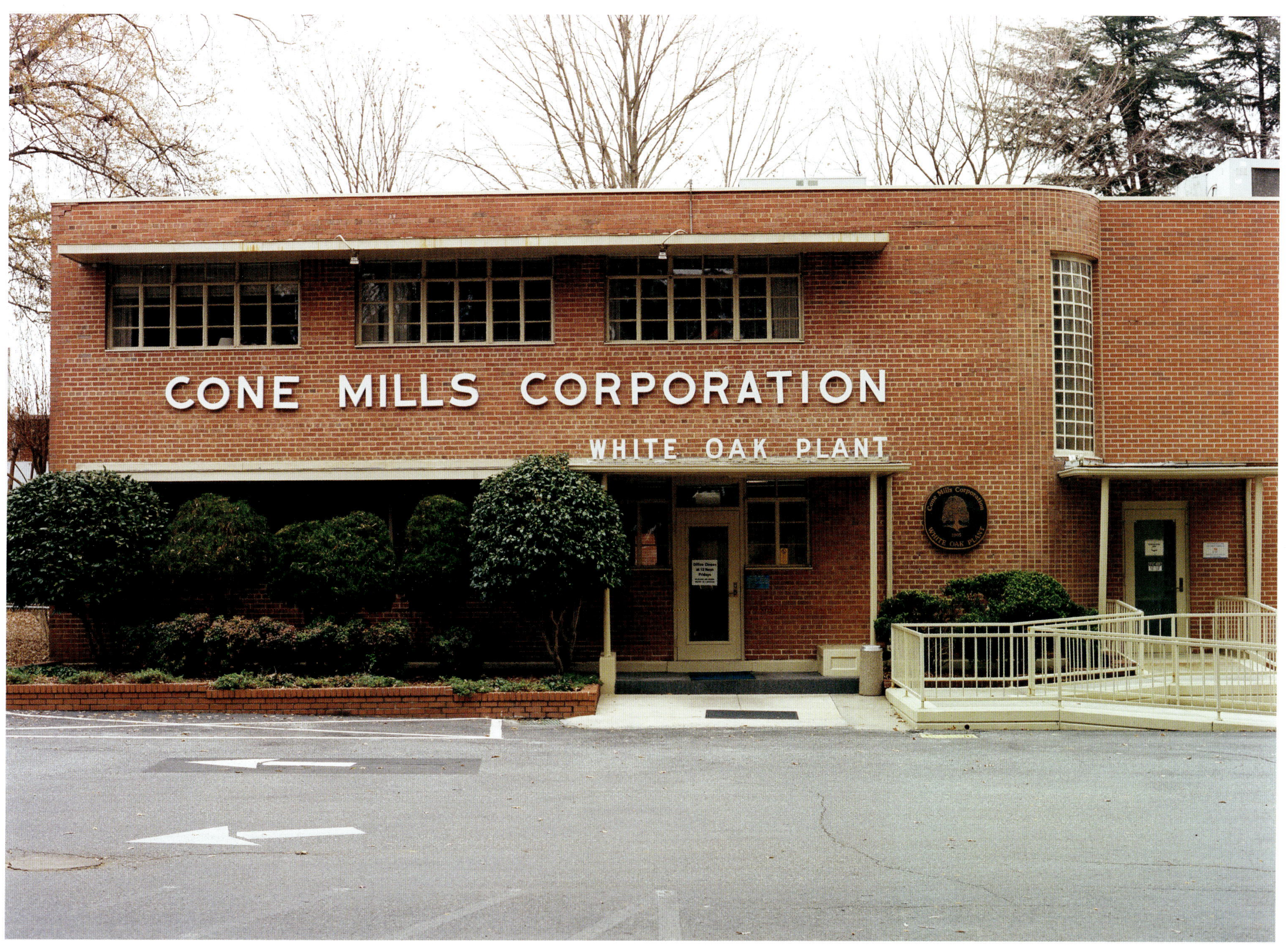

Front Office, Cone Mills White Oak Plant

PREFACE

In 2009 I started dreaming of the idea of opening a menswear store that focused on brands that were made in America and engineered to outlive whoever purchased them. The United States had once been one of the major textiles manufacturers in the world, but by the turn of the twenty-first century, it had become hard to fill even a 2,000-square-foot retail space with brands still made stateside. In 2016, this dream was realized when my wife, our two sons, and I opened our store in Petaluma, California. We exclusively carried denim made by Tellason, using fabric woven at Cone Mills' White Oak plant in Greensboro, North Carolina, which was then cut and sewn in San Francisco. A year prior, I visited White Oak for the first time and stood on the wooden floors that housed the storied Draper X-3 looms, which were famous for making selvage denim. For a country that's still relatively young, I can't quite describe the feeling of standing in a space that had been in existence for more than a hundred years and had made such an impact on the world.

My analog interest in shooting film photography is not unlike weaving selvage denim on shuttle looms. While mechanical in nature, they both require technicians to manually operate them. They take time and finesse. They're not always perfect, but that's what gives them their distinct character. Late in the summer of 2017, I started hearing rumblings of White Oak closing. Knowing how tremendous the impact that mill had made, not only to the local economy but to global fashion at large, I wanted to do what I could to capture its final days and help preserve its legacy. I reached out and sent a letter to the powers that be, asking if I could fund my own trip from California to North Carolina to capture the final days of the mill in operation. In a few short weeks, I received a response letting me know that I was one of dozens who had made the same request, and that if I wished to submit a proposal, they would be selecting two finalists in November who would be invited out during the final weeks of production. I was one of the two.

I've sorted this collection of images into three chapters, each from three different trips to Greensboro between 2017 and 2022. During that first trip I was able to see the mill in operation firsthand and walked tens of thousands of its more than 1 million square feet. During that same trip, I visited Hudson's Hill in downtown Greensboro, a retail store very similar to the one my wife and I owned back in California. It was there that I met one of its proprietors, Evan Morrison. To this day, I don't know anyone more steeped in American textile history.

We became fast friends and have since stayed in touch about developments at White Oak. In January 2018, the lights went out forever, or so we thought . . .

Through my own research, I would learn that the property was purchased by J.W. Demolition, a company that knew from experience the value of the building materials used throughout the plant, and had the expertise to extract and preserve them to be reused in new construction projects. I was granted permission by the new owner to return, and in February 2018 I was on a plane headed back out to Greensboro. This time, I intended to walk every inch of the buildings that, for years, claimed the title of largest producer of denim in the world. That was a jarring experience. Something that had lived and breathed for multiple generations was so abruptly vacant. Workstations with drinks still at them. Denim still hanging from conveyers. Hundreds of people worked there every day, and now the only sounds were my work boots on the wooden floor, echoing throughout the empty rooms. Multiple contributing factors would ultimately lead up to the day of the last timecard being punched, and those stories could fill the pages of this book. Those stories are to be told by those who were there. What I will say that it was ultimately a combination of corporate greed, government malfeasance, and the American consumer's obsession with fast fashion that led to the lights being turned off.

I would spend three days walking and photographing the entire property. Evan joined me for two of them. We ended up having lunch with the owner of J.W. Demolition one of those days. He and Evan would go on to quickly become friends. Over the course of the next two years, they would continue to talk about the potential of weaving denim once again at White Oak. Evan connected with former Cone employee and denim industry veteran Ralph Tharpe, and what started as conversations eventually led to the founding of the White Oak Legacy Foundation (W.O.L.F), a nonprofit organization dedicated to the history and preservation of American textiles, focusing on Greensboro and the role it played in the history of denim production. W.O.L.F. established a subsidiary, Proximity Manufacturing Company, which began weaving denim again on some of the same Draper X-3 looms still inside White Oak in early 2021. During my most recent trip in September 2022, I got to capture their production in action. They're carrying on that very legacy, now for three consecutive centuries, weaving selvage denim on shuttle looms at the very same site in Greensboro, North Carolina, where it all started. This is not an exit.

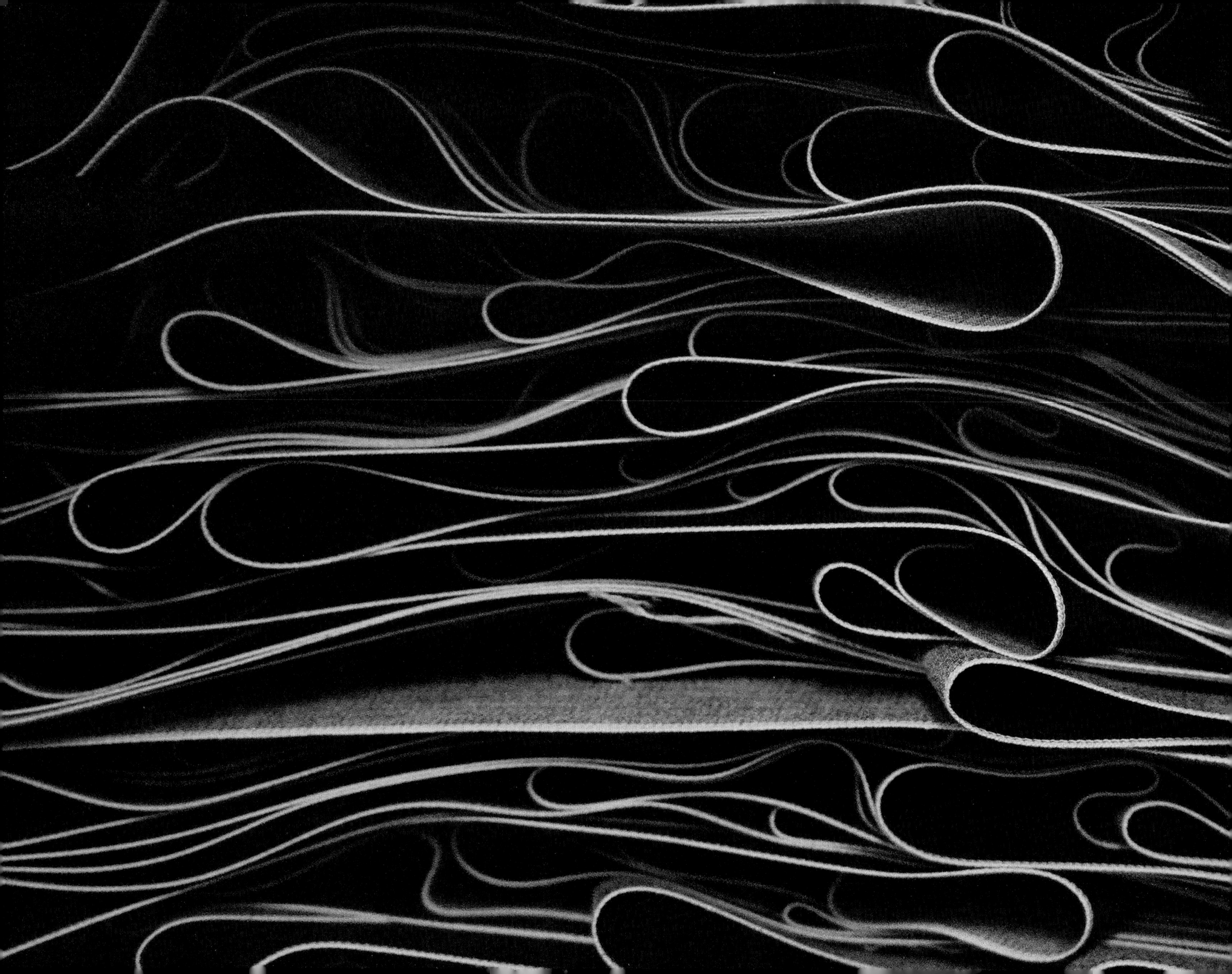

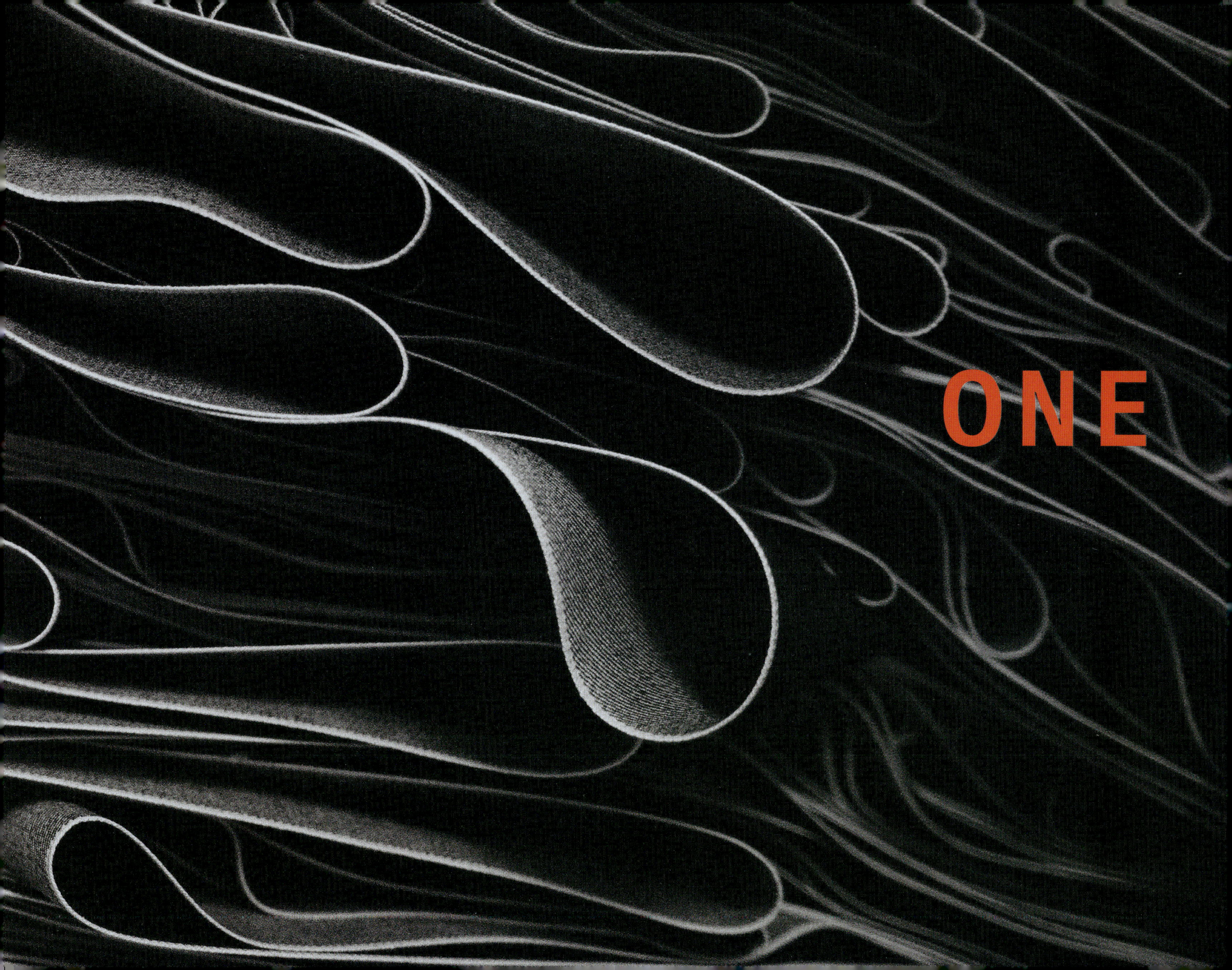

ONE

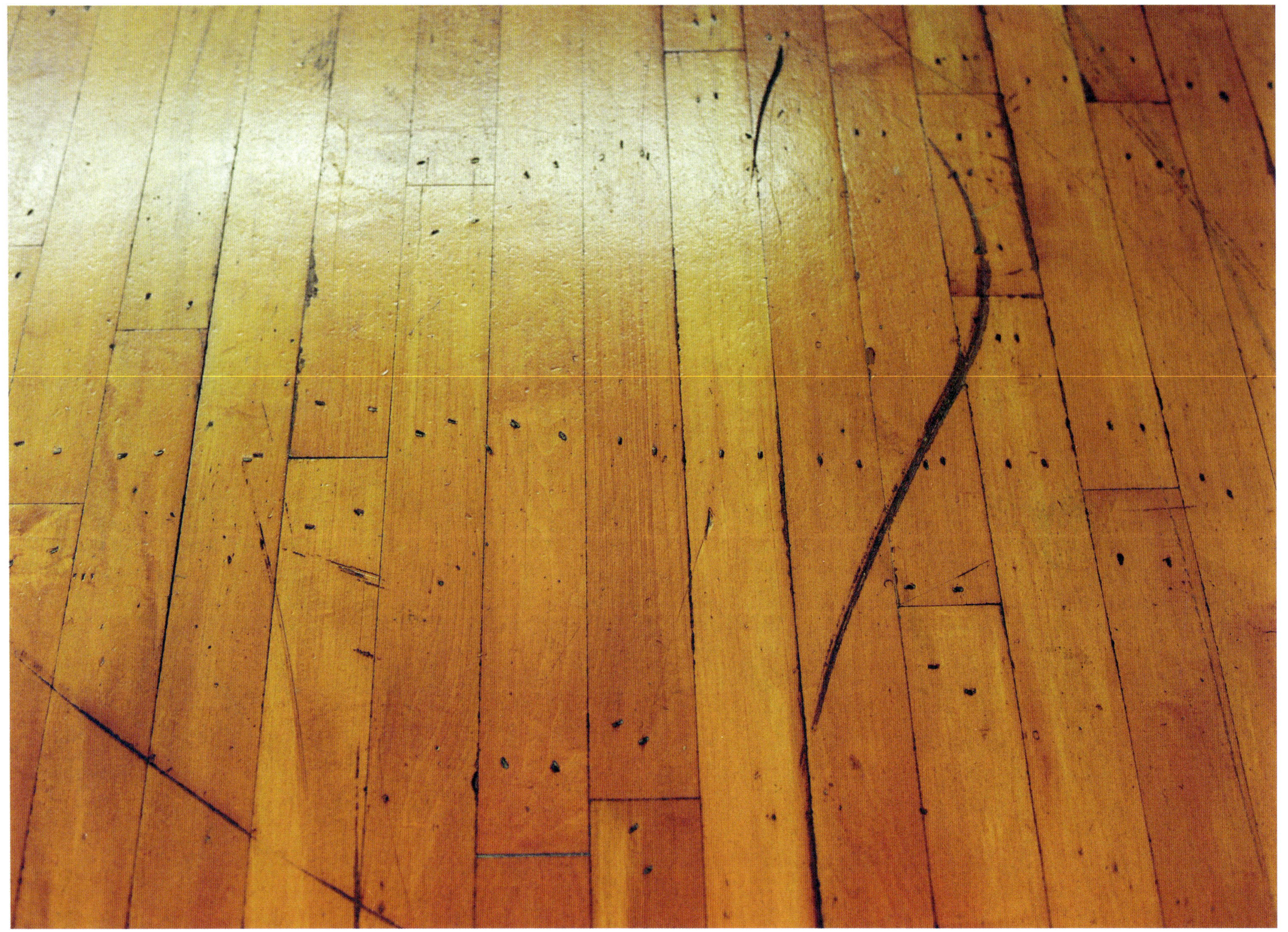

Maple flooring in the Weaving Department

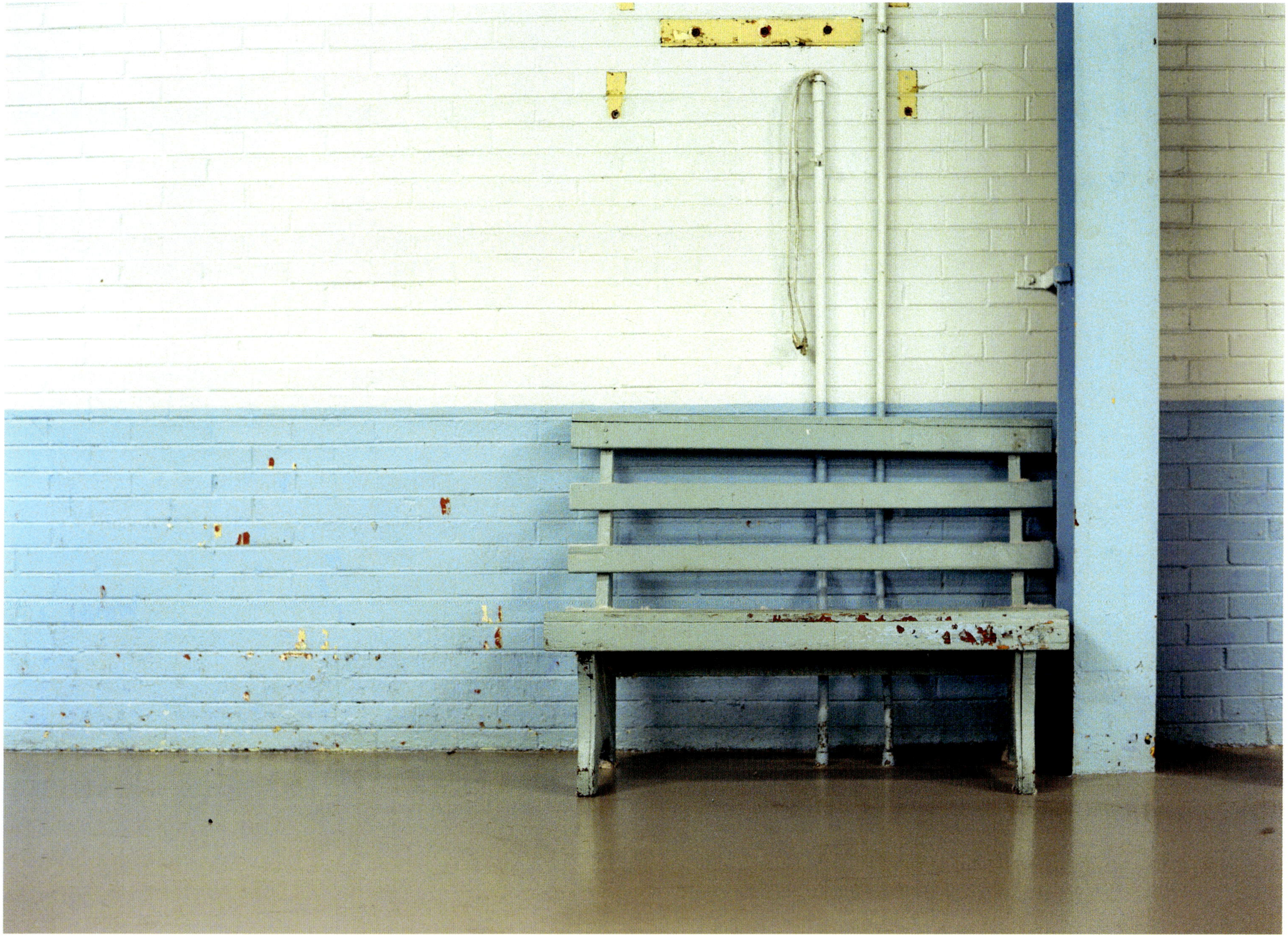

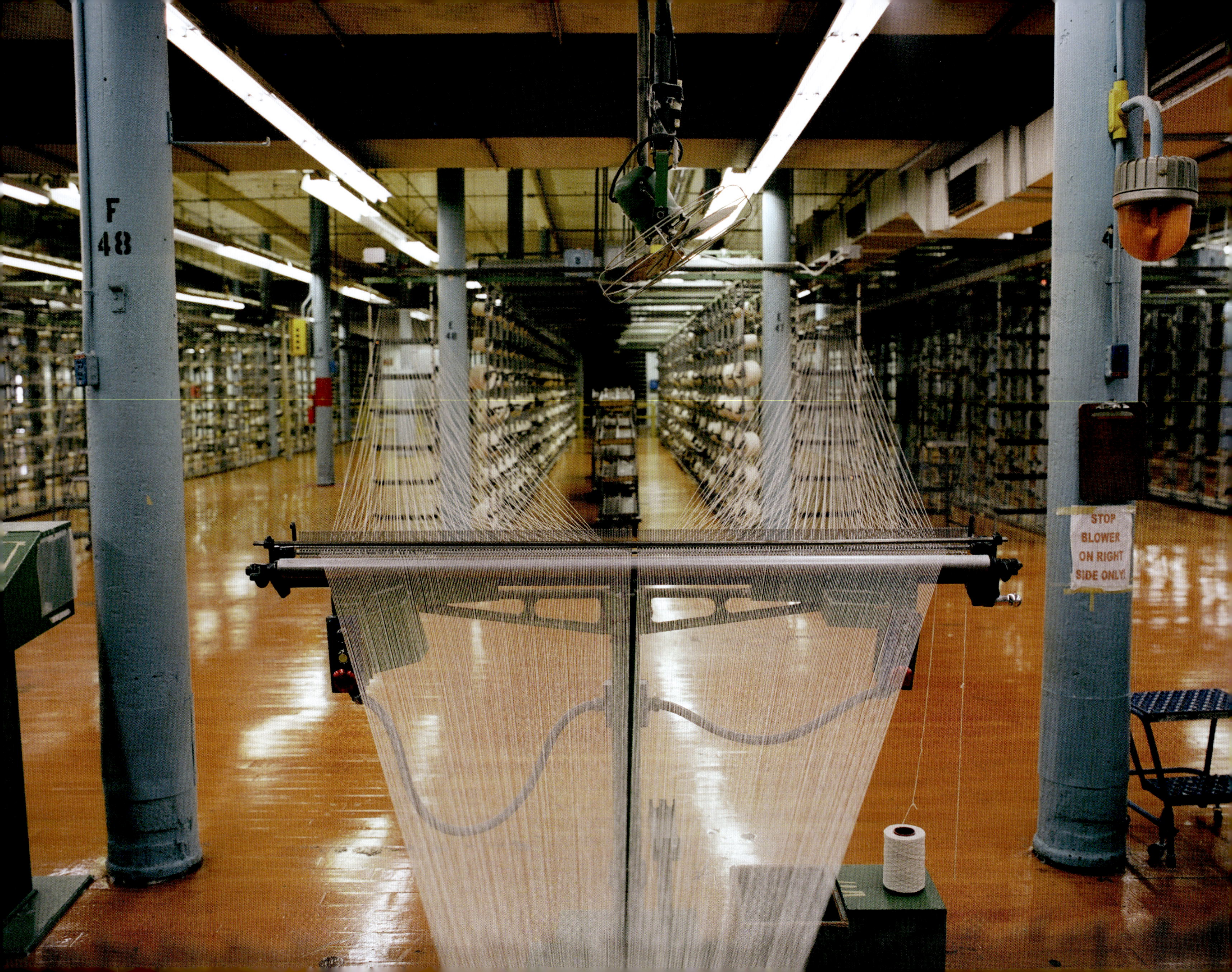

F
48
STOP
BLOWER
ON RIGHT
SIDE ONLY!

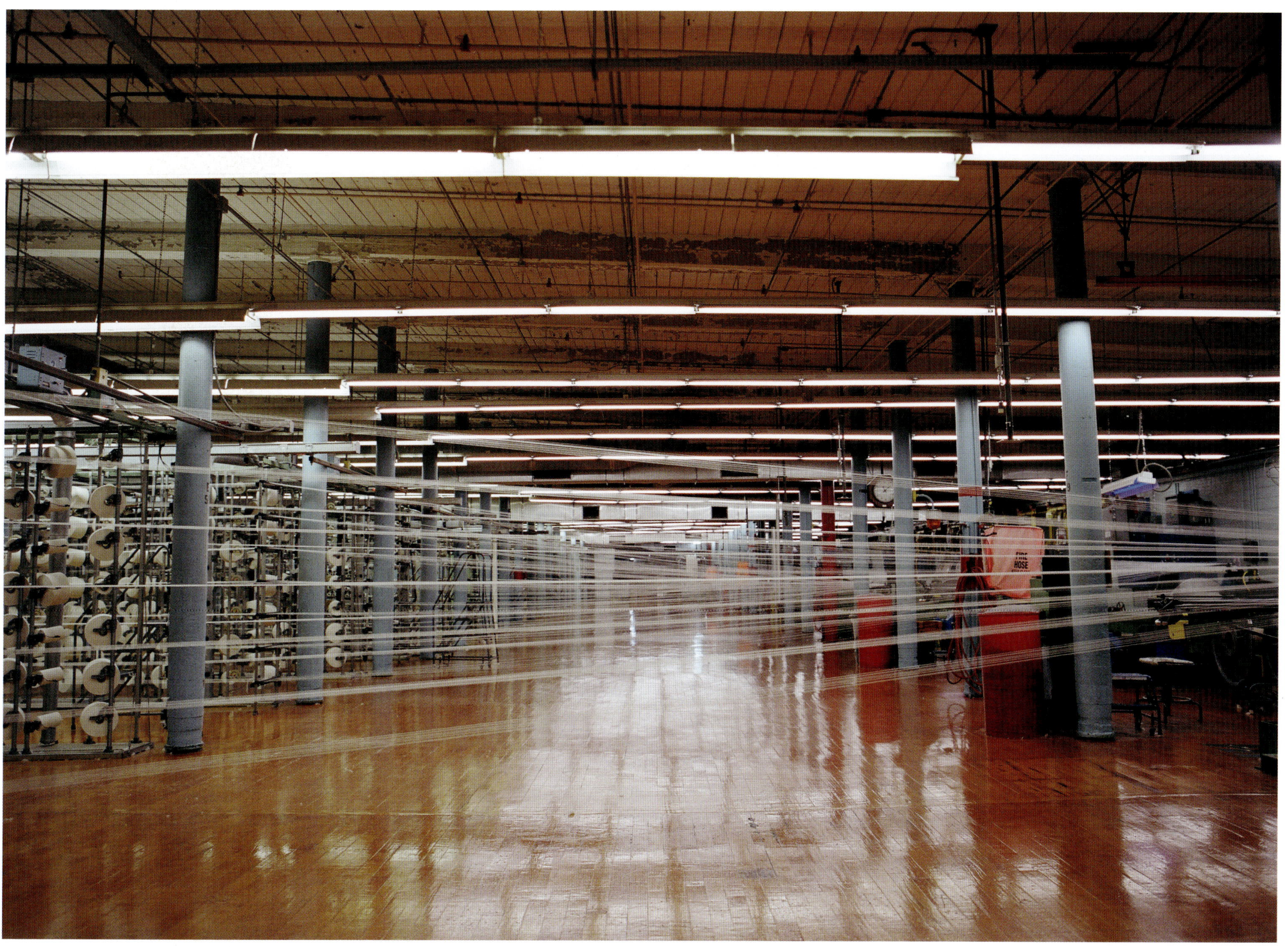

Ball warper // Warping room

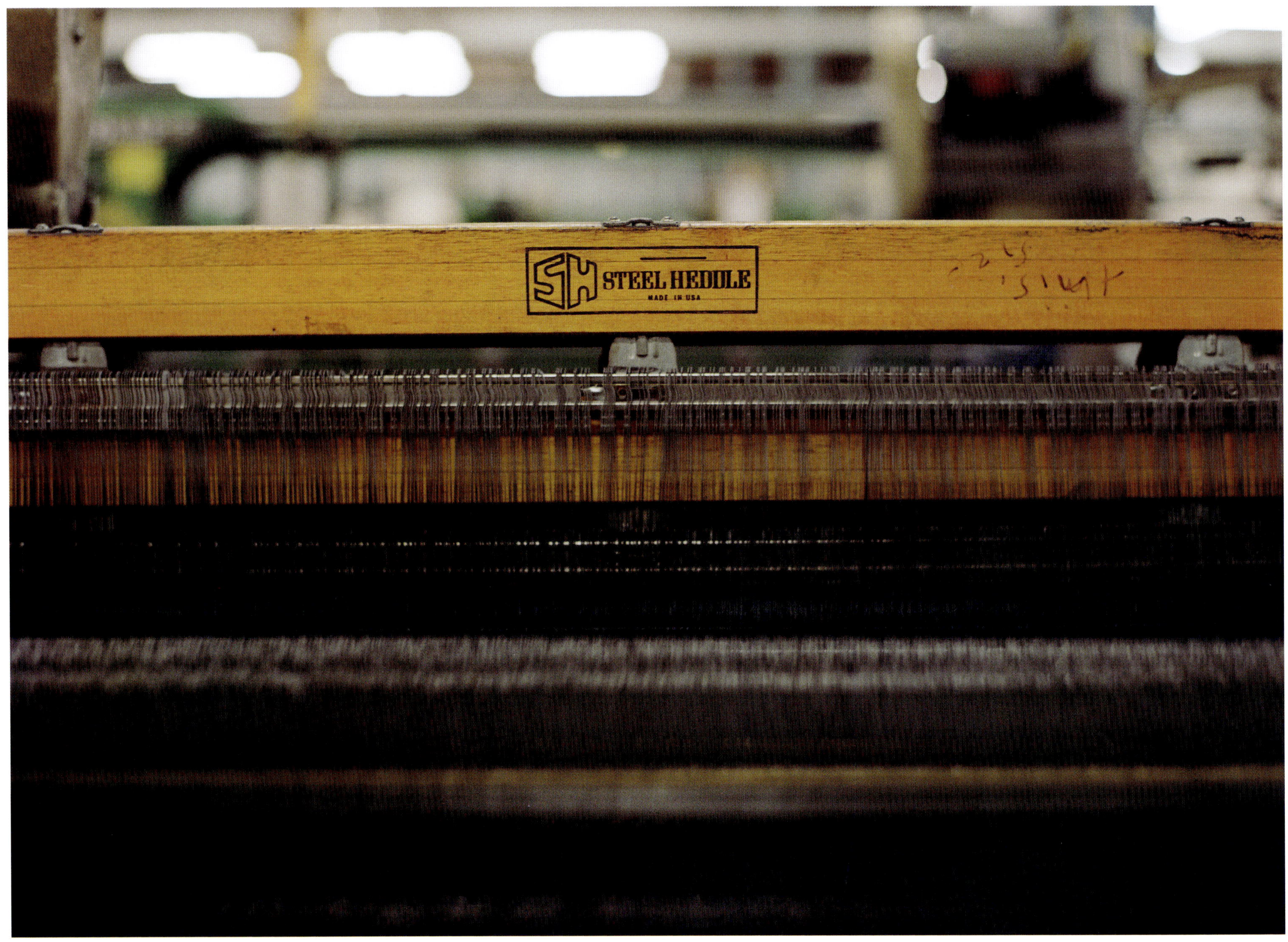

Draper X-3 loom harness // Selvage yarn being added to freshly dyed indigo yarn in the Slashing Department

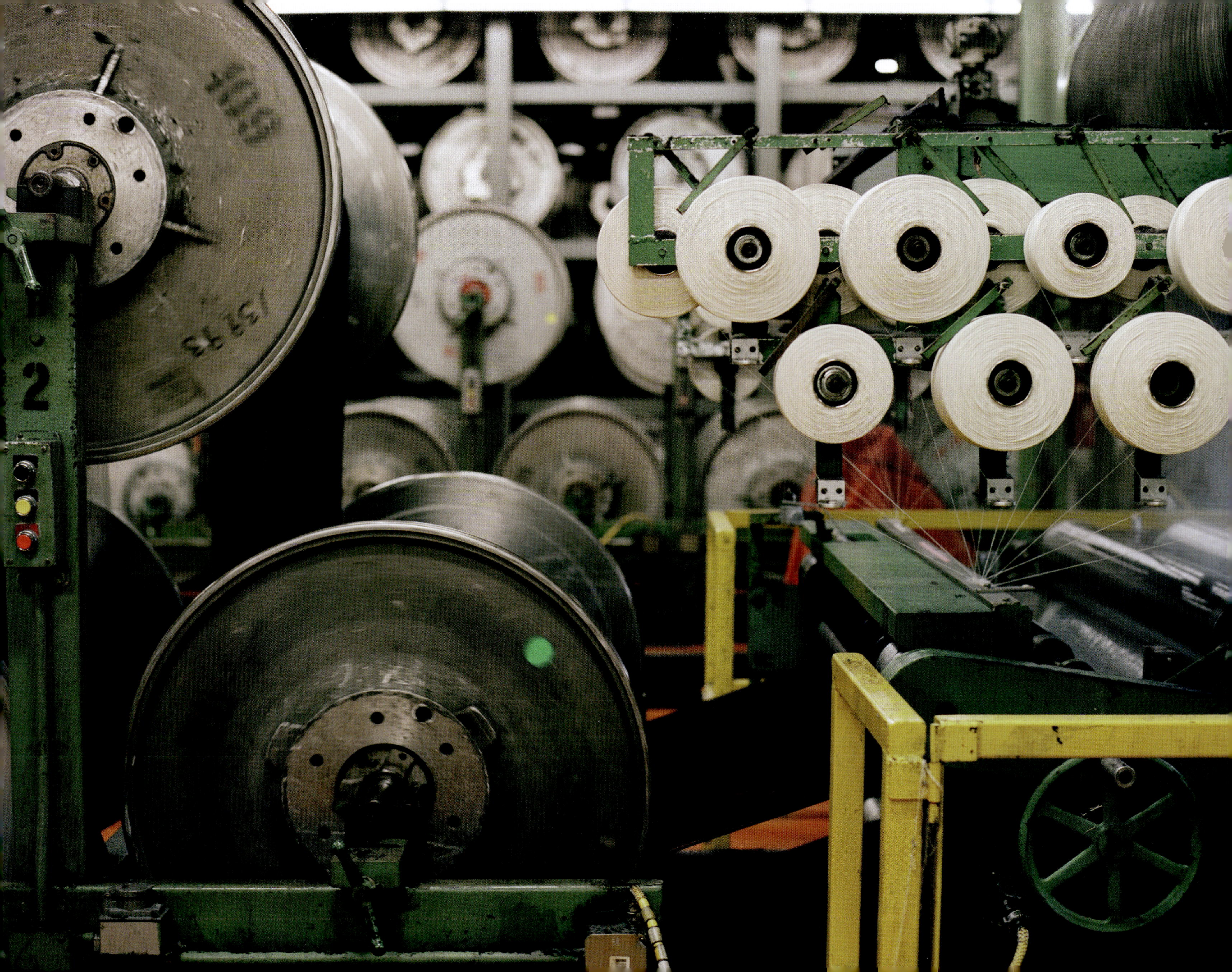

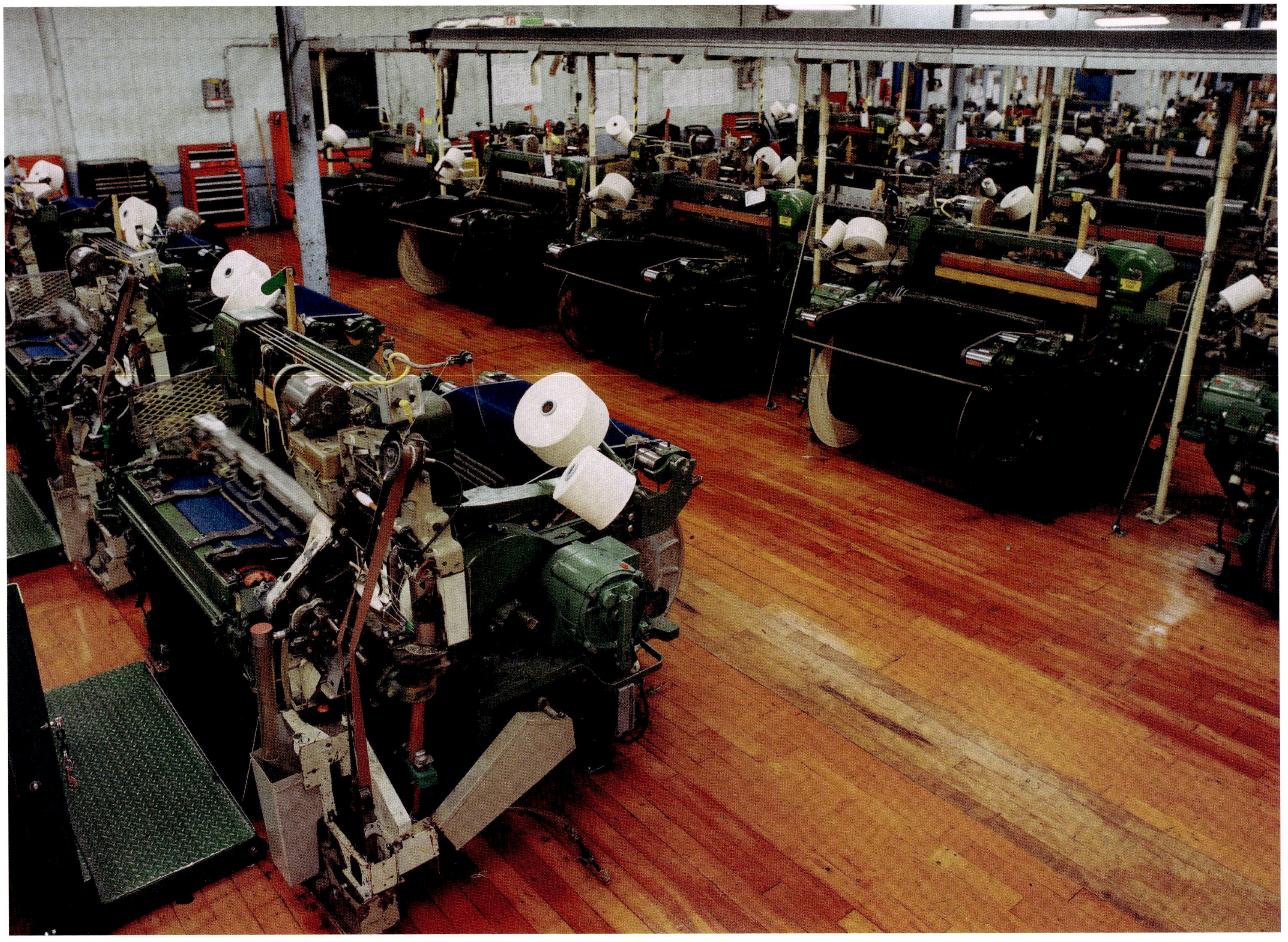

Shuttle looms in the X-3 Weave Room

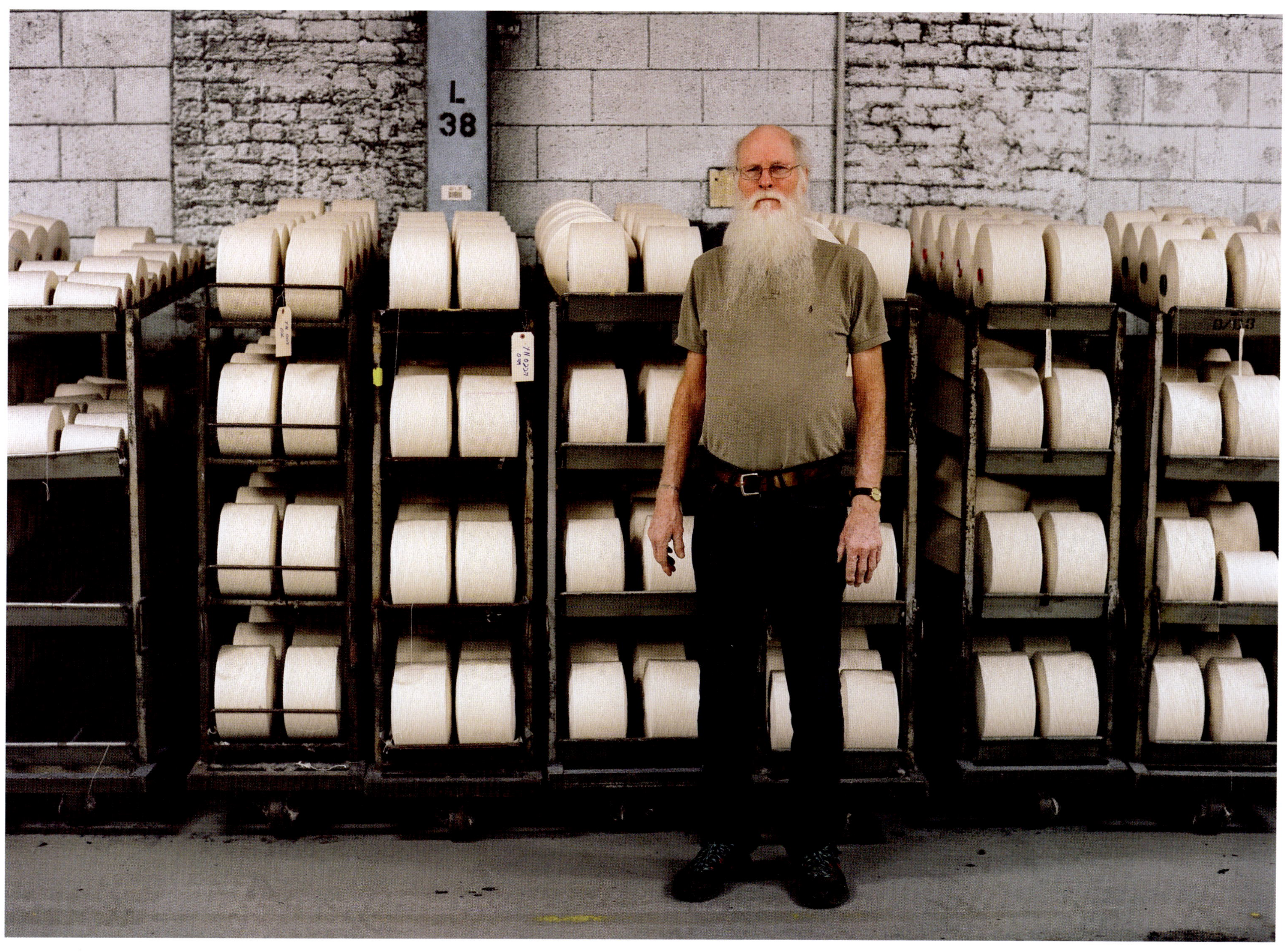

George "Red" Westmorland Jr., Warper Tender, filling yarn carts in the Weave Room

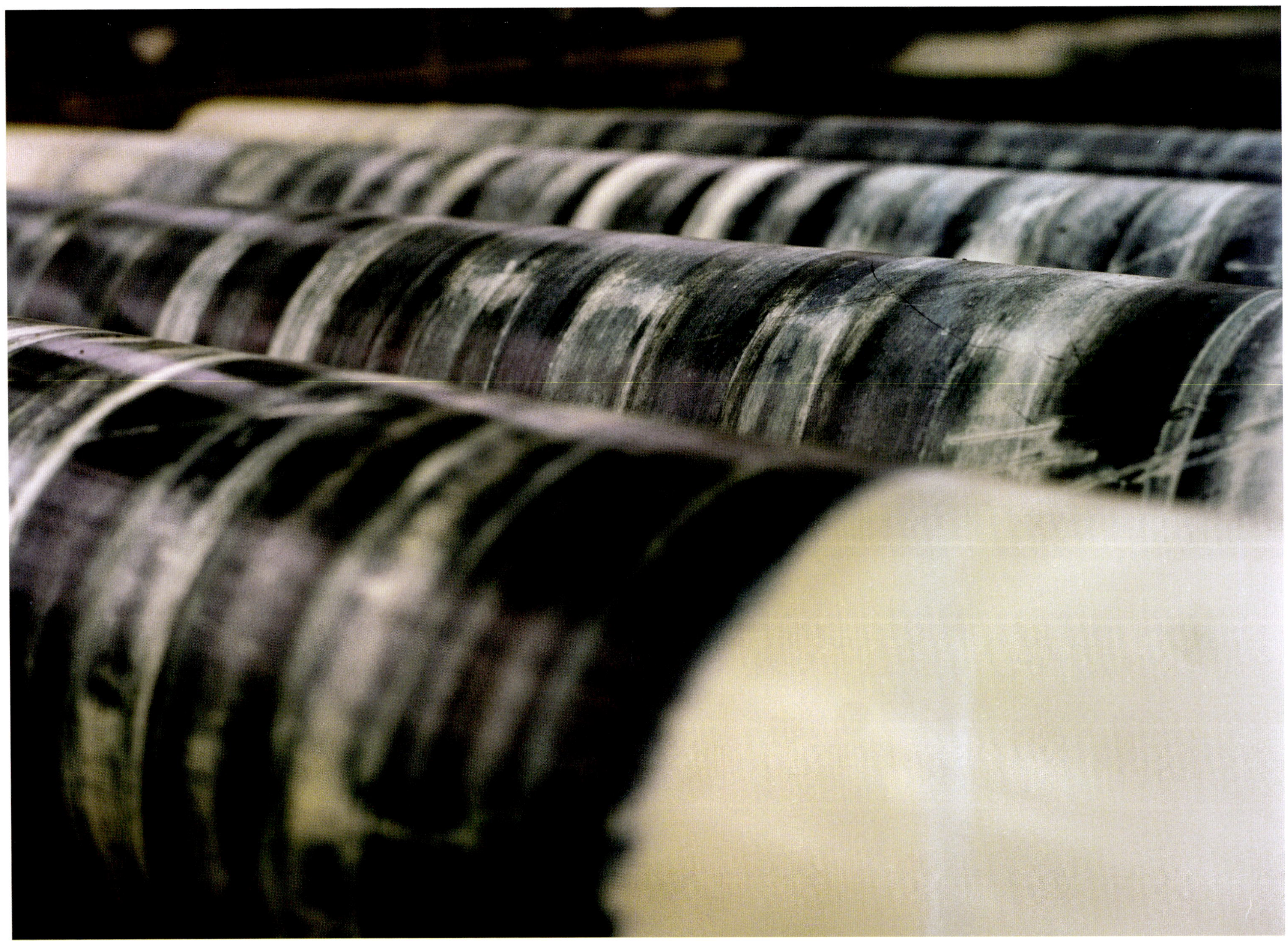

Rollers stained with indigo in the Dye House

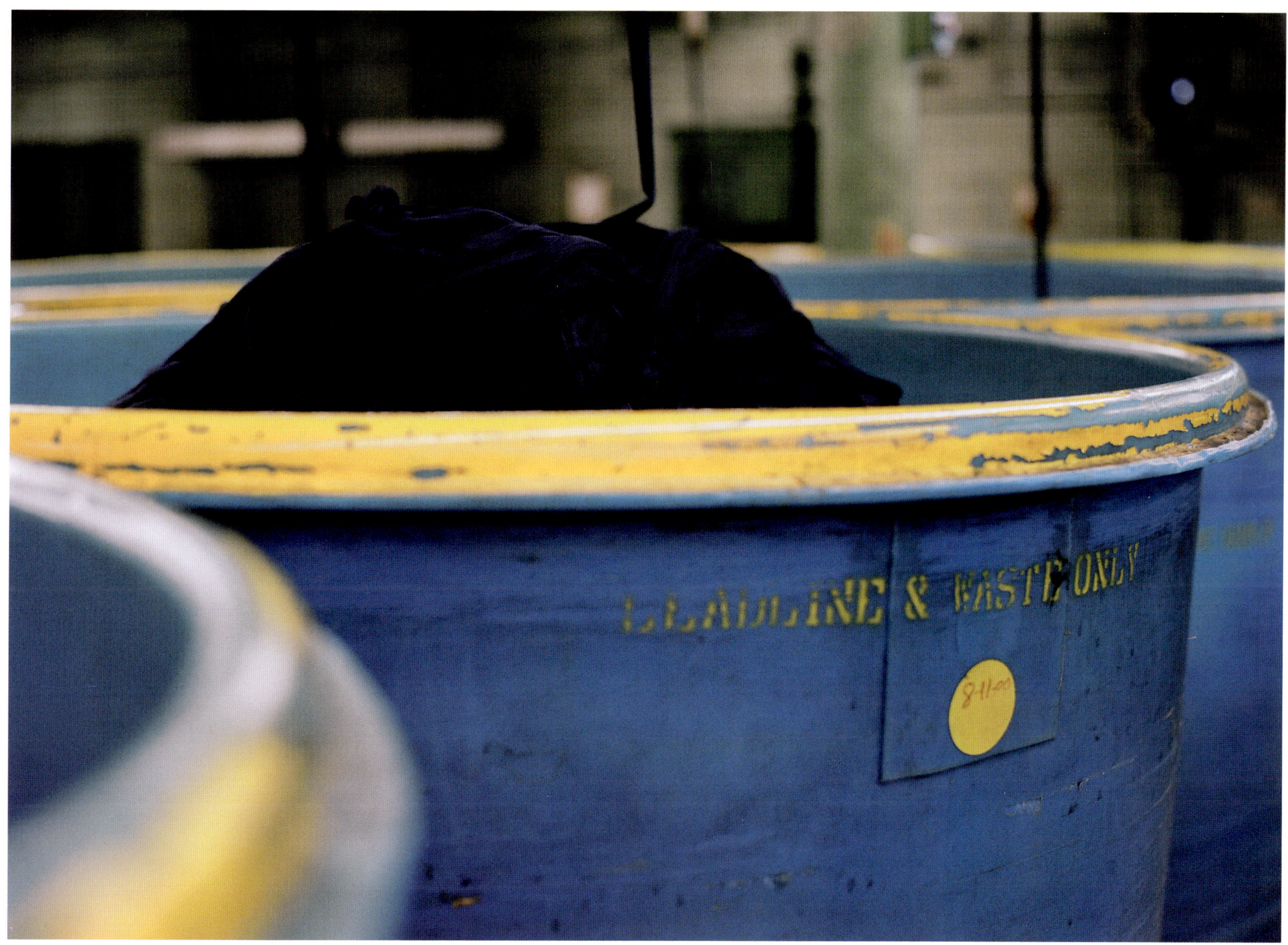

Rope-dyed indigo warp in the Dye House

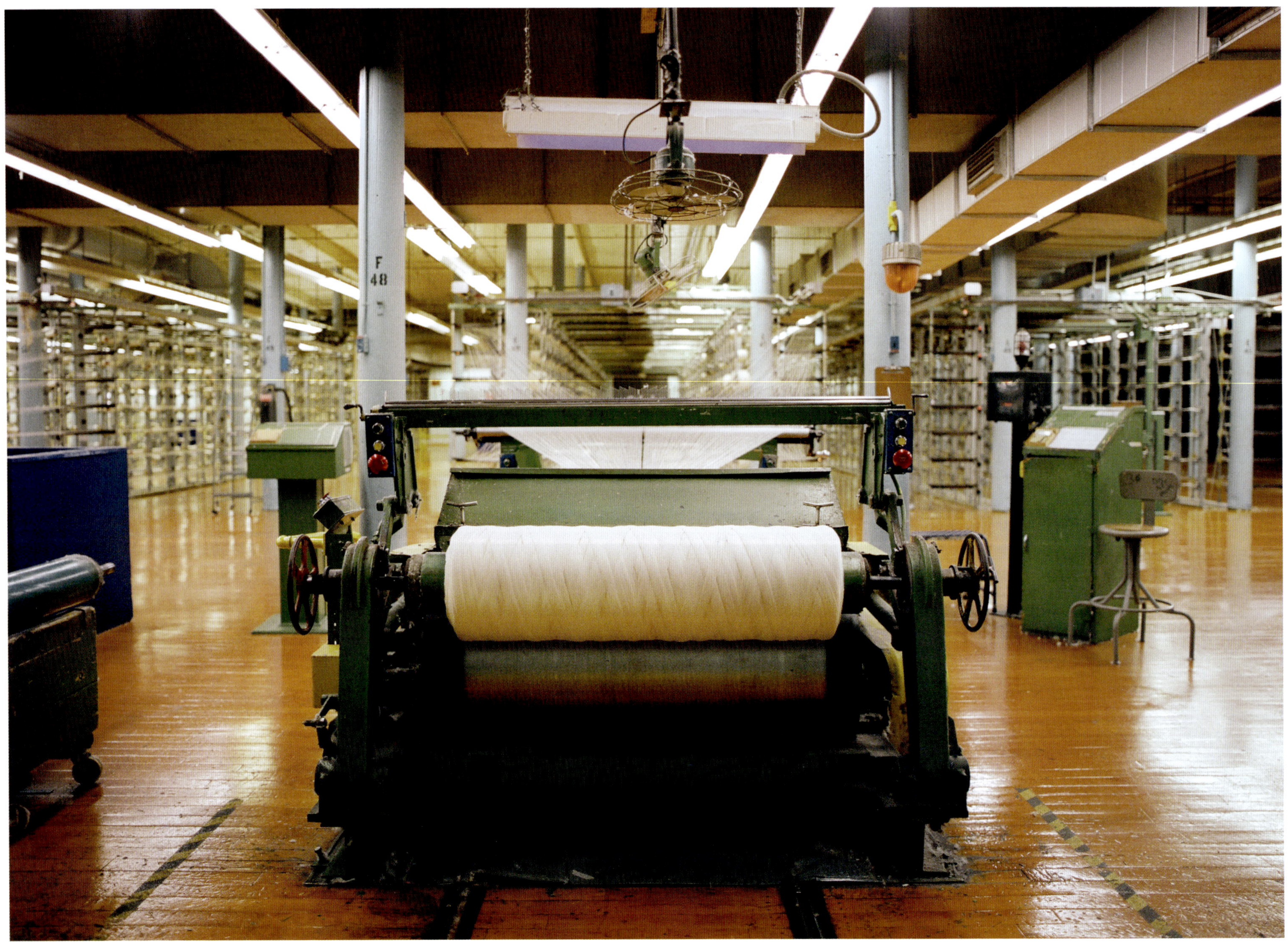

Ball warper, Warping Department

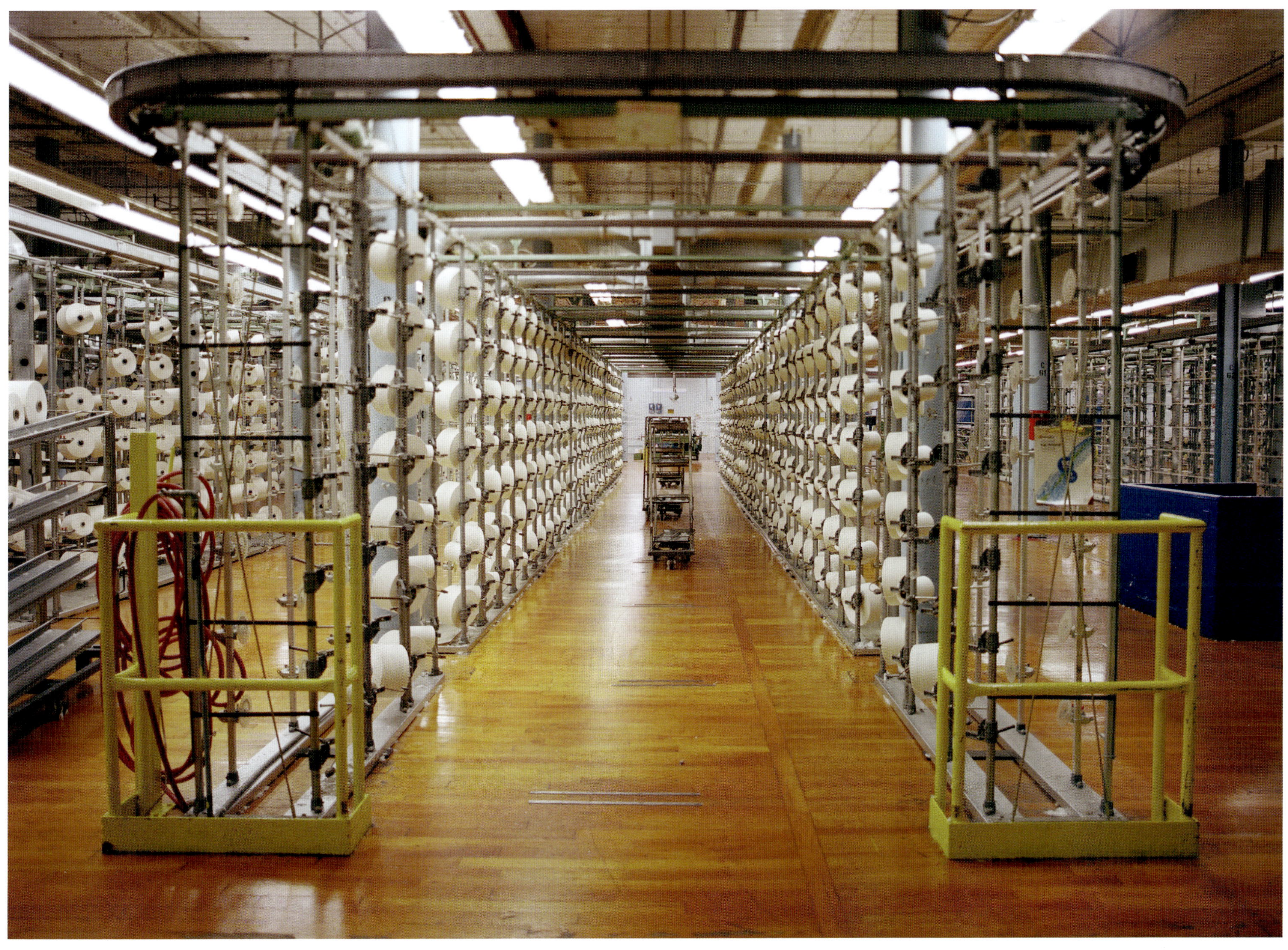

Undyed yarn on creels, Warping Department

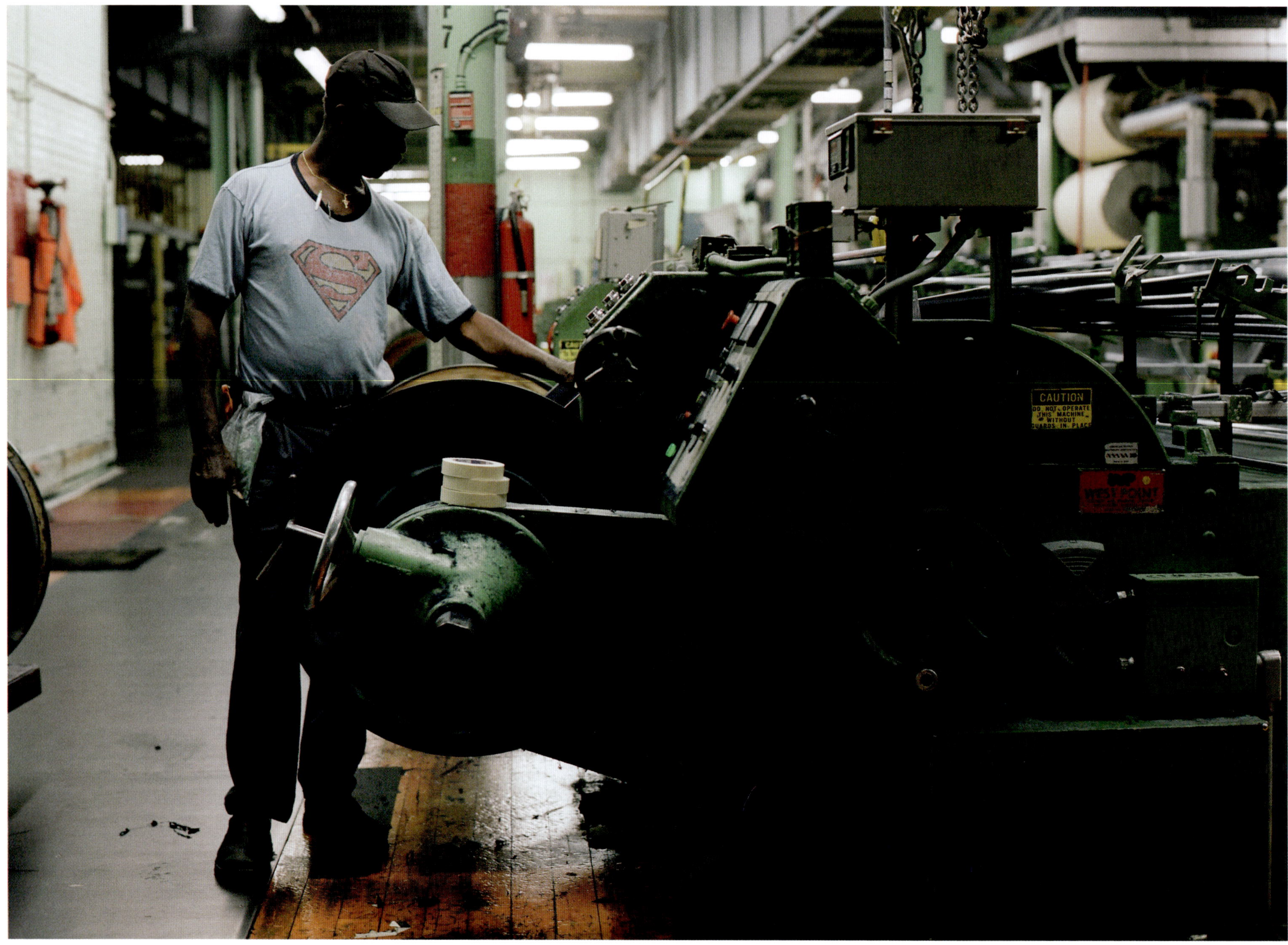

Willie Capers Jr., Slasher Tender, tending a beaming frame in the Slashing Department // Finishing Department

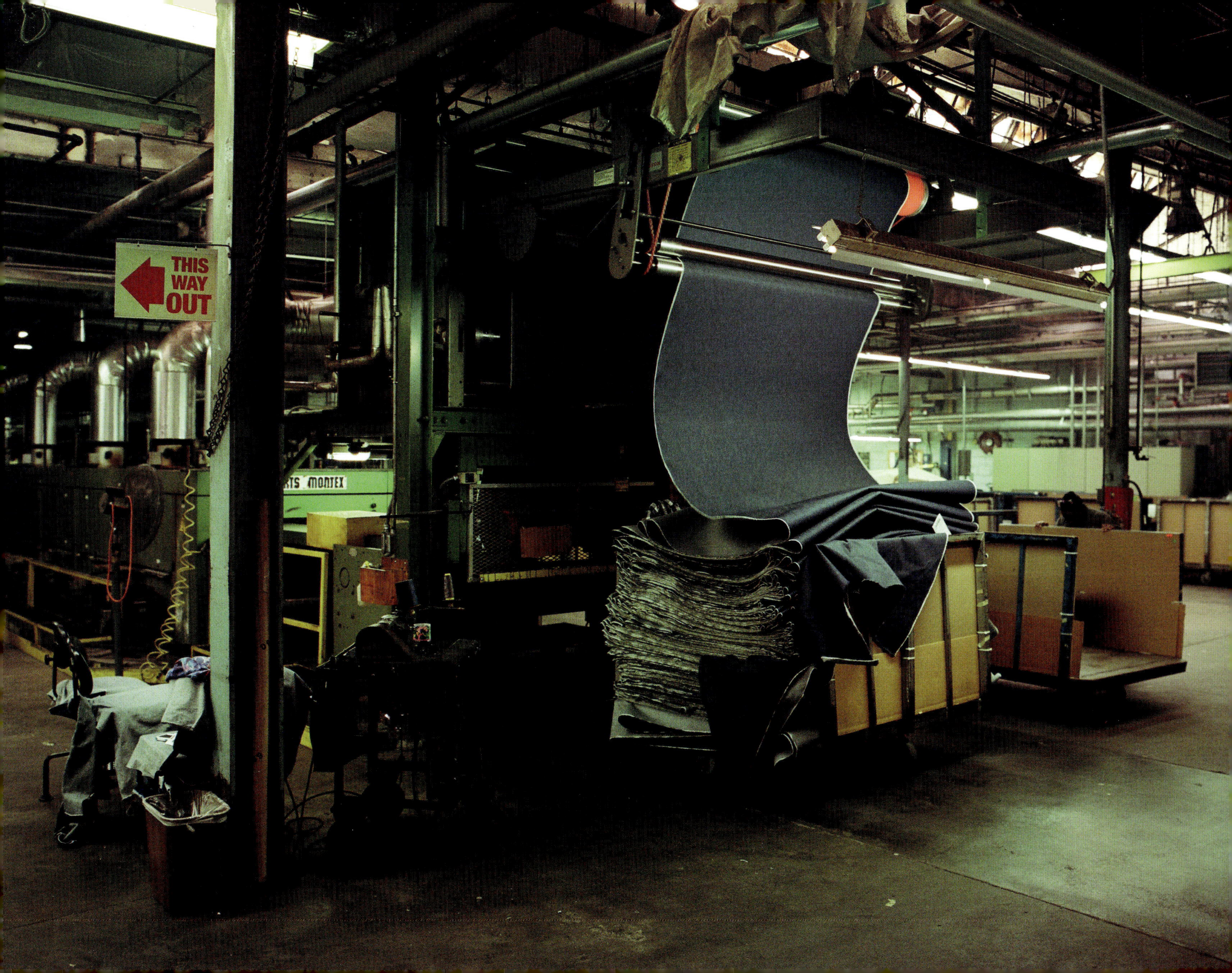
THIS
WAY
OUT

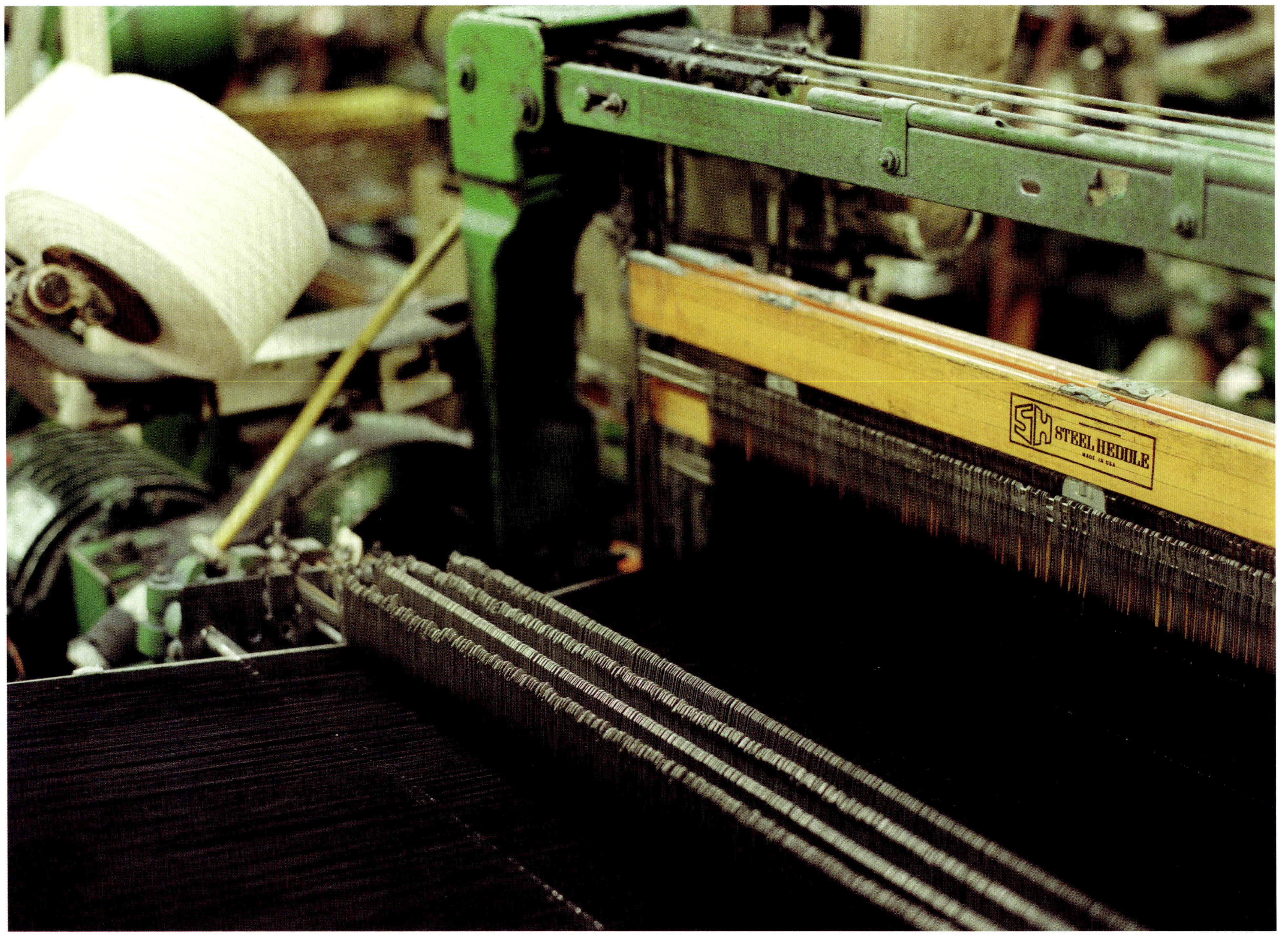

Let-off side of a Draper X-3 loom in the Weave Room

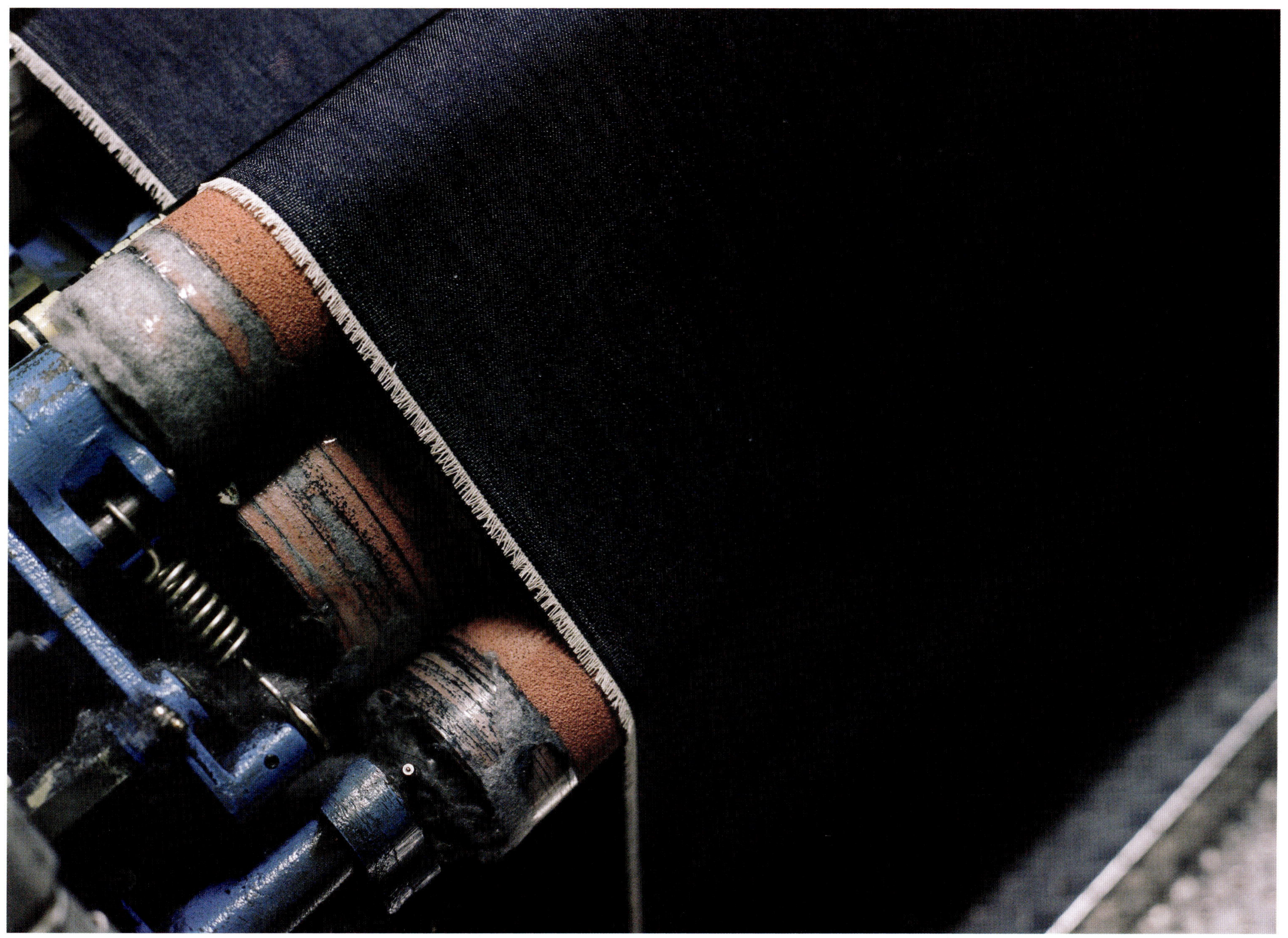

3 x 1 non-selvage denim in the Weave Room

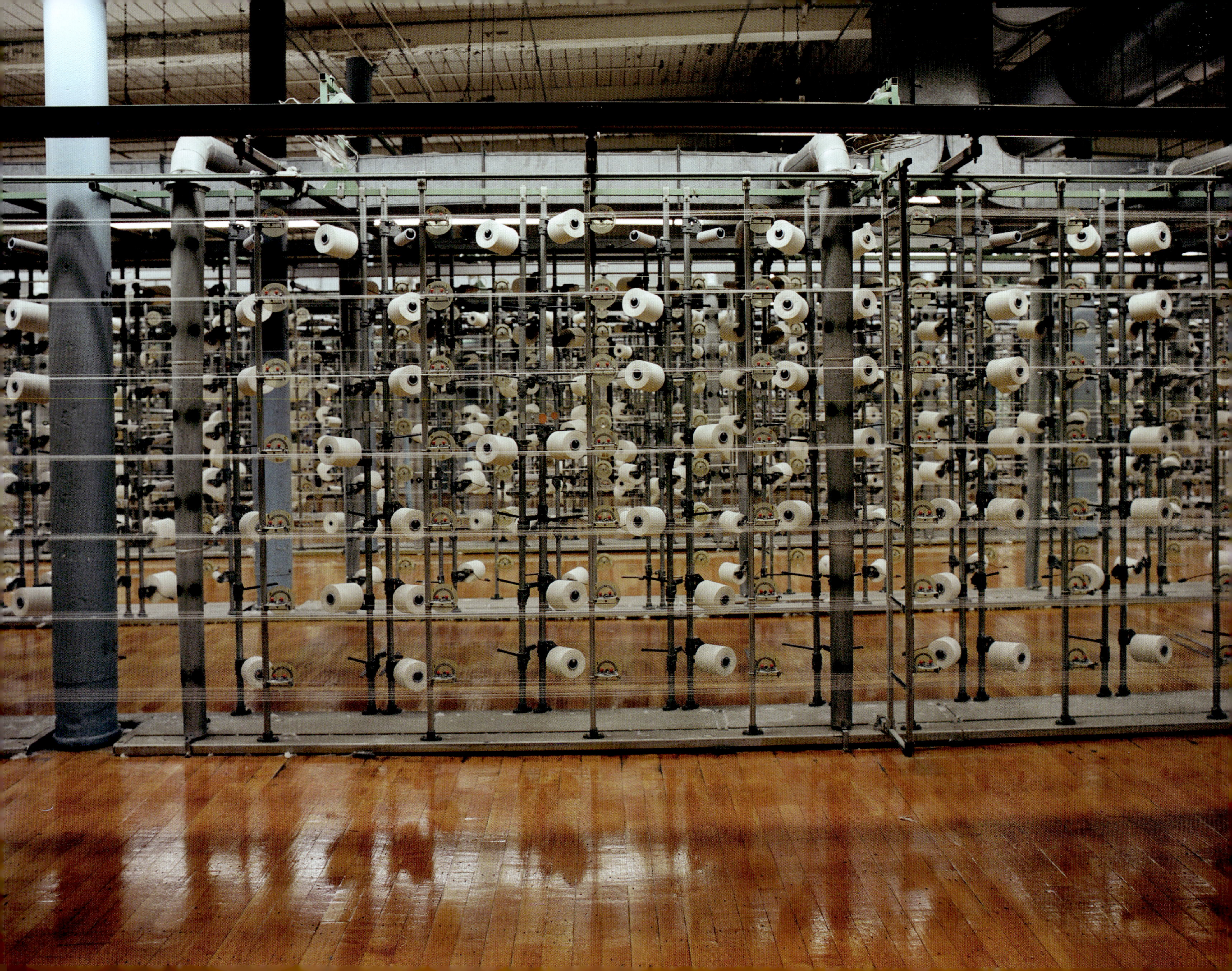

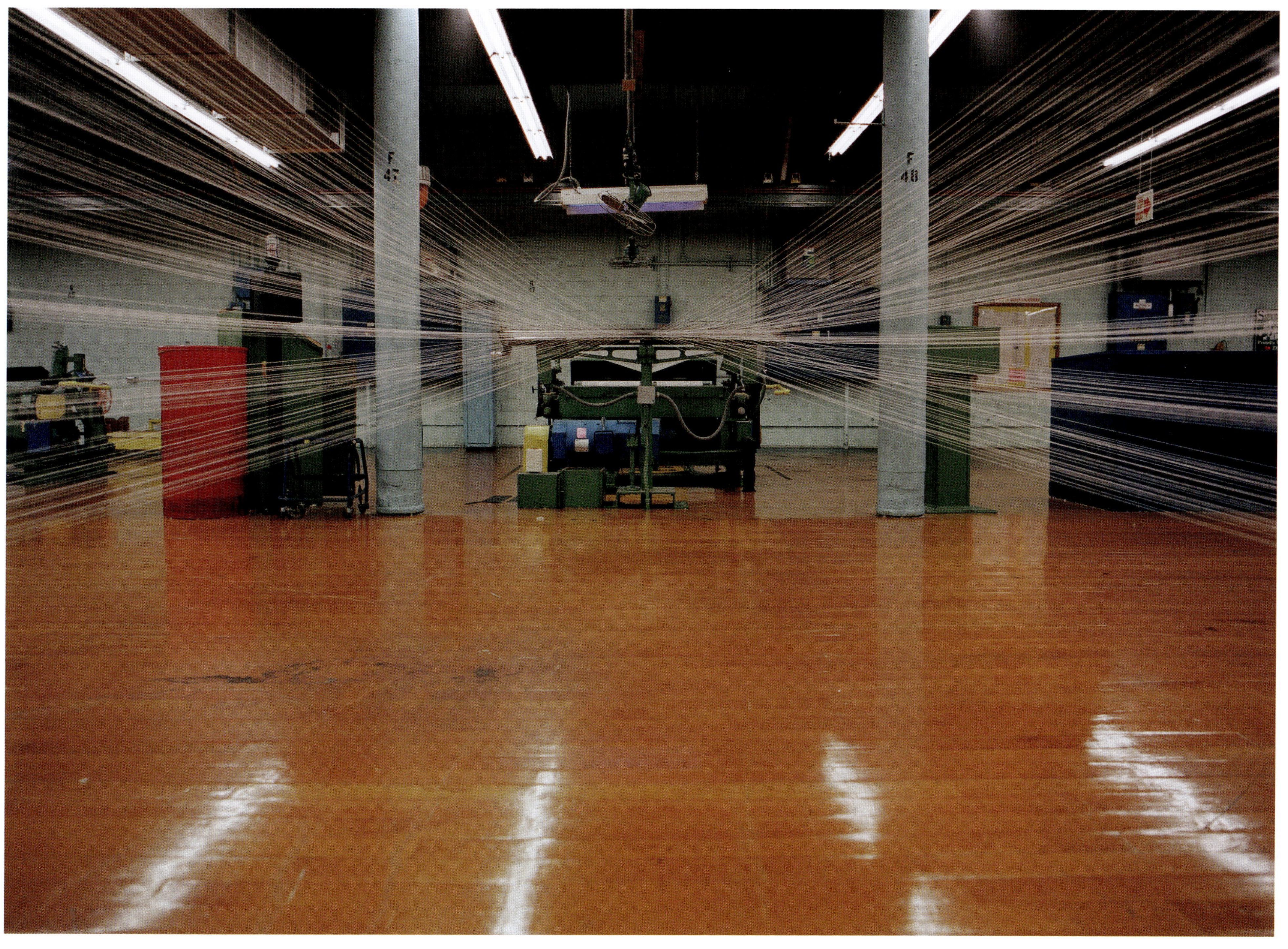

Undyed yarn on creels in the Warping Department   //   Ball warper in the Warping Department

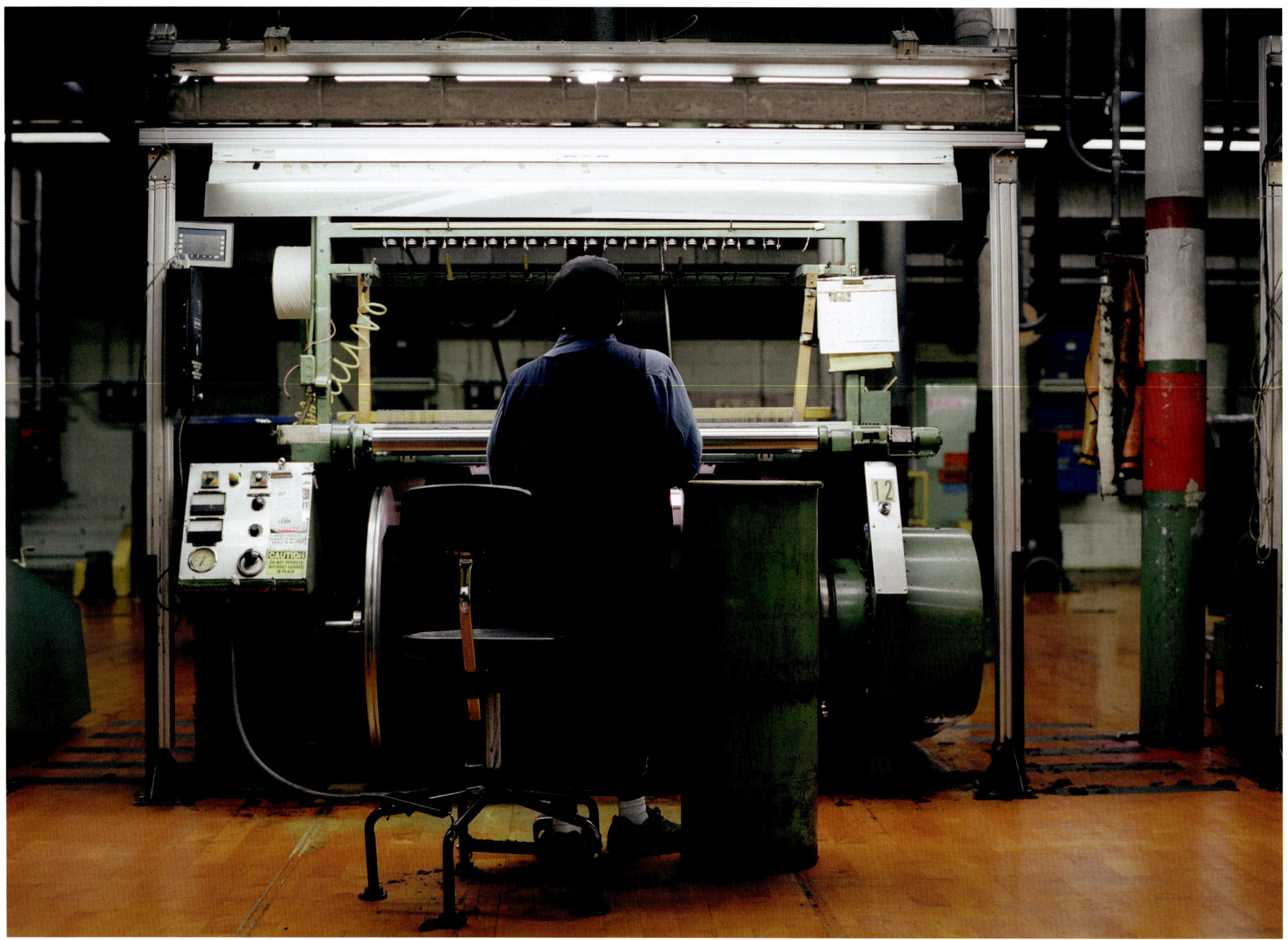

Lillie Neal, Beamer Tender, tending a rebeaming frame in the Beaming Department

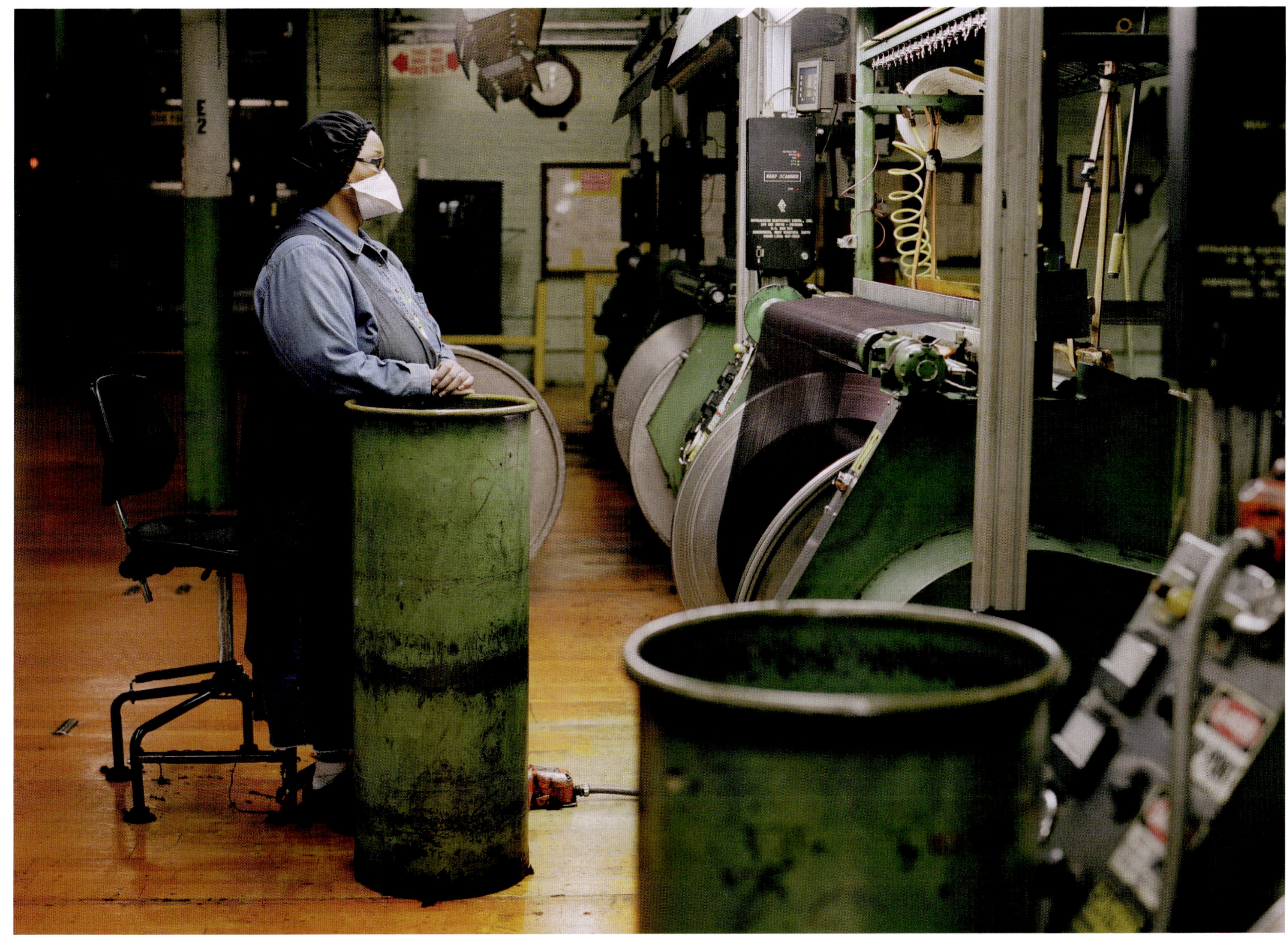

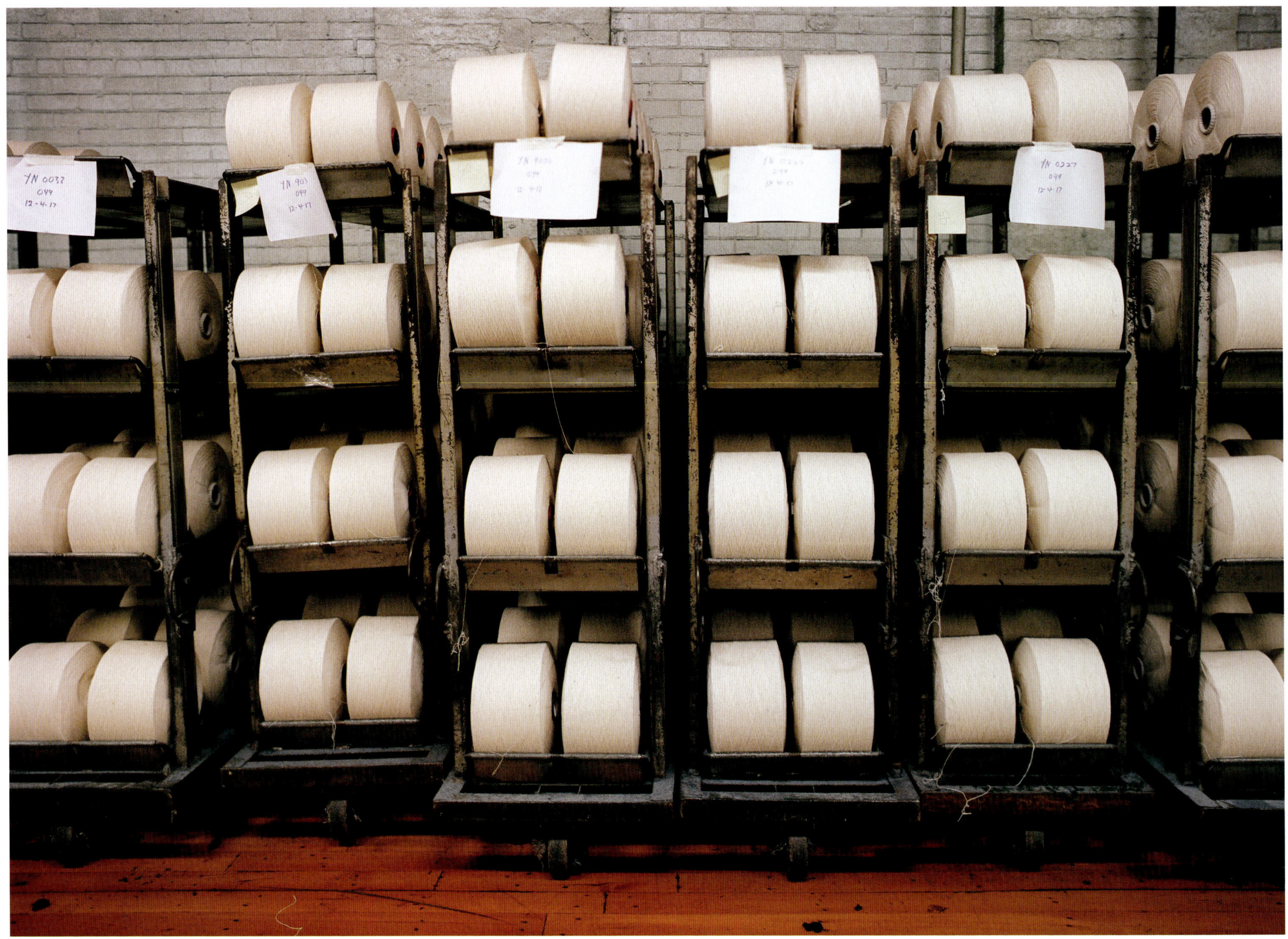

Yarn carts in the Weave Room in the Weaving Department // Signage in the Weave Room

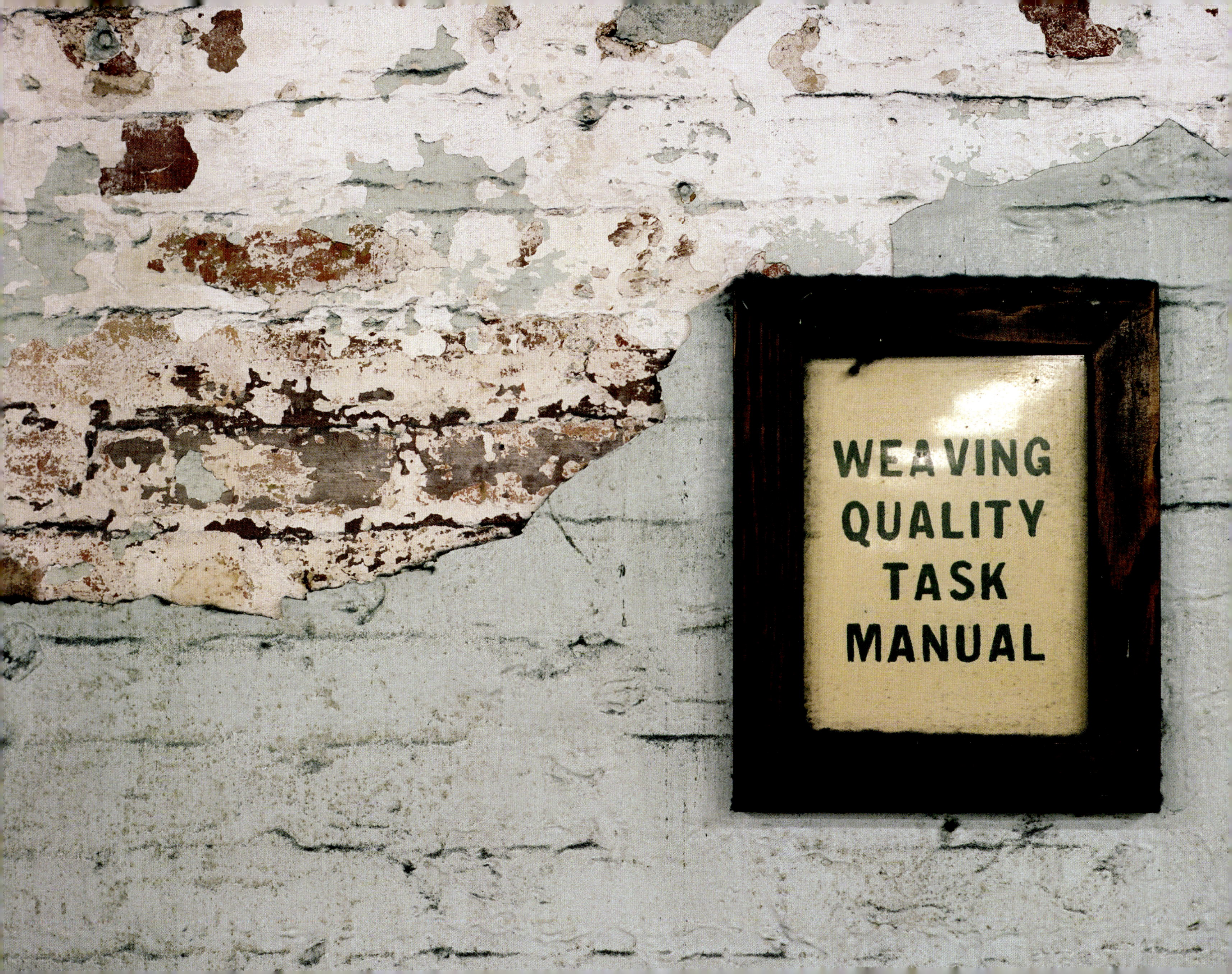

WEAVING
QUALITY
TASK
MANUAL

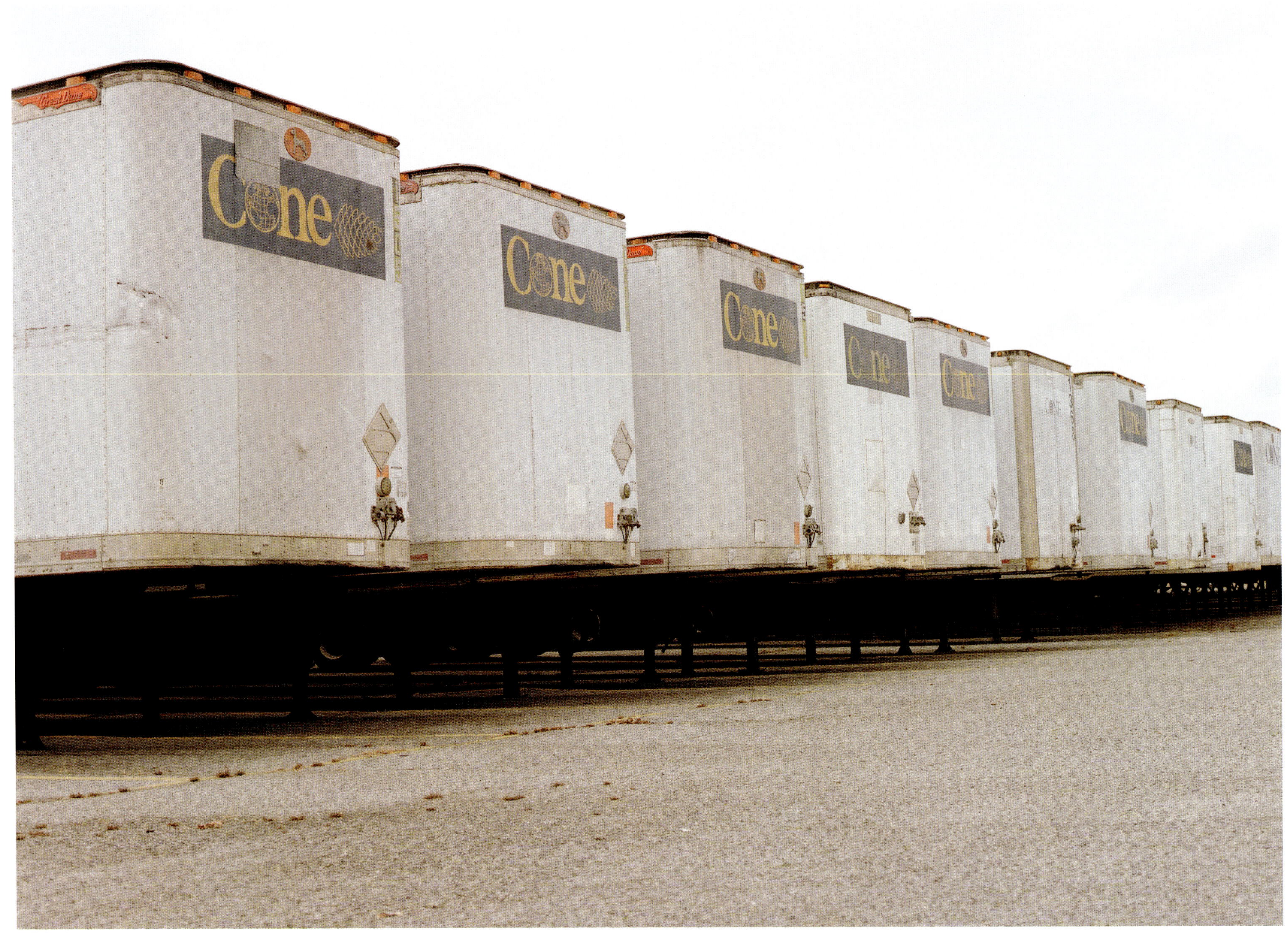

Fleet of eighteen-wheeler drydock trailers in parking lot

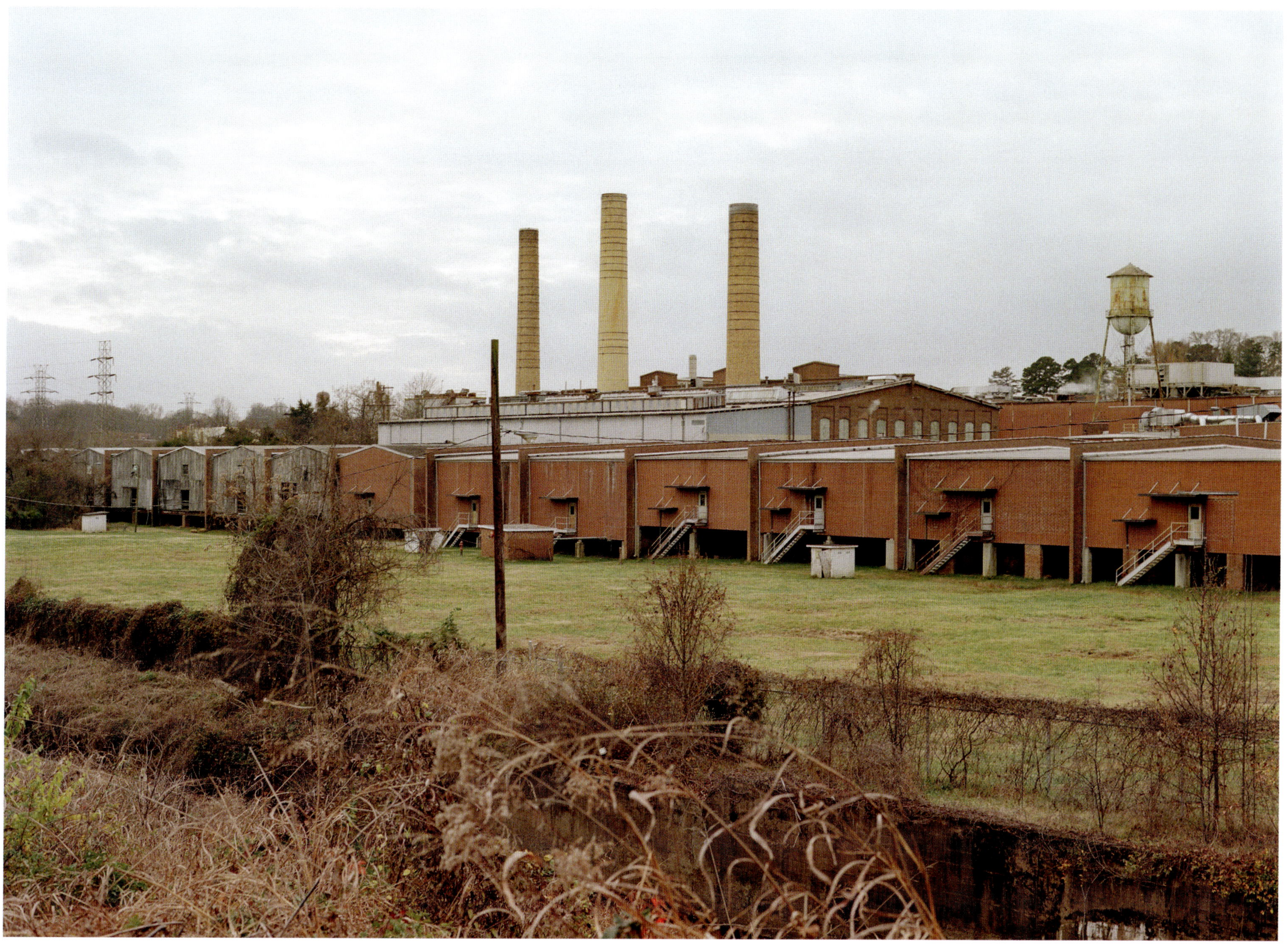

Cotton Warehouses in the foreground, with the White Oak Cotton Mills Powerhouse
in the background

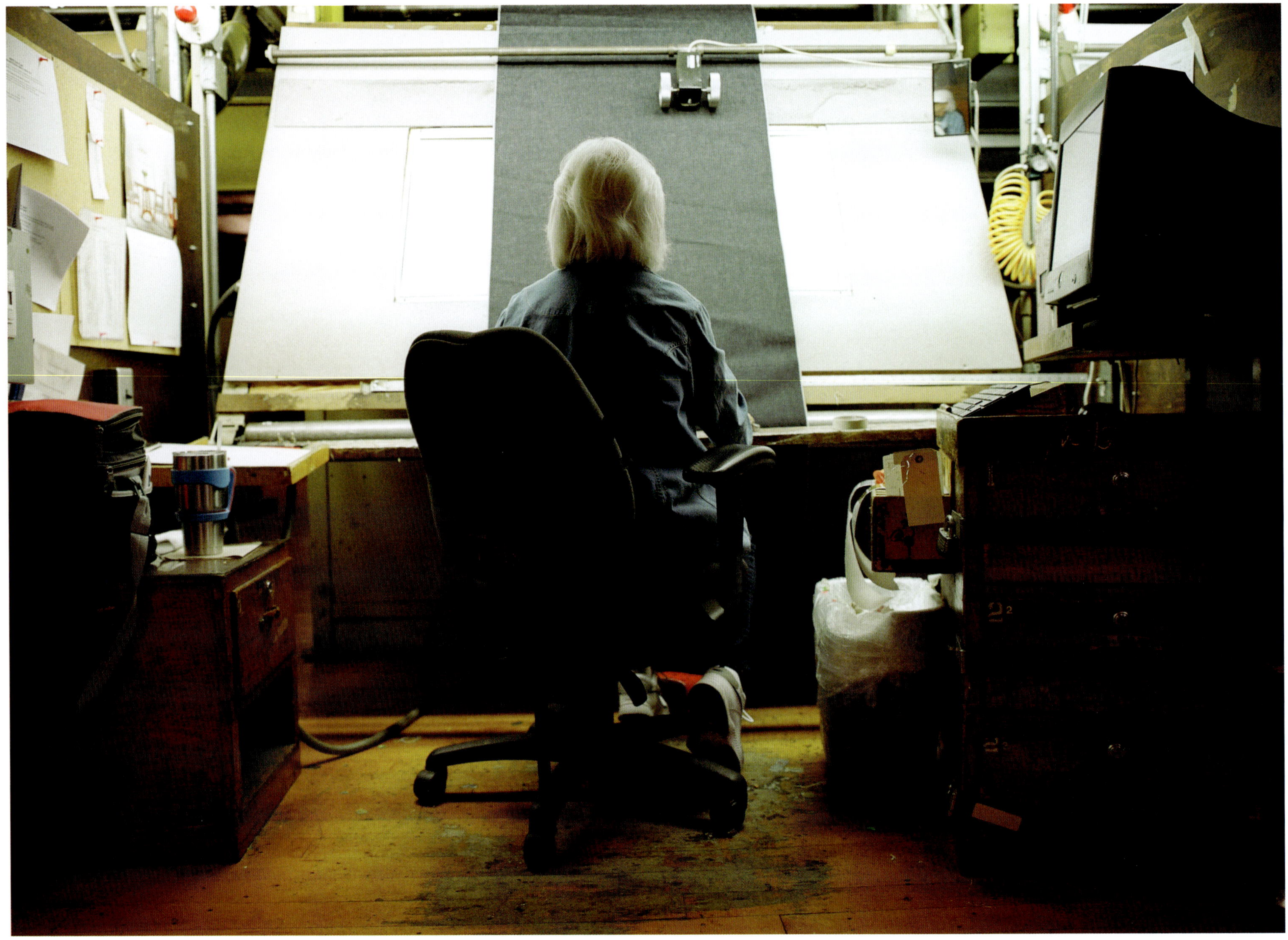

Sara Wyrick inspecting cloth from an X-3 loom in the Quality Control Department

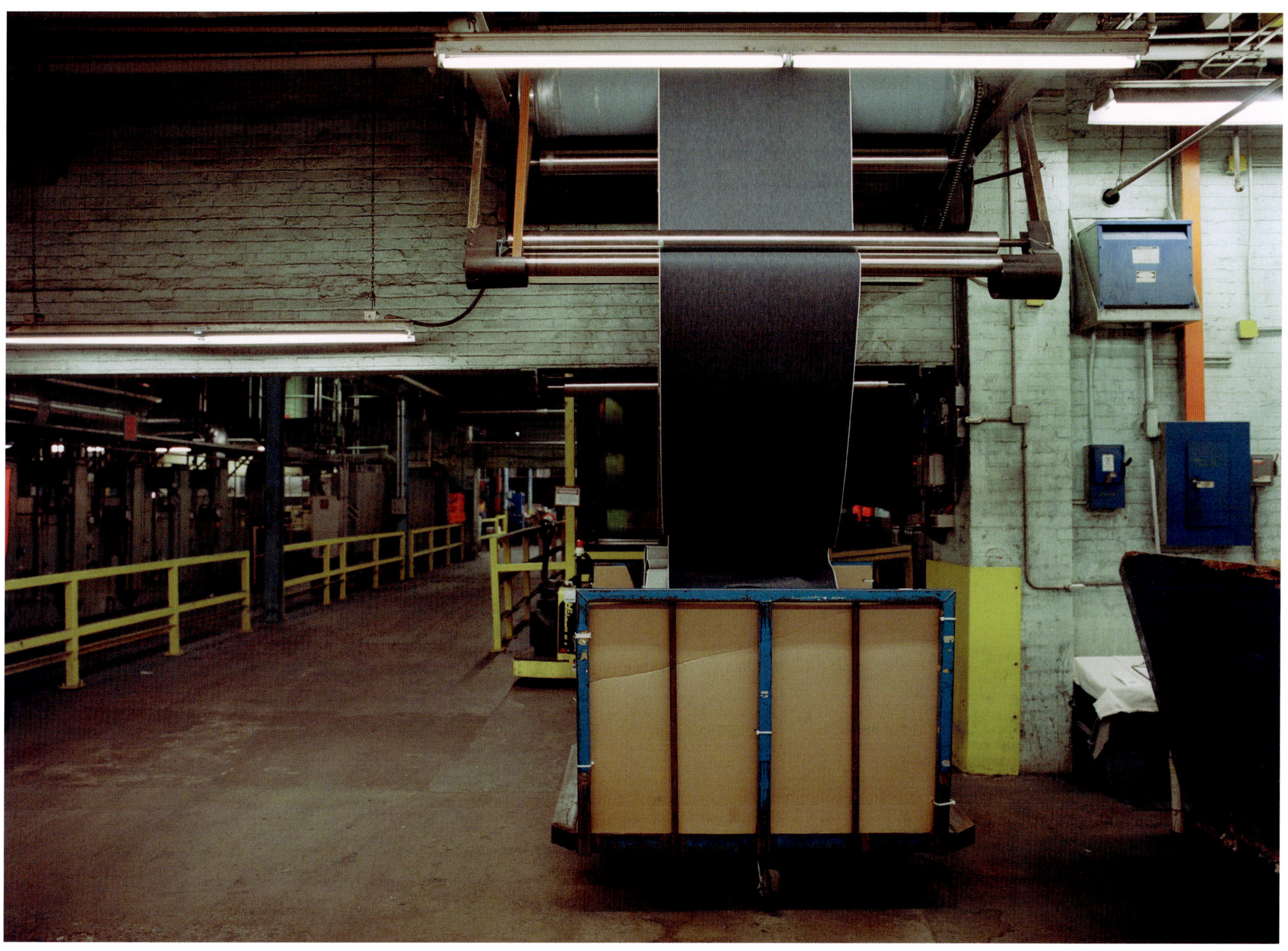

Cloth tentering in the Finishing Department

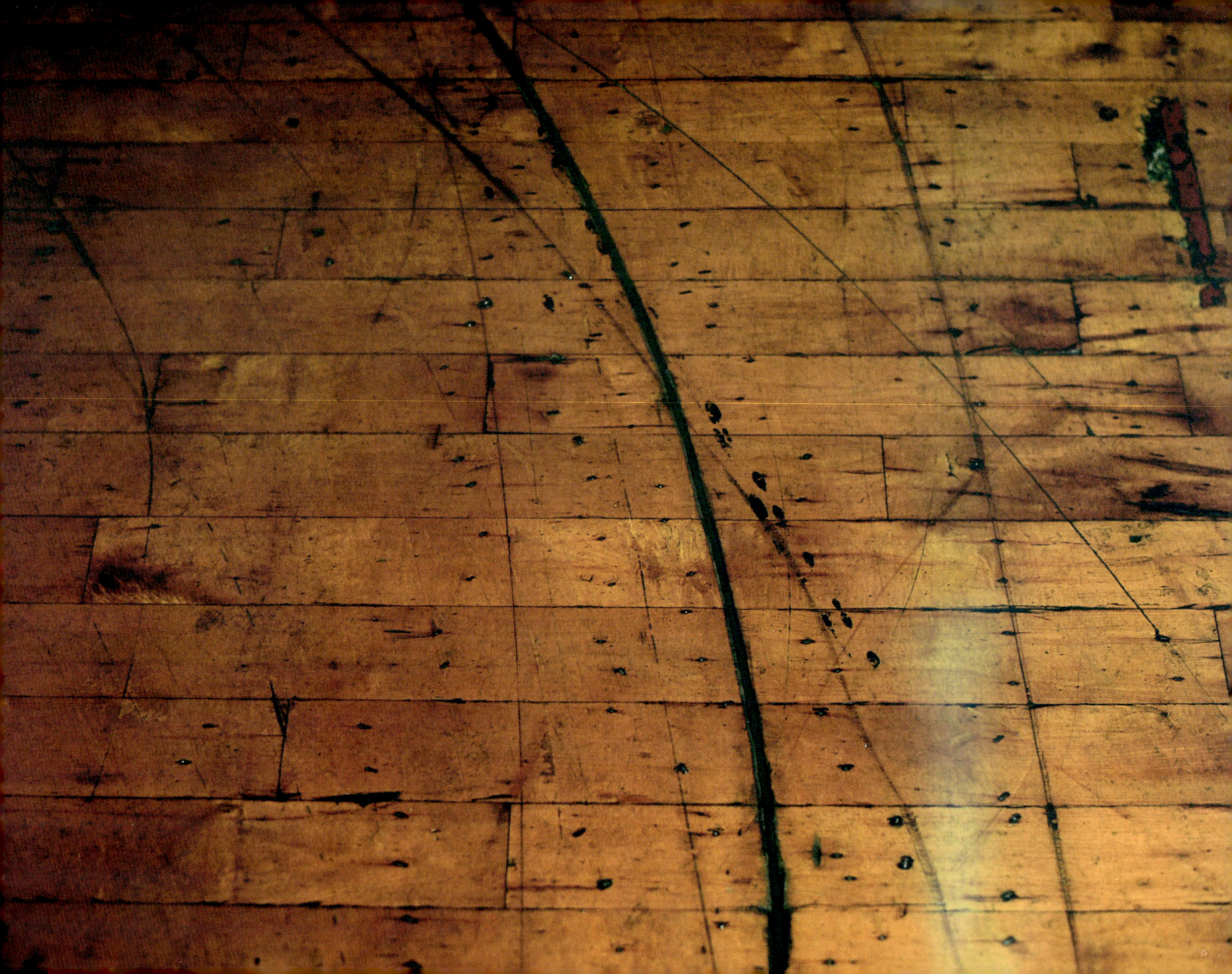

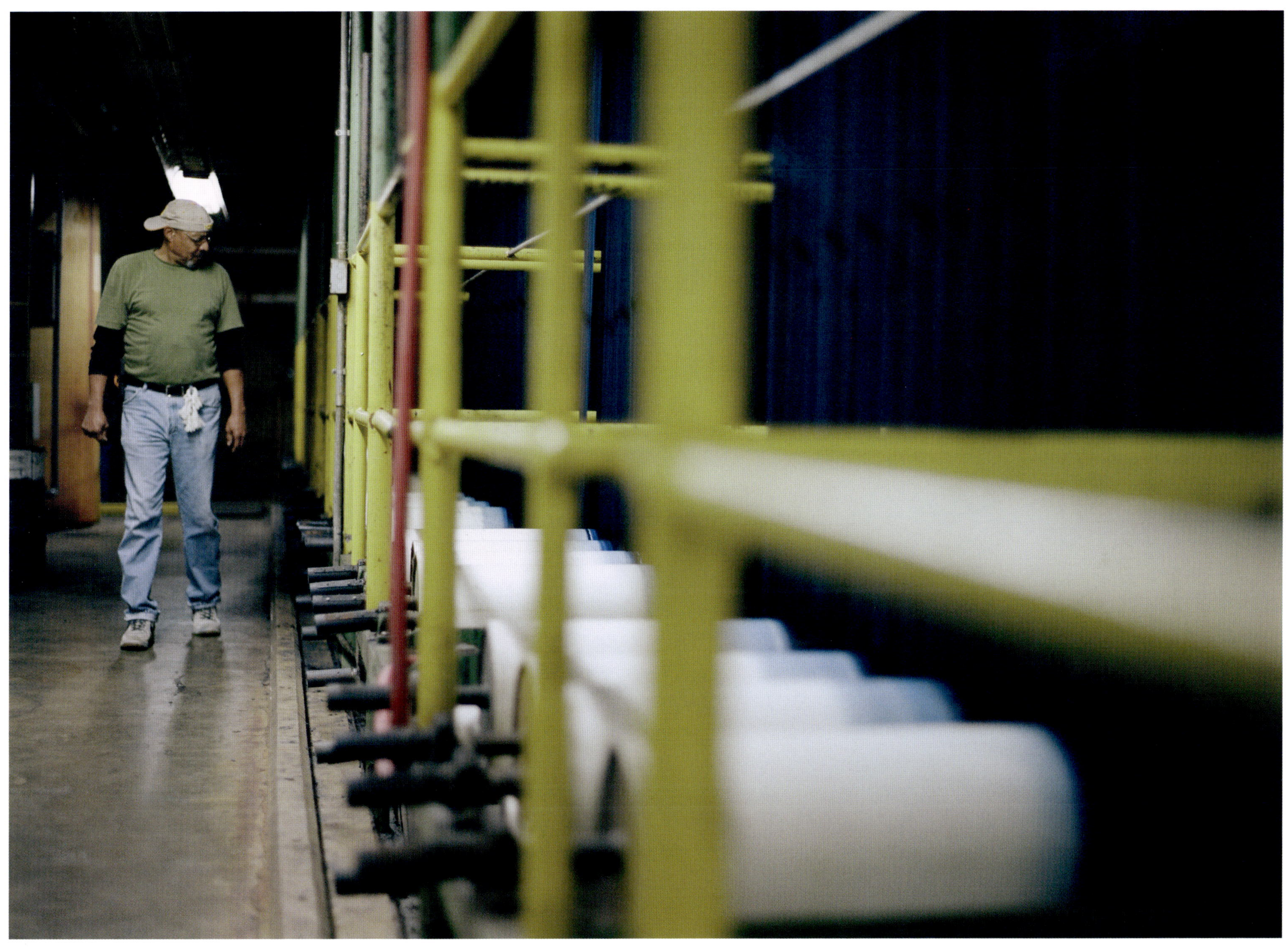

Maple floors, White Oak Cotton Mills // Clarence Goins, Sky Tender, walking the Skying Range in the Dye House

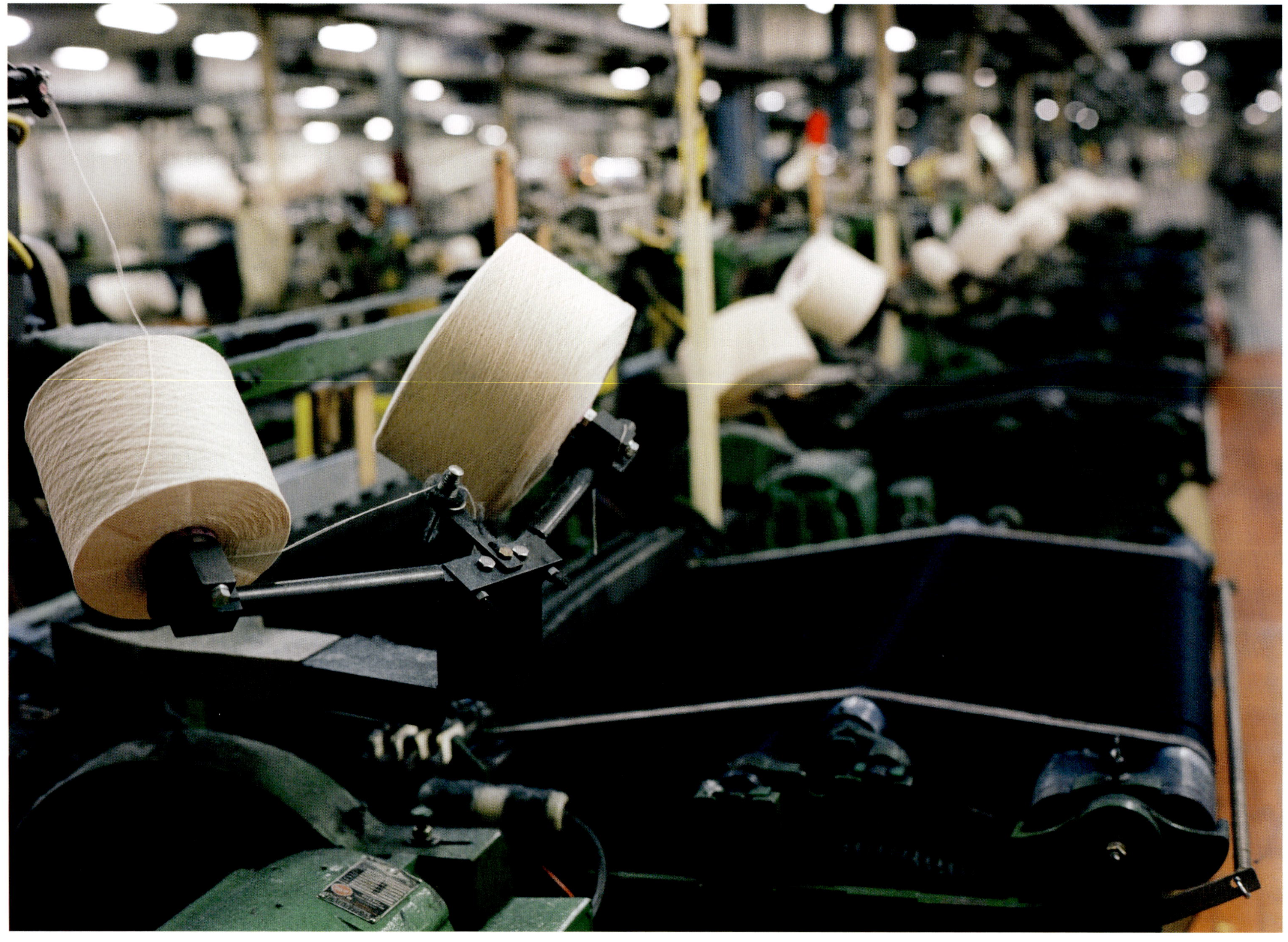

Filling yarn on X-3 shuttle looms in the Weave Room   //   Debbie Lindsey operating an X-3 shuttle loom in the Weave Room

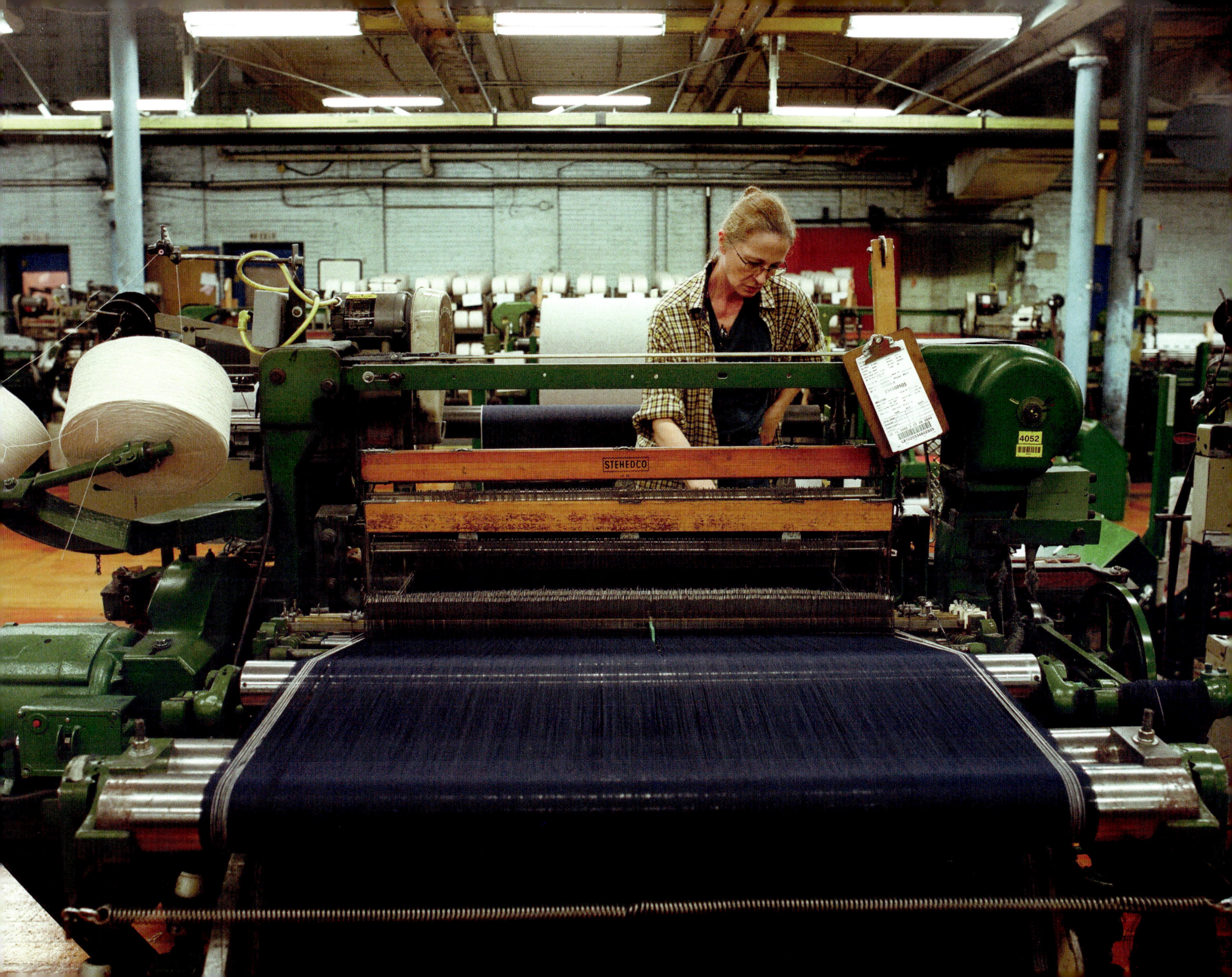

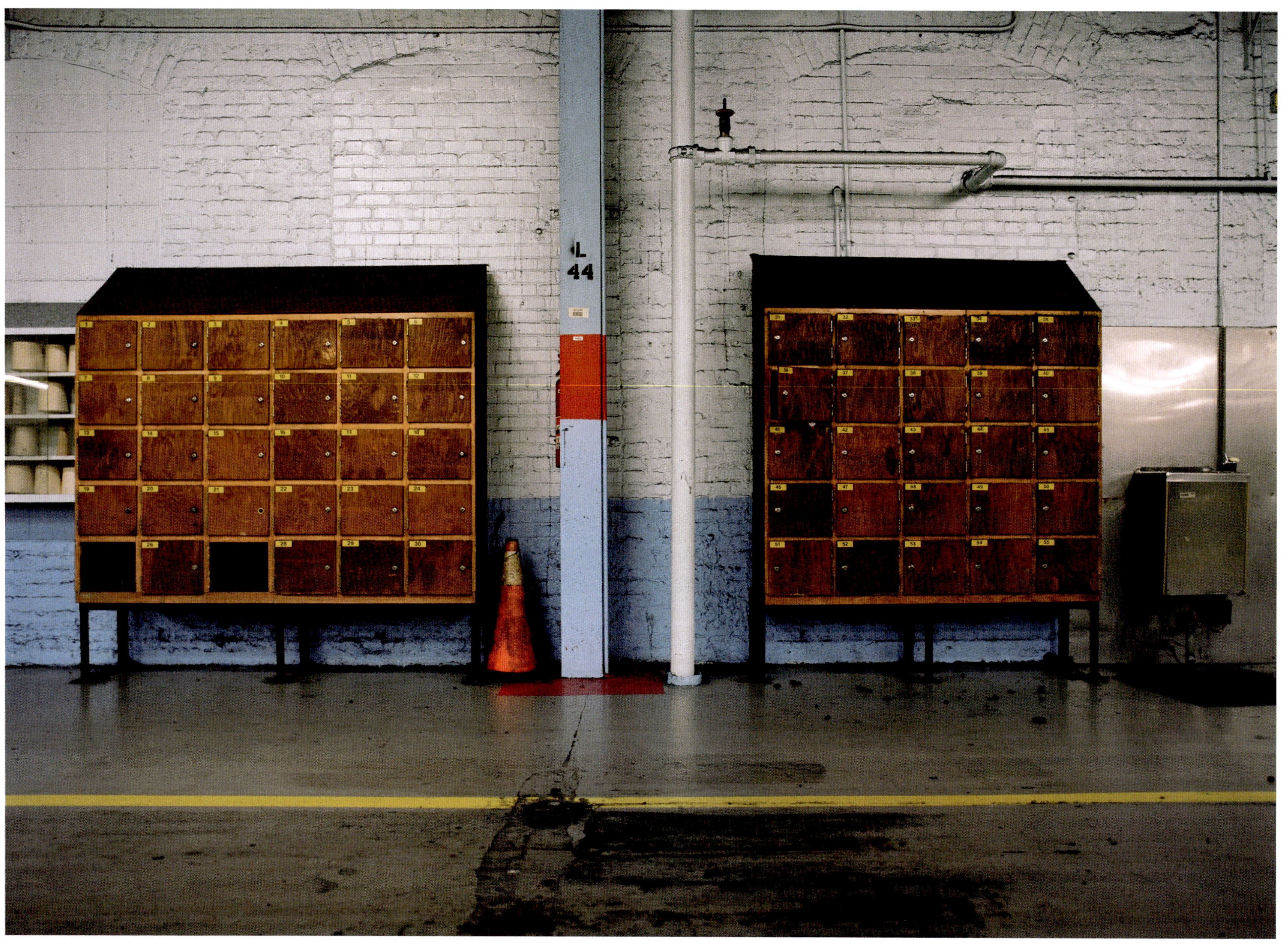

Employee lockers in the Weave Room

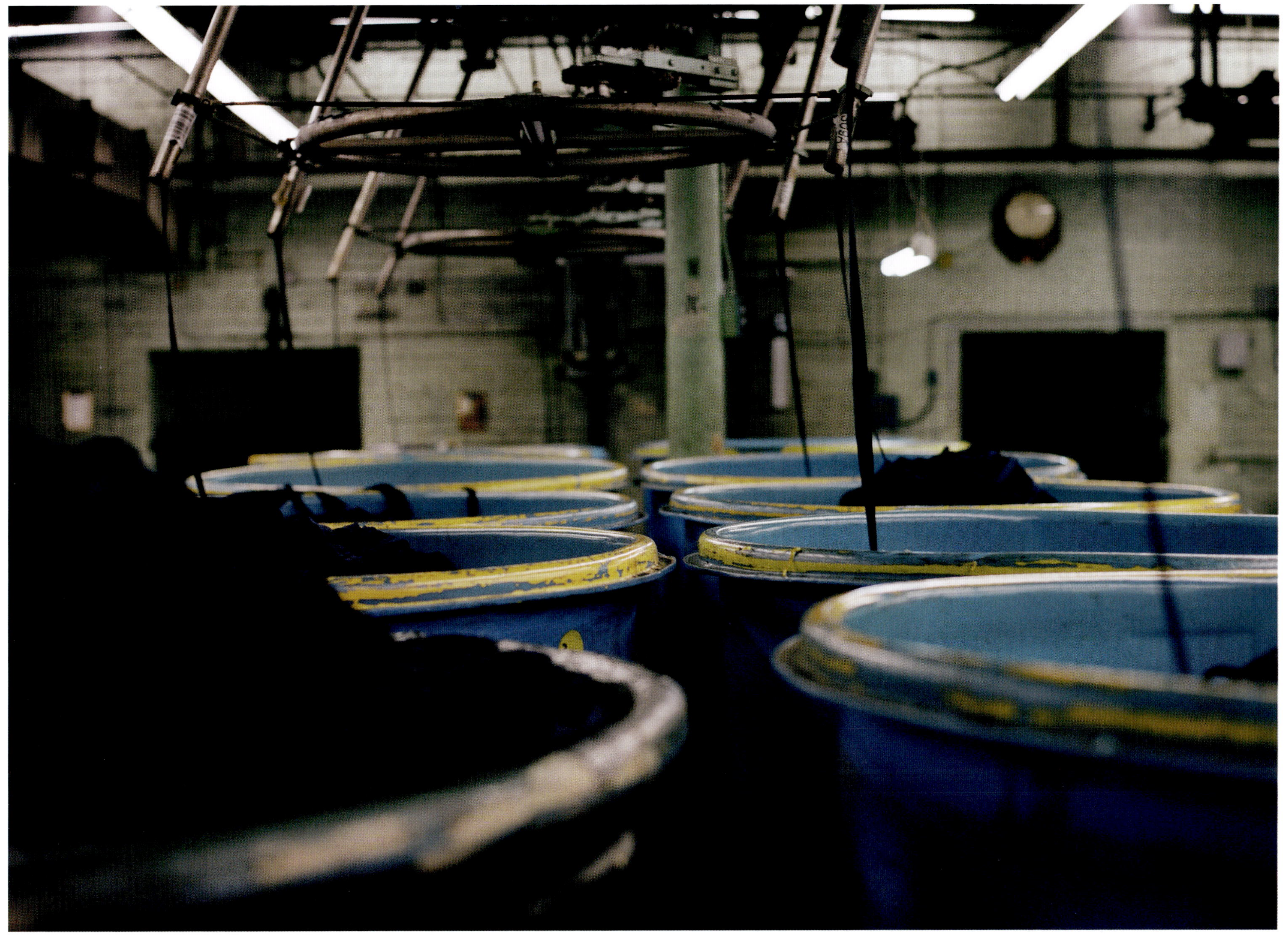

Indigo-dyed balled warp loading into tubs for rebeaming in the Dye House

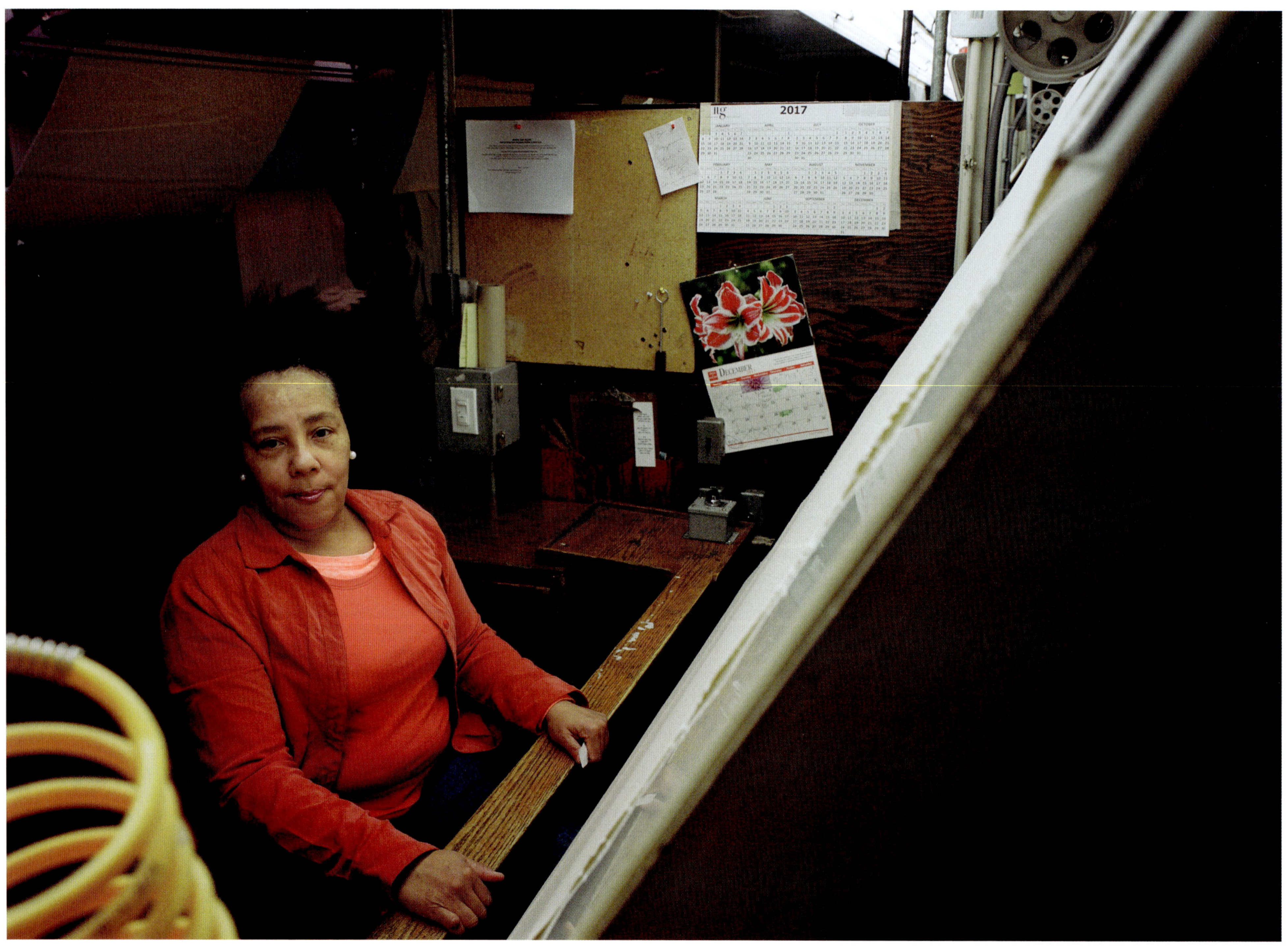

Dawn Whitted, Inspector of Finished Goods, inspecting cloth in the
Quality Control Department

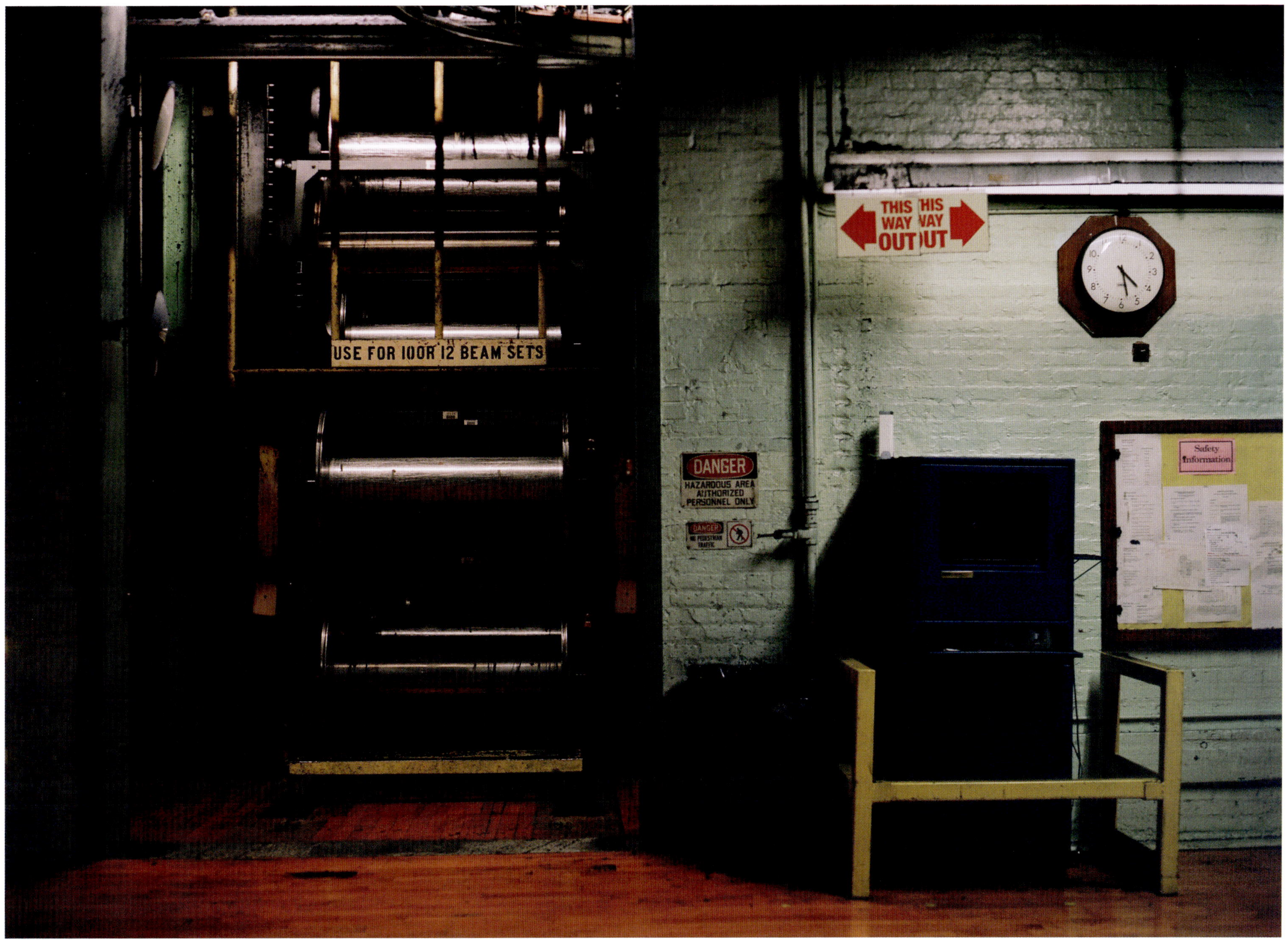
USE FOR 10 OR 12 BEAM SETS
THIS
WAY
OUT
THIS
WAY
OUT
DANGER
HAZARDOUS AREA
AUTHORIZED
PERSONNEL ONLY
DANGER
NO PEDESTRIAN
TRAFFIC
Safety
Information

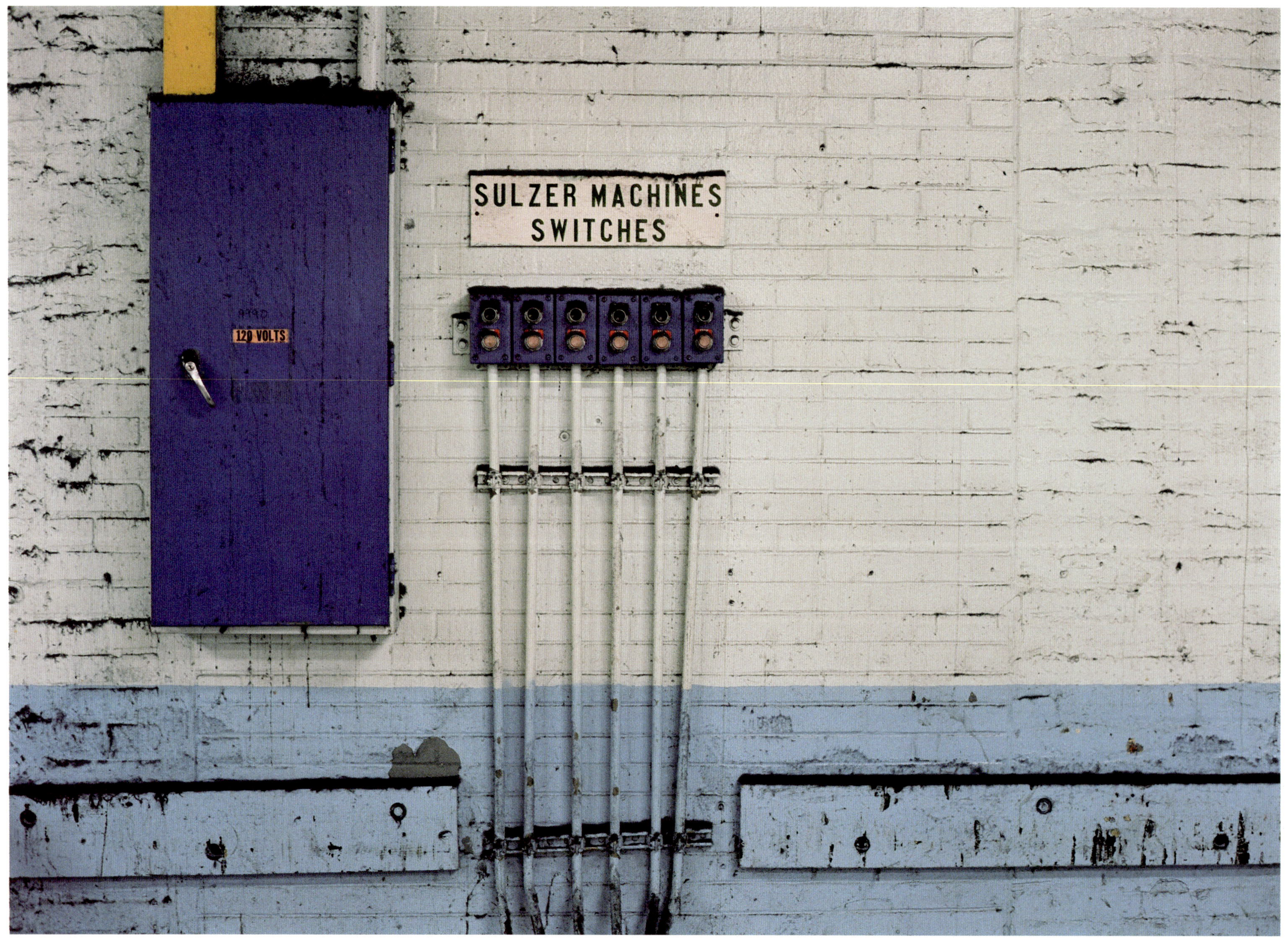

48

Electrical controls in the Weave Room // Freshly dyed indigo balled yarns in the Dye House

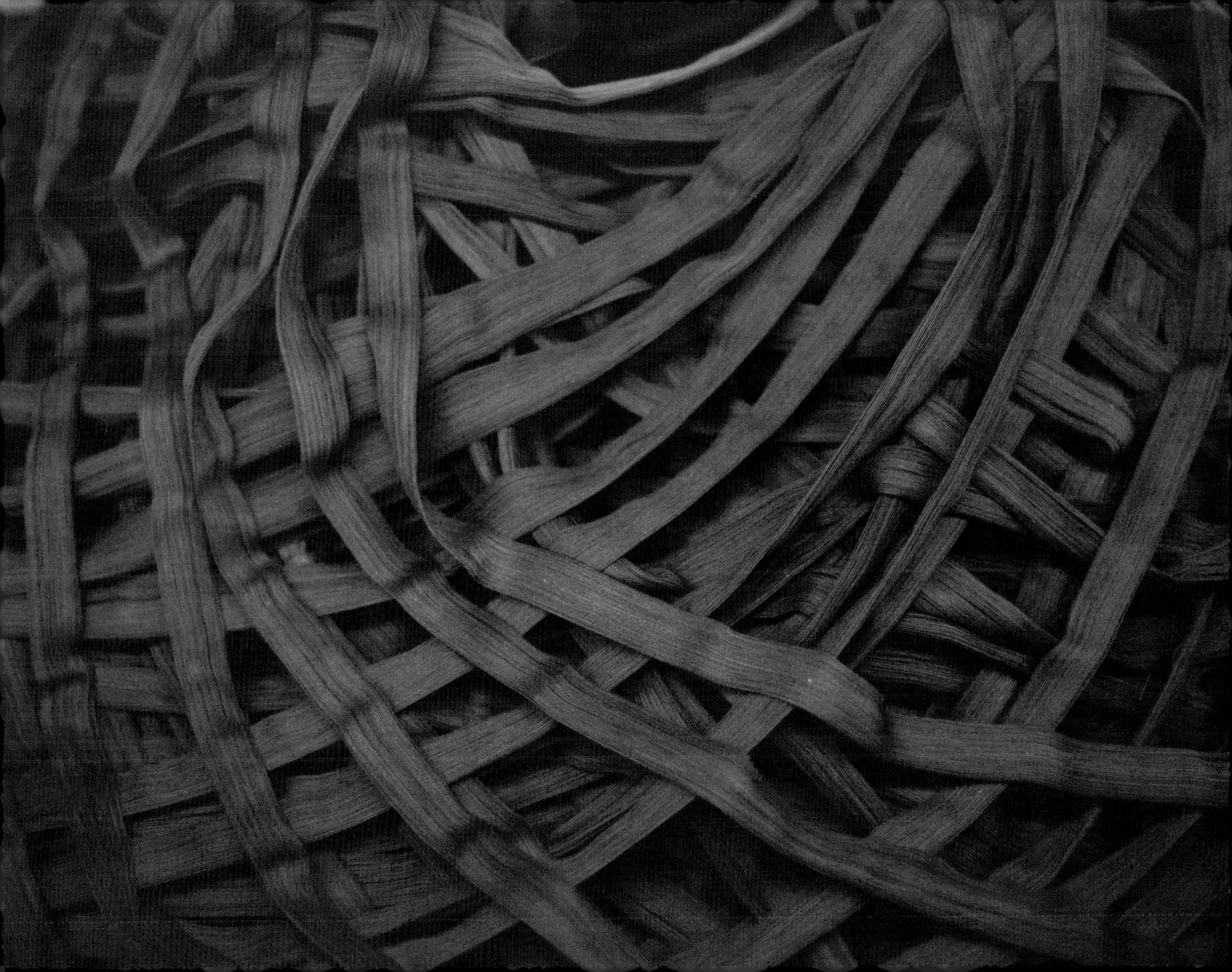

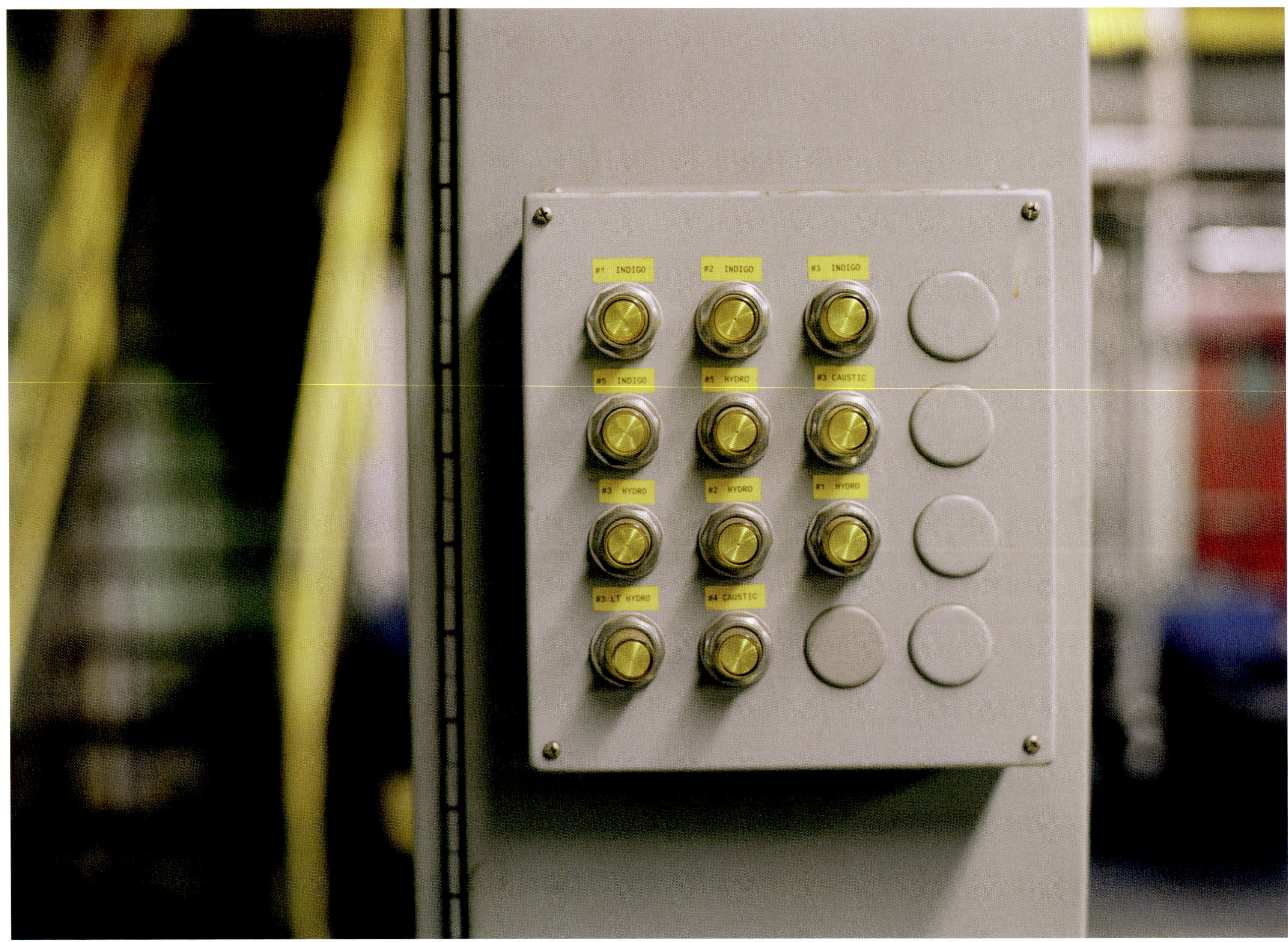

Dye range control panel in the Dye House

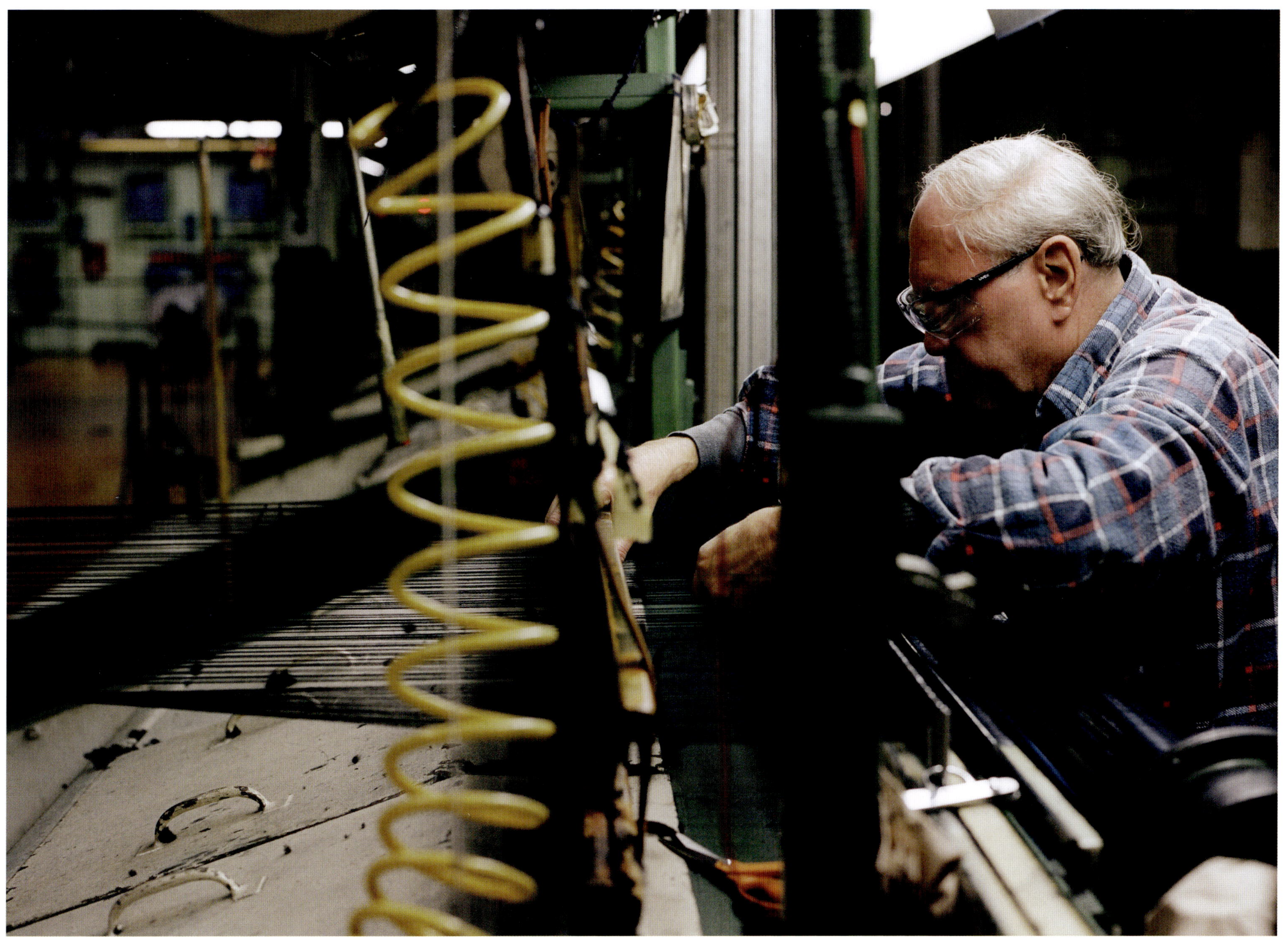

Jake Miller, Lead Beamer, working a rebeaming frame in the Beaming Department

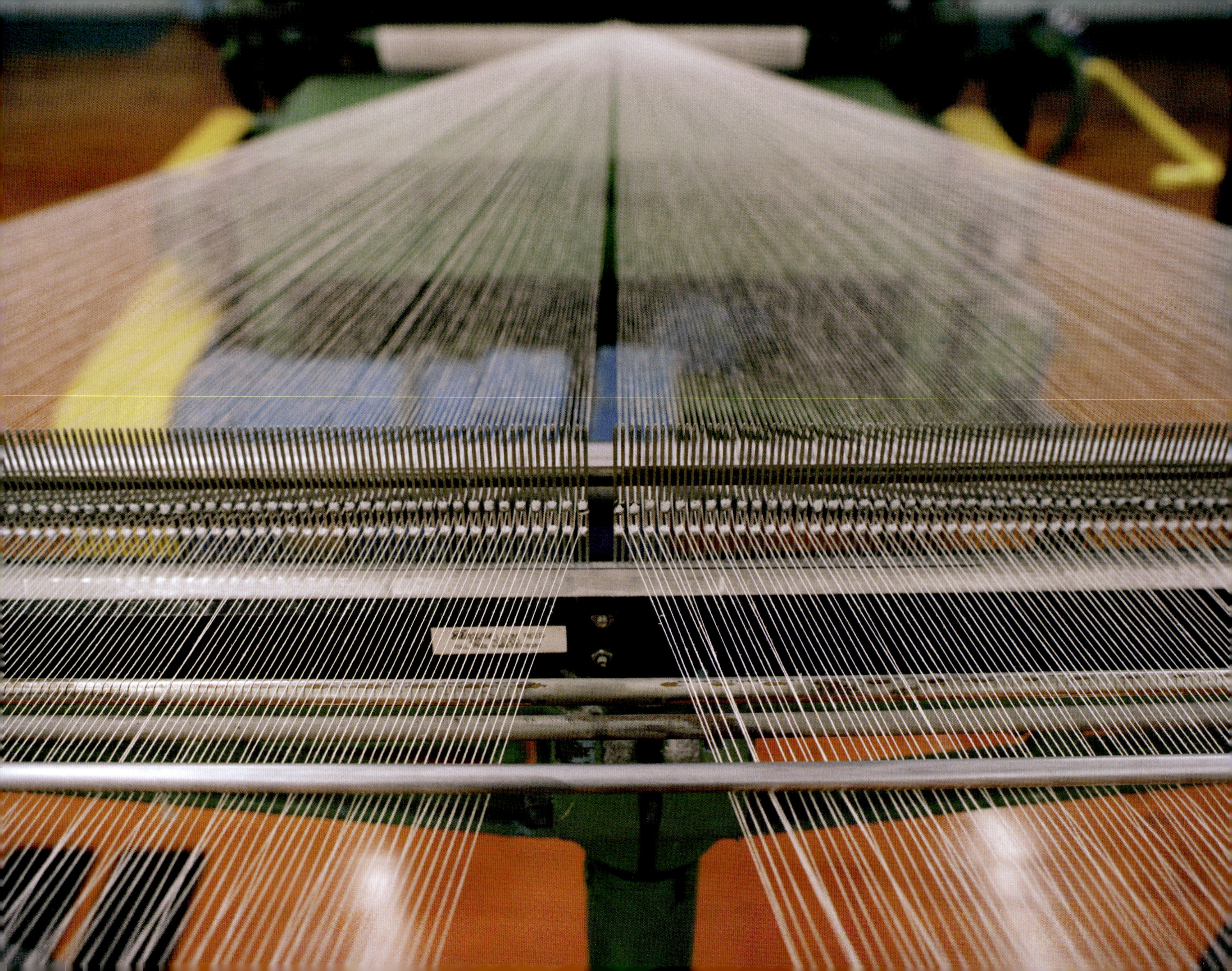

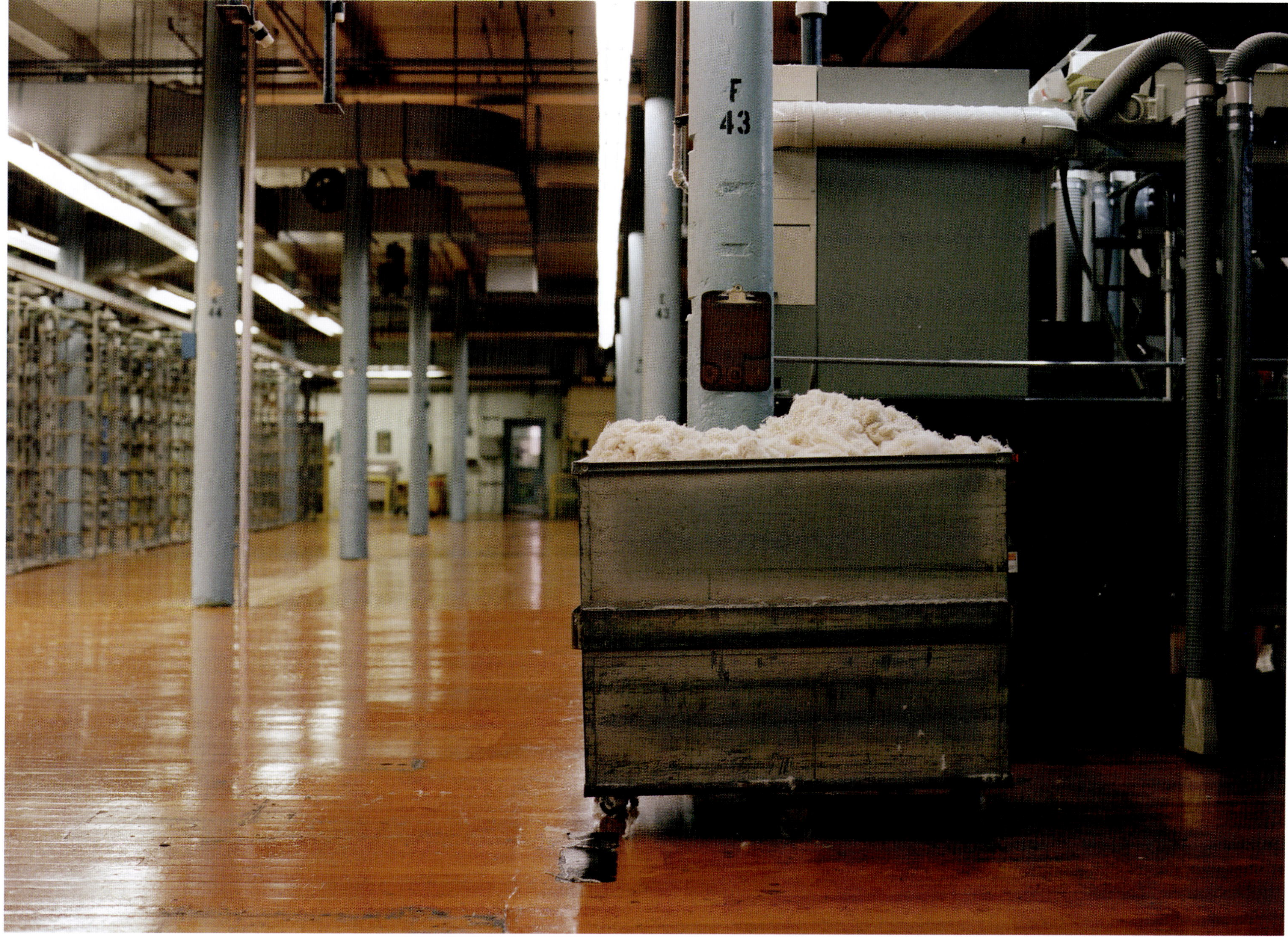

Ball warping in the Warping Department   //   Cotton scrap bin in the Warping Department

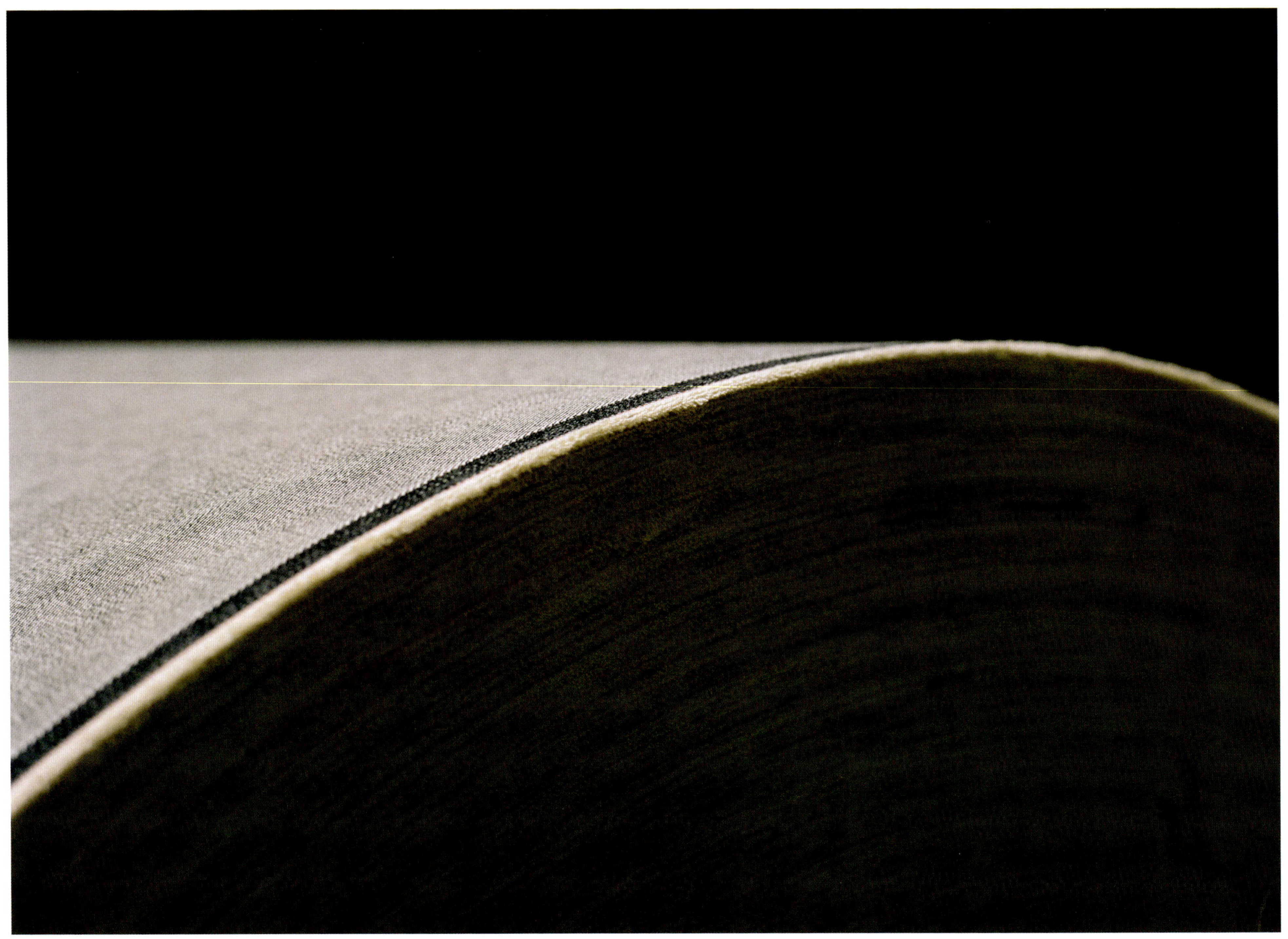

54

3 x 1 denim on large roll

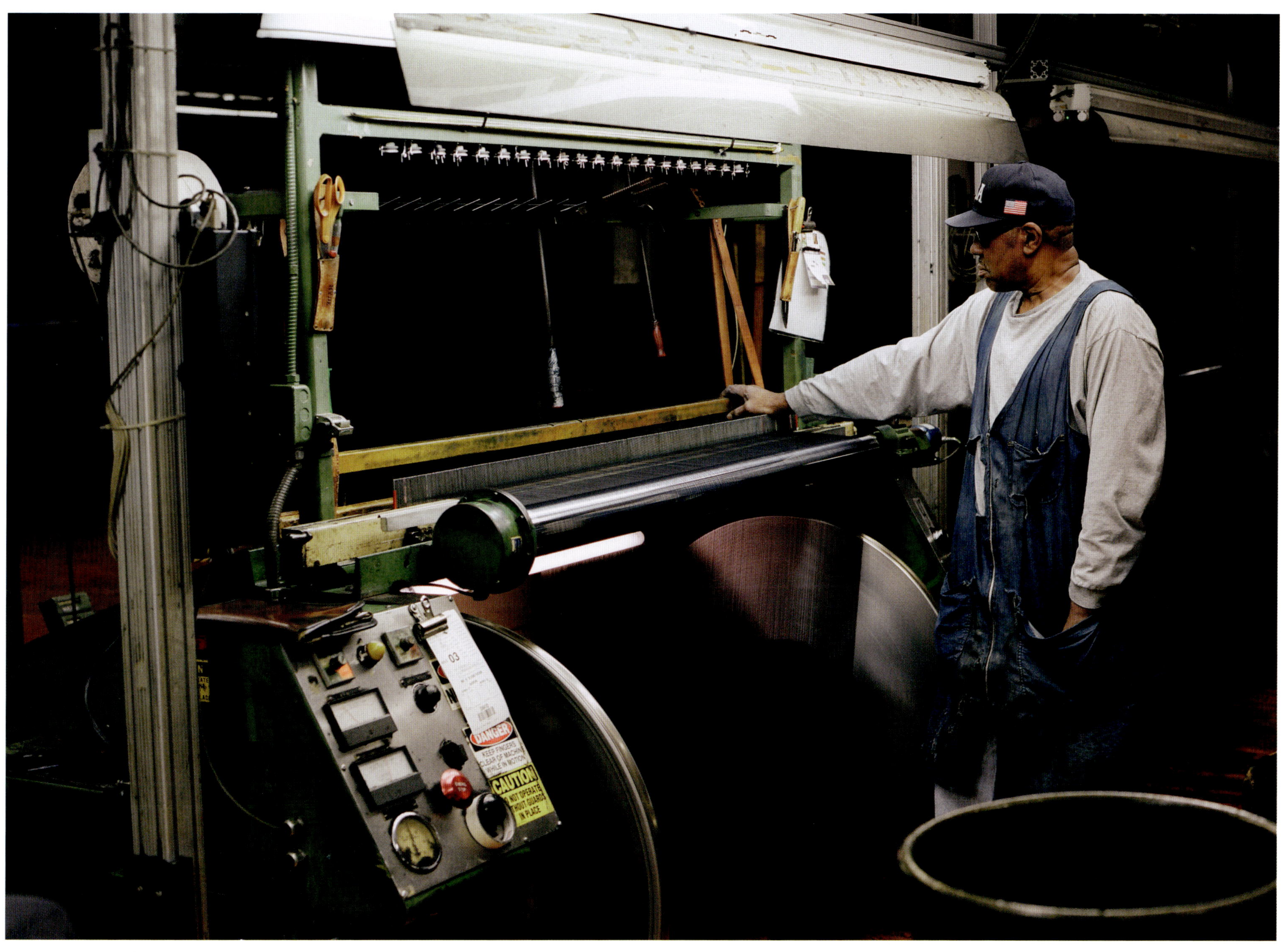

Bobby Hamilton, Beamer Tender, tending a rebeaming frame in the Beaming Department

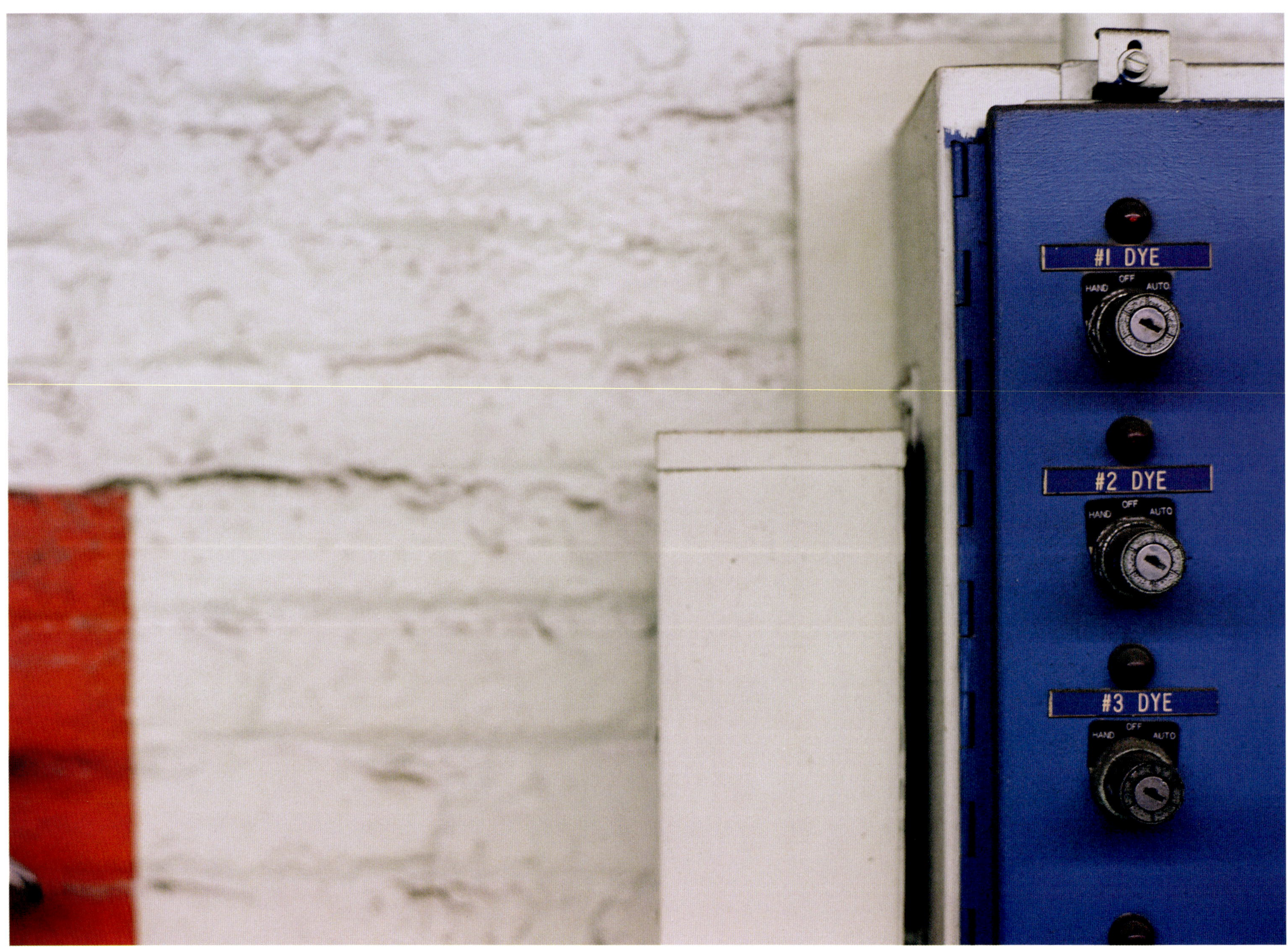

Control panel in the Dye House

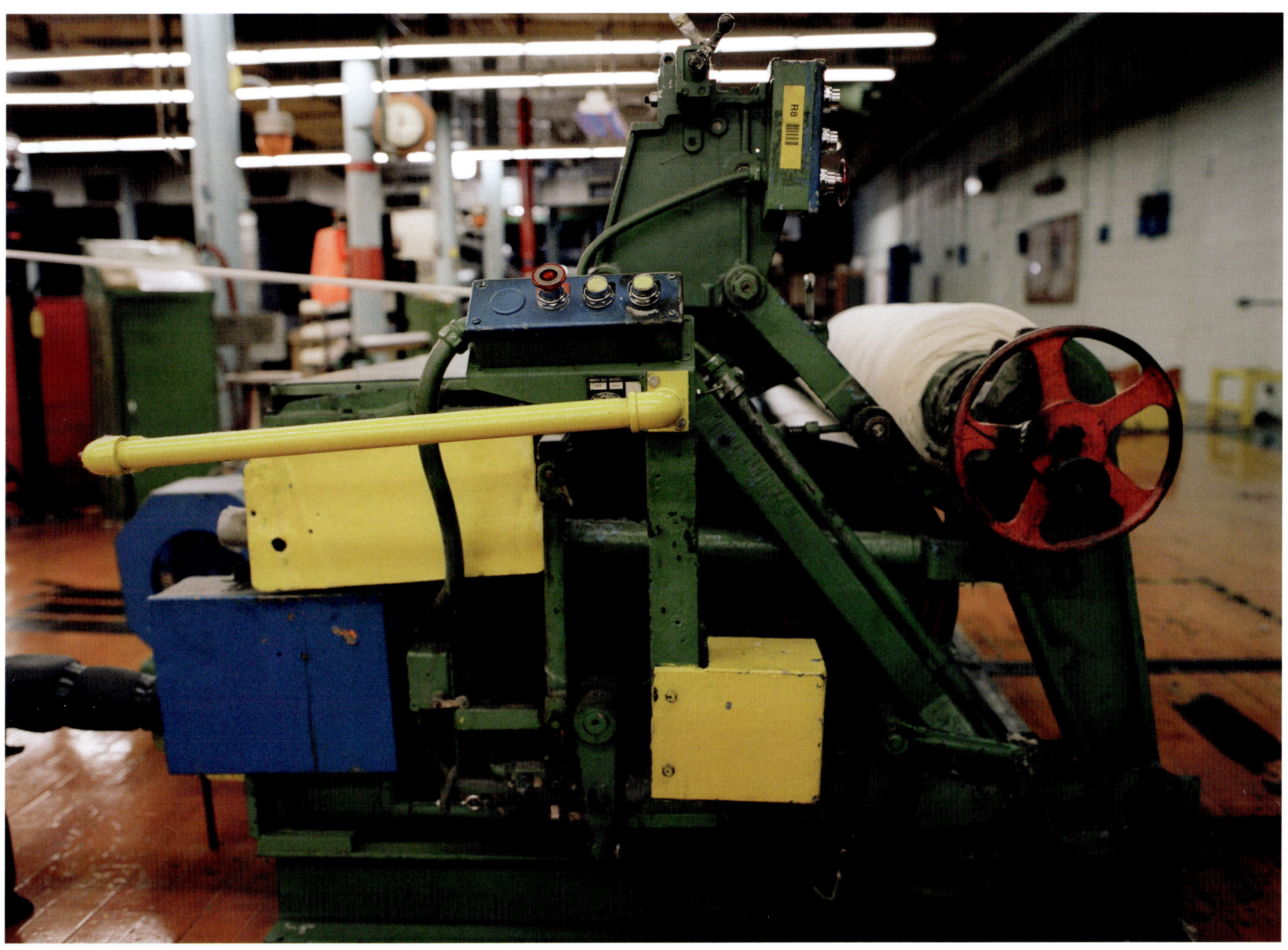

Ball Warper in the Warping Department

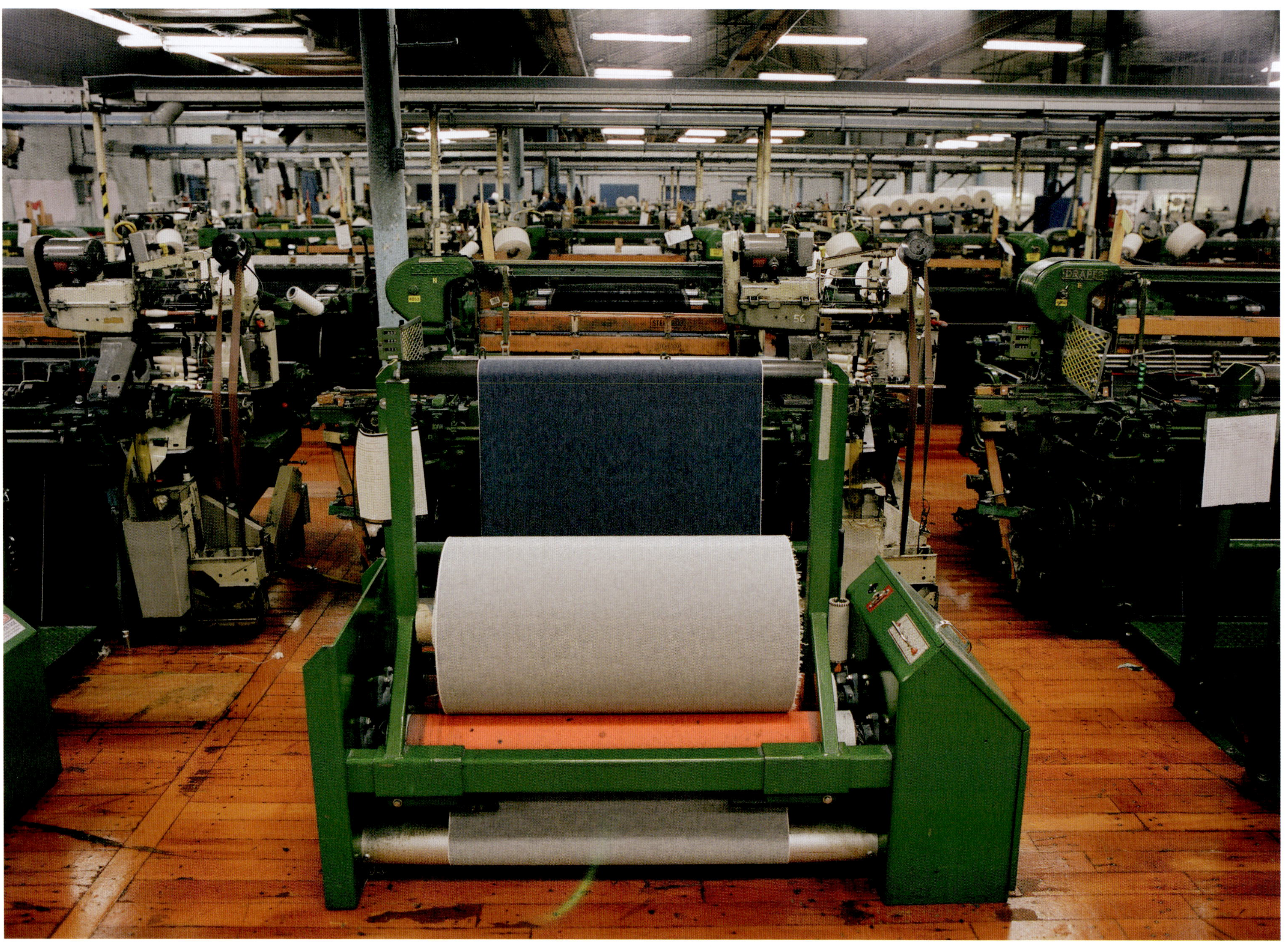

A veiw of the X-3 weave floor in the Weave Room // David "Rabbit" George, Fixer, in a weaver's alley on the X-3 weave floor in the Weave Room

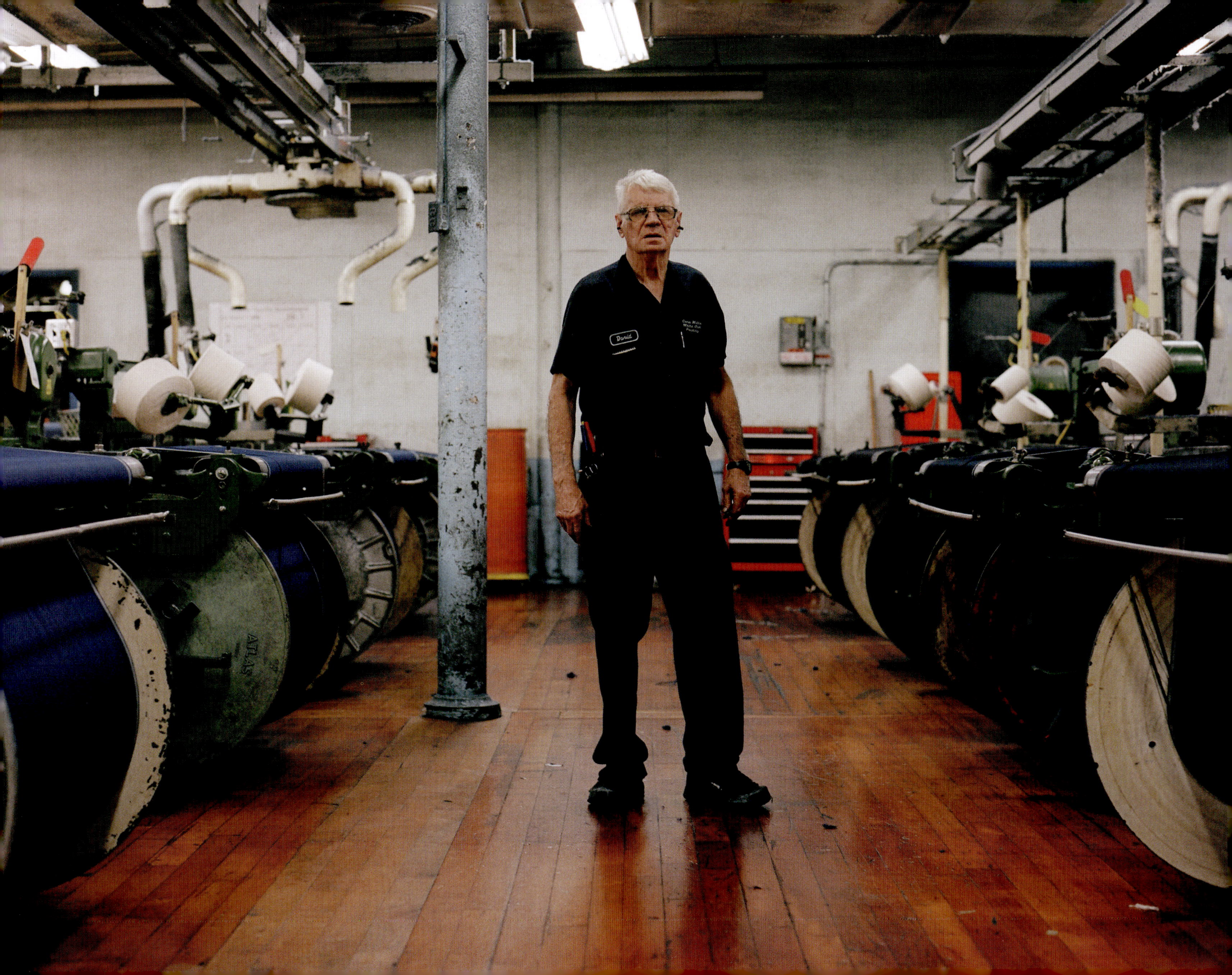

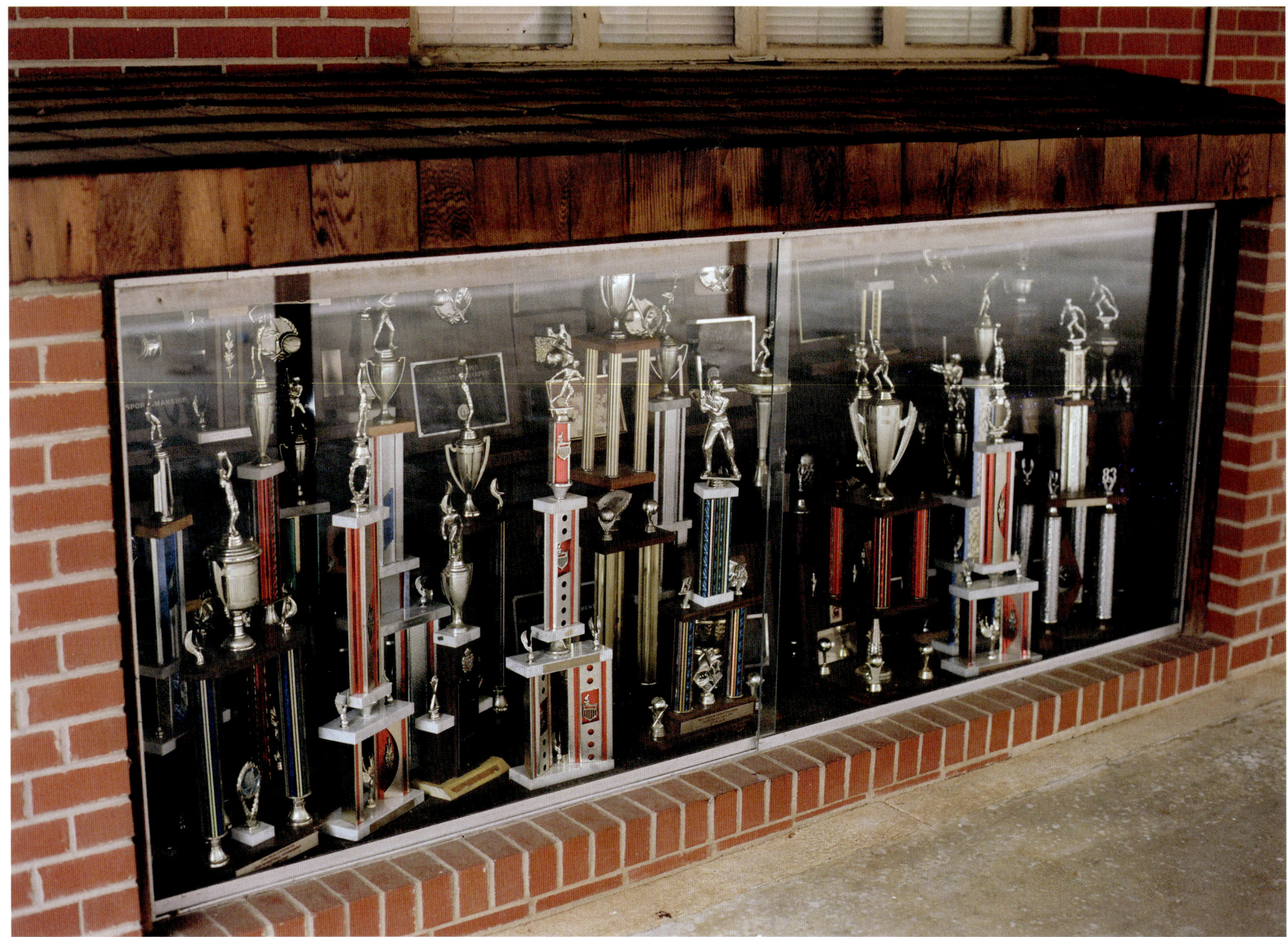

Intramural trophy case at the Front Office

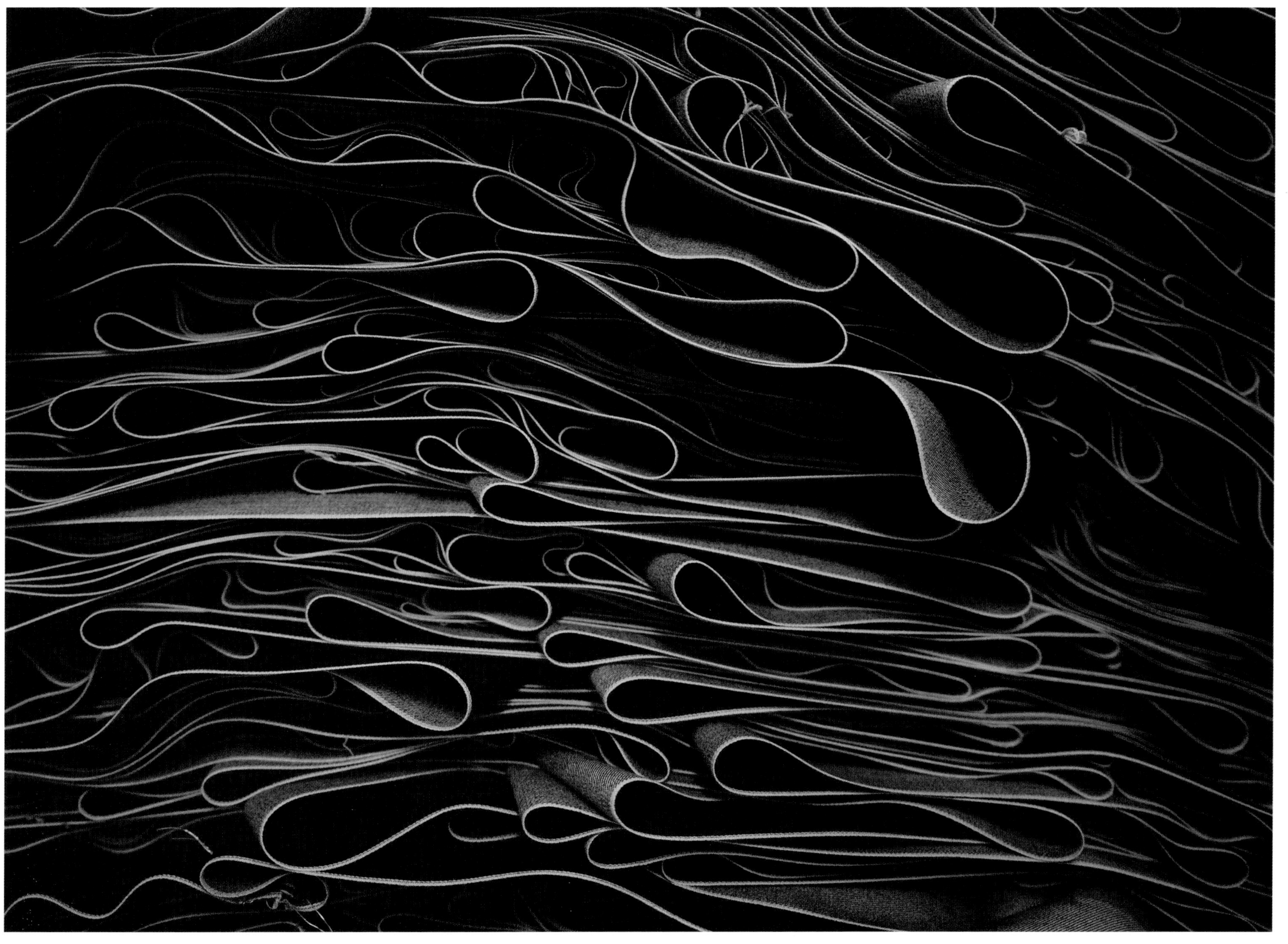

Freshly finished denim fabric being rolled up in the Finishing Department

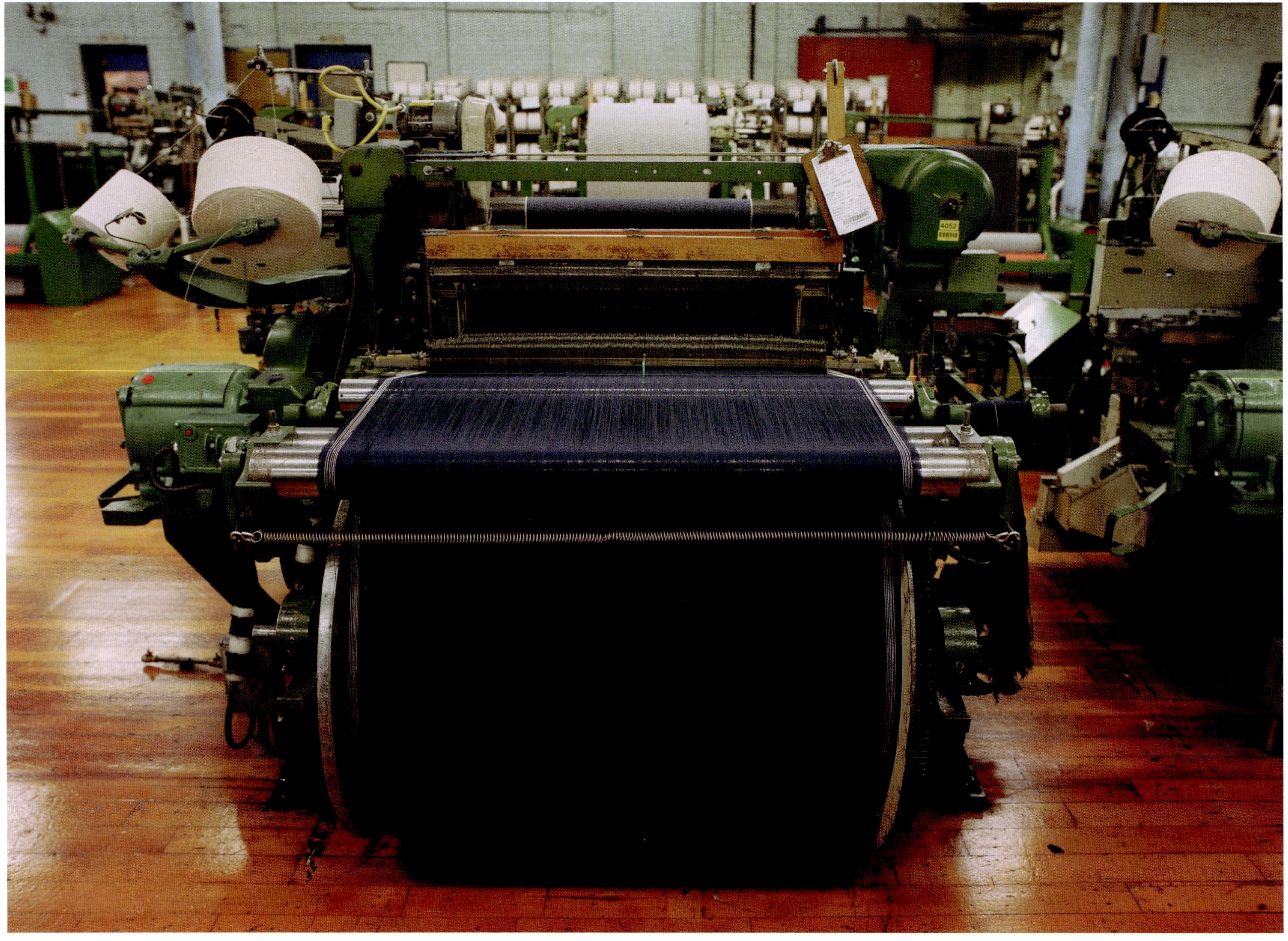

Draper X-3 shuttle loom in the Weave Room // Deborah Graves, Weaver, on the X-3 weave floor in the Weave Room

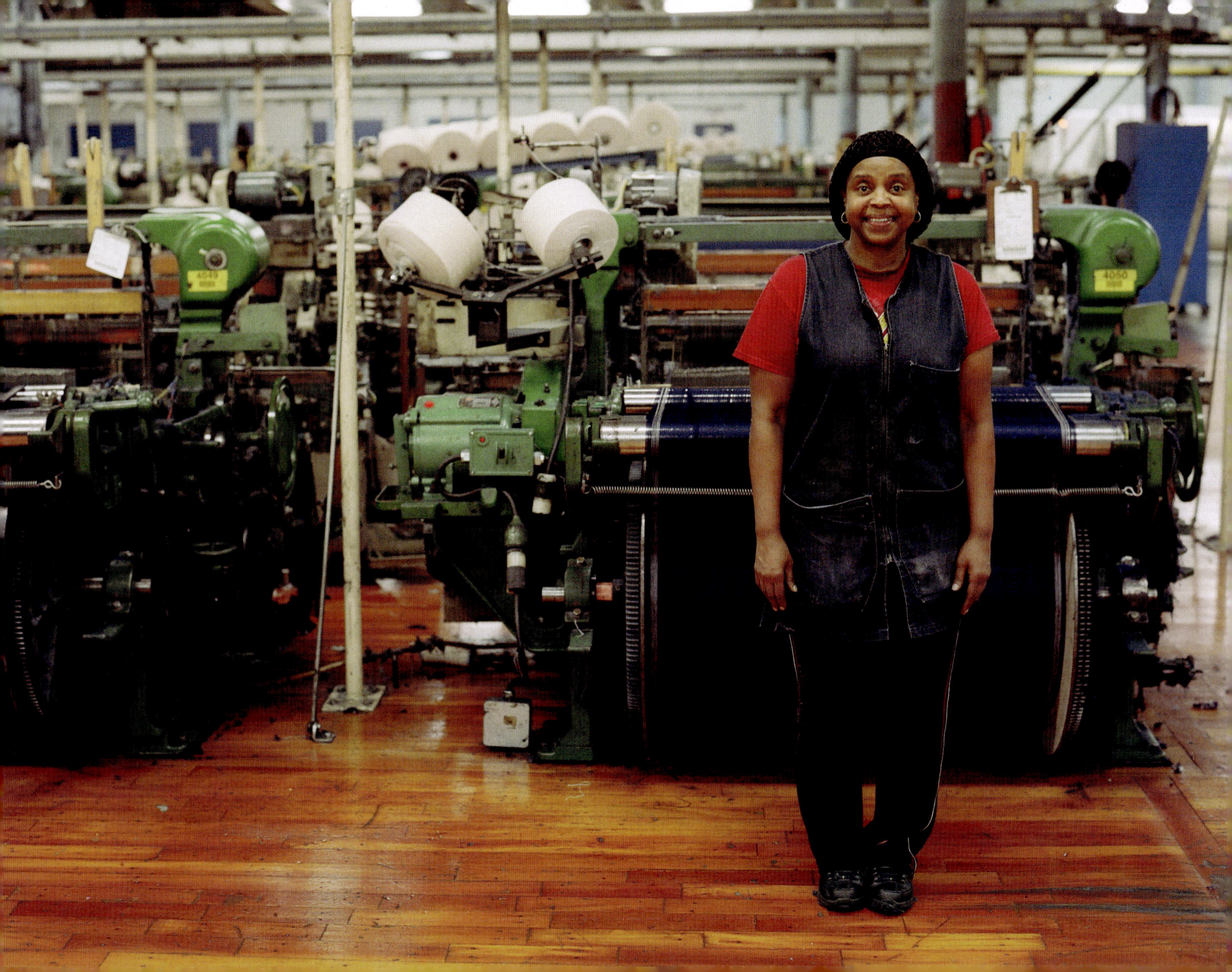

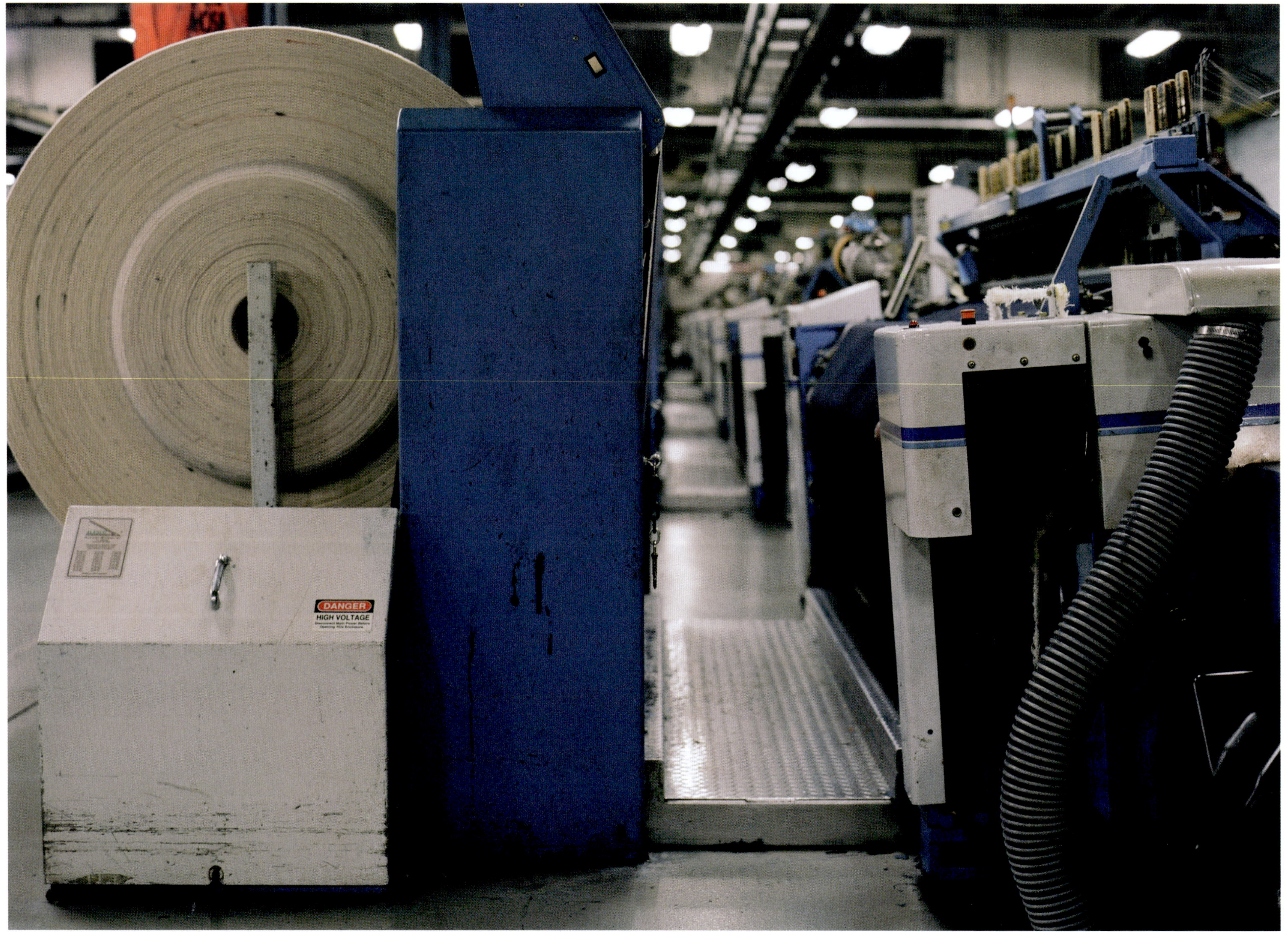

Modern air jet loom in the Weaving Department

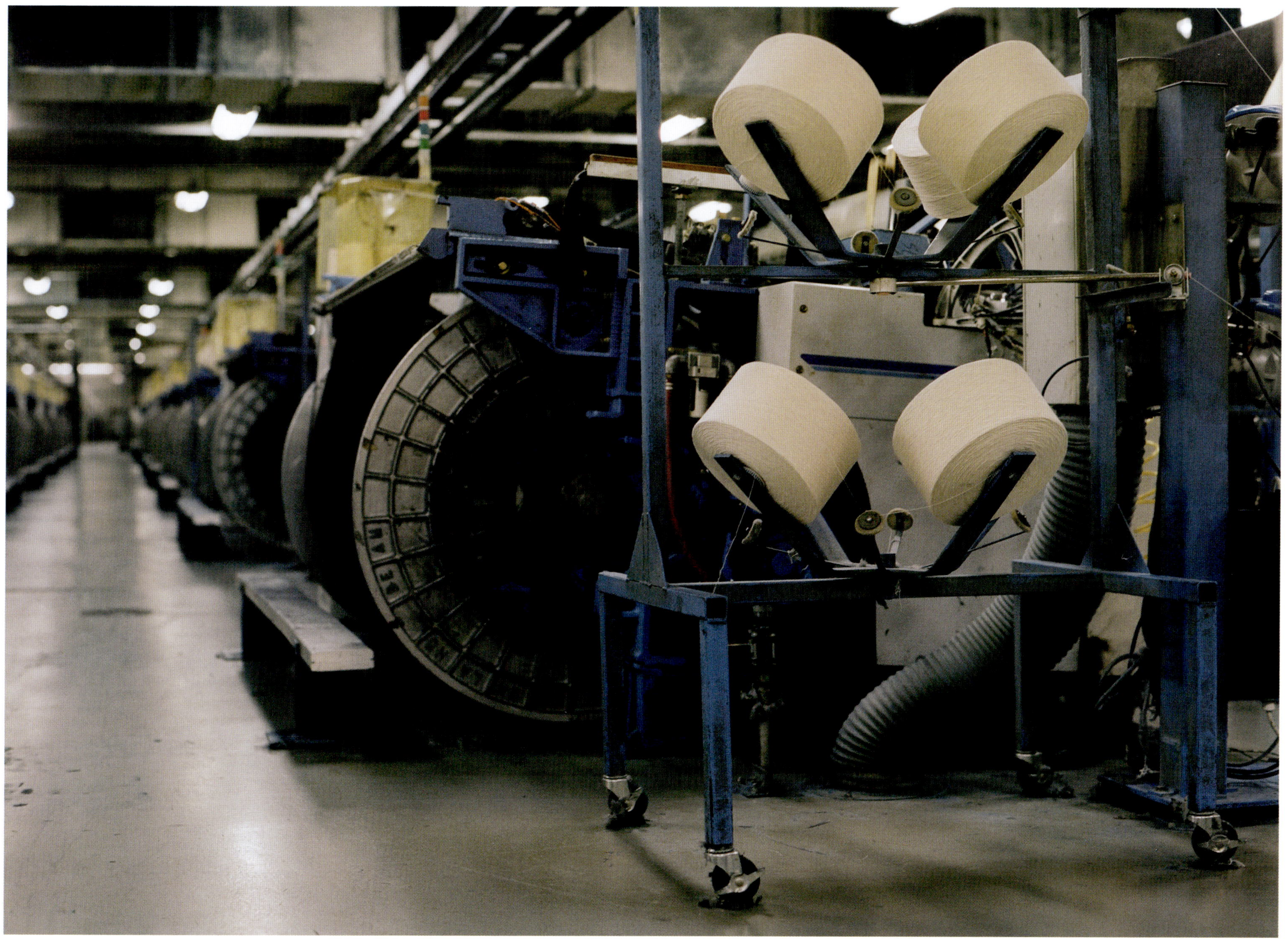

Filling yarn beside a modern loom in the Weaving Department

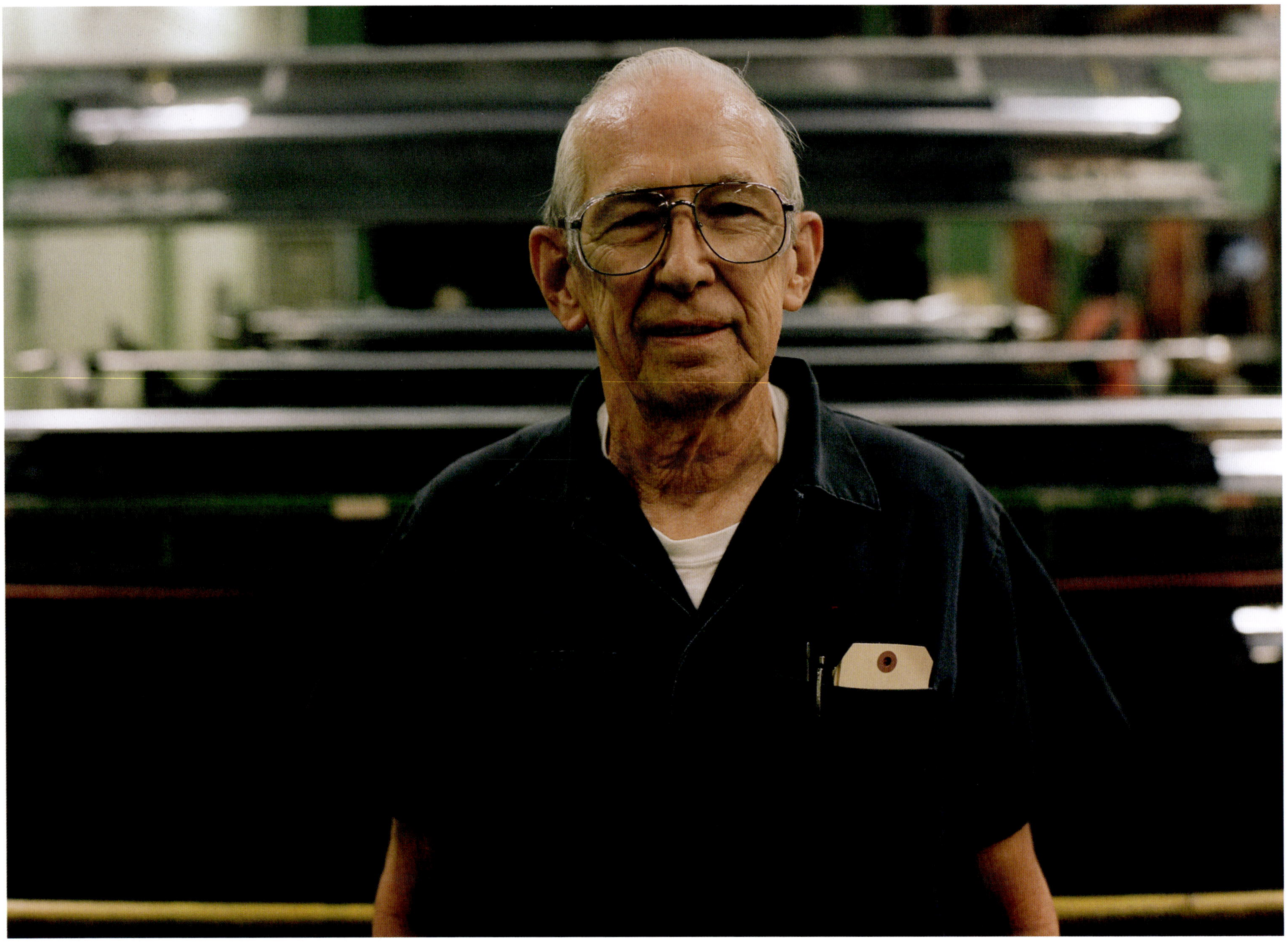

Elbert "Frank" Williams, Slasher Fixer in the Slashing Department.
Employeed at White Oak for sixty-two years.

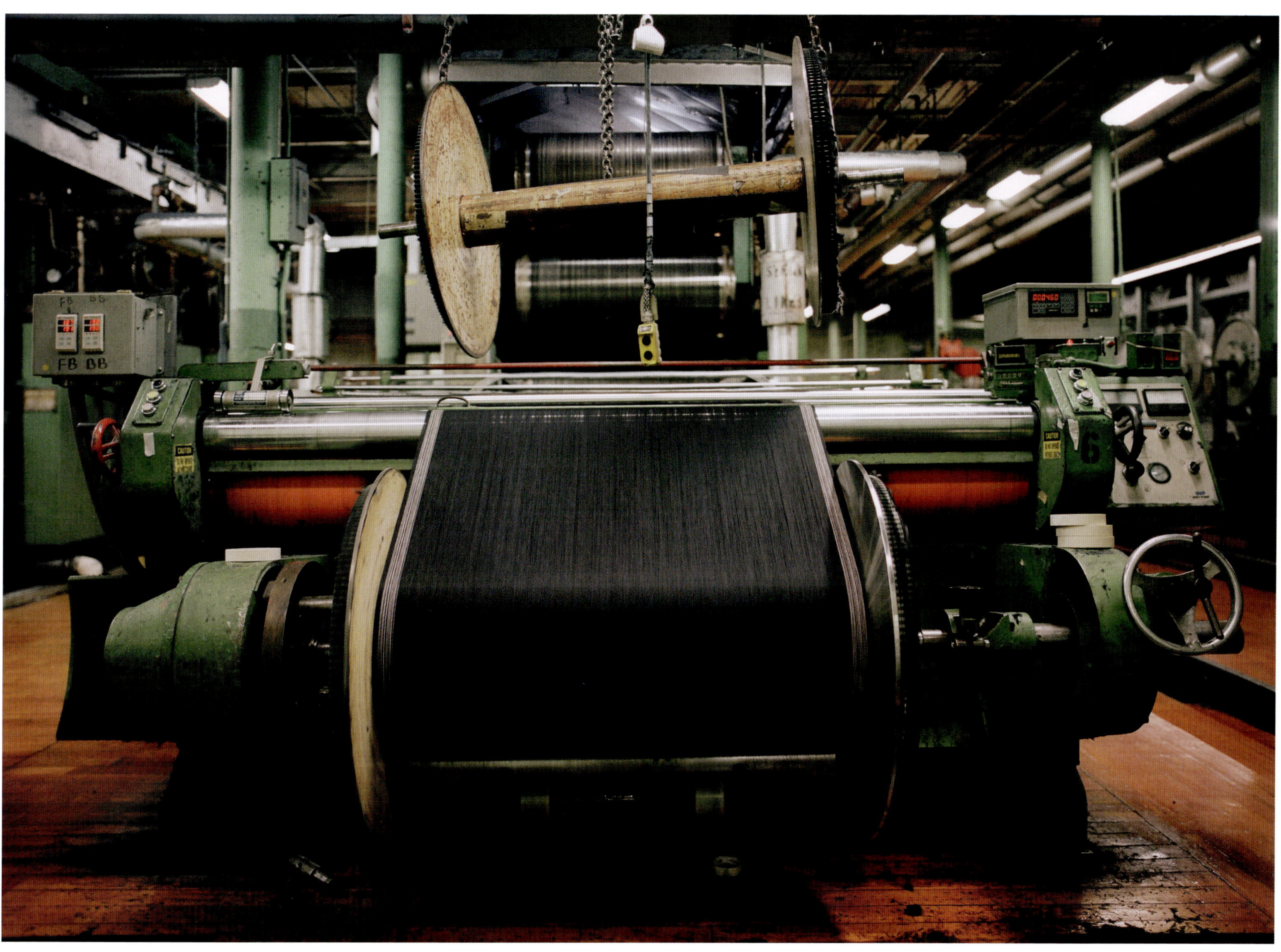

Winding warp yarns onto a loom beam for an X-3 in the Slashing Department

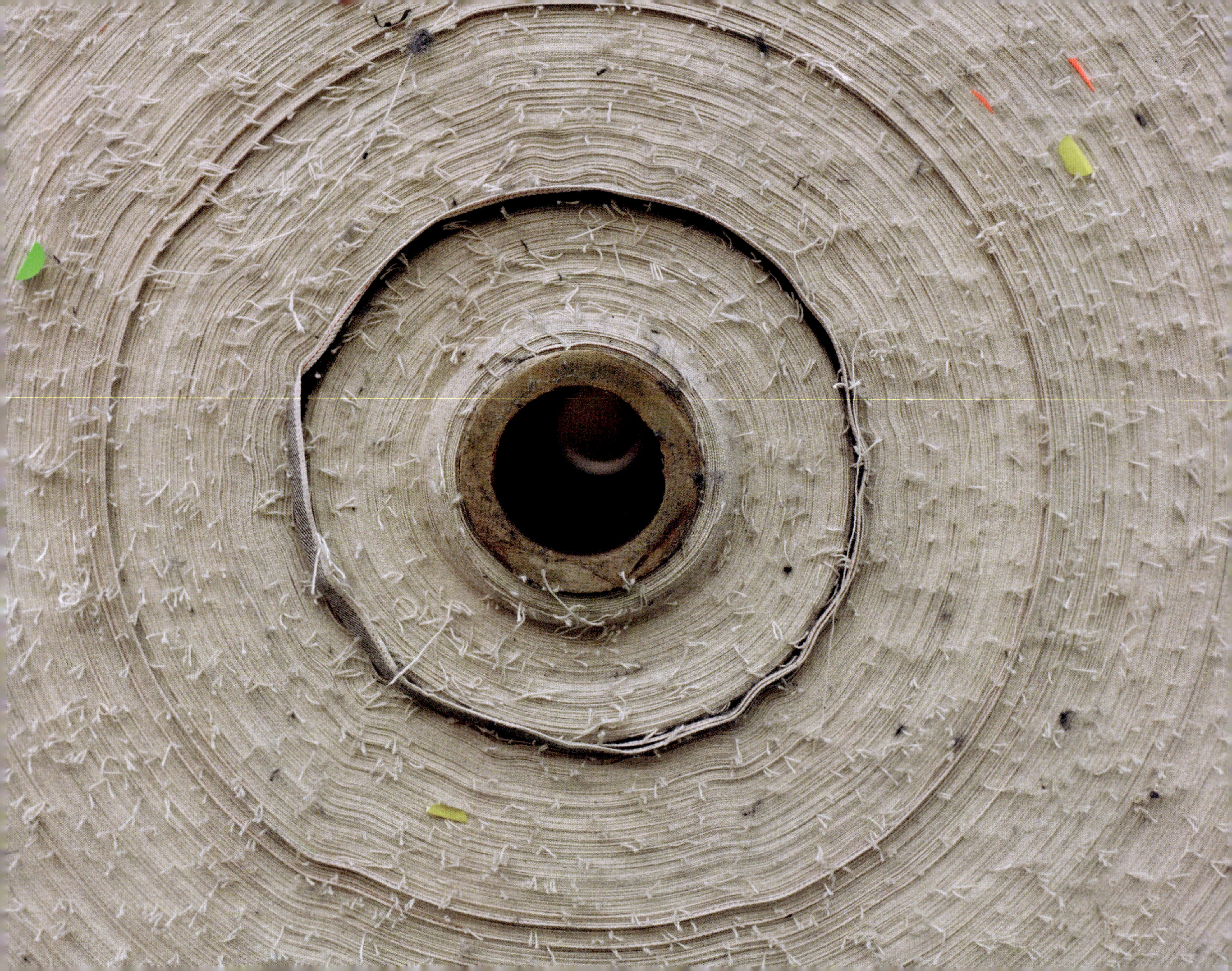

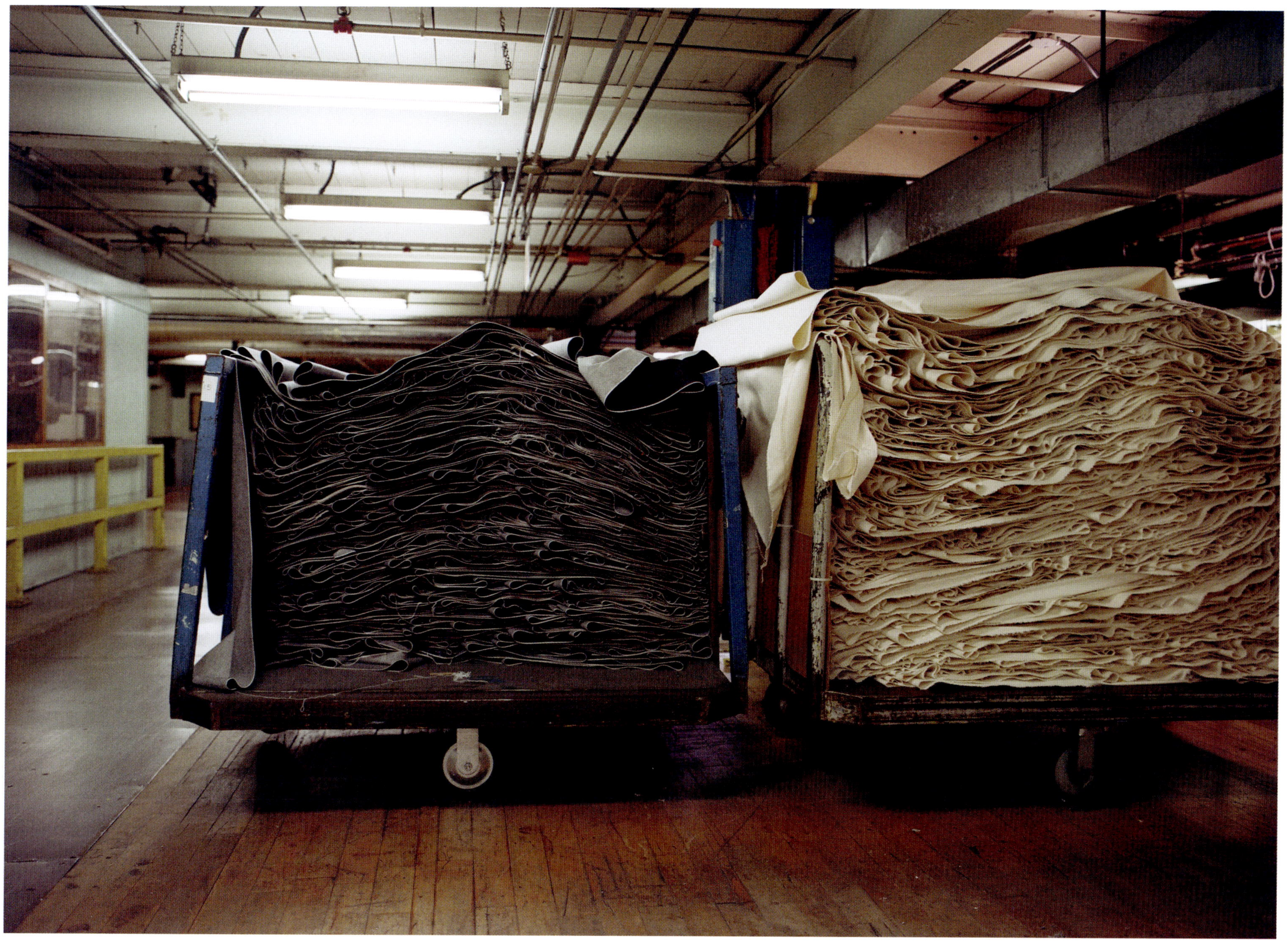

Selvage denim on large roll // Denim in the finishing process on carts in the Finishing Department

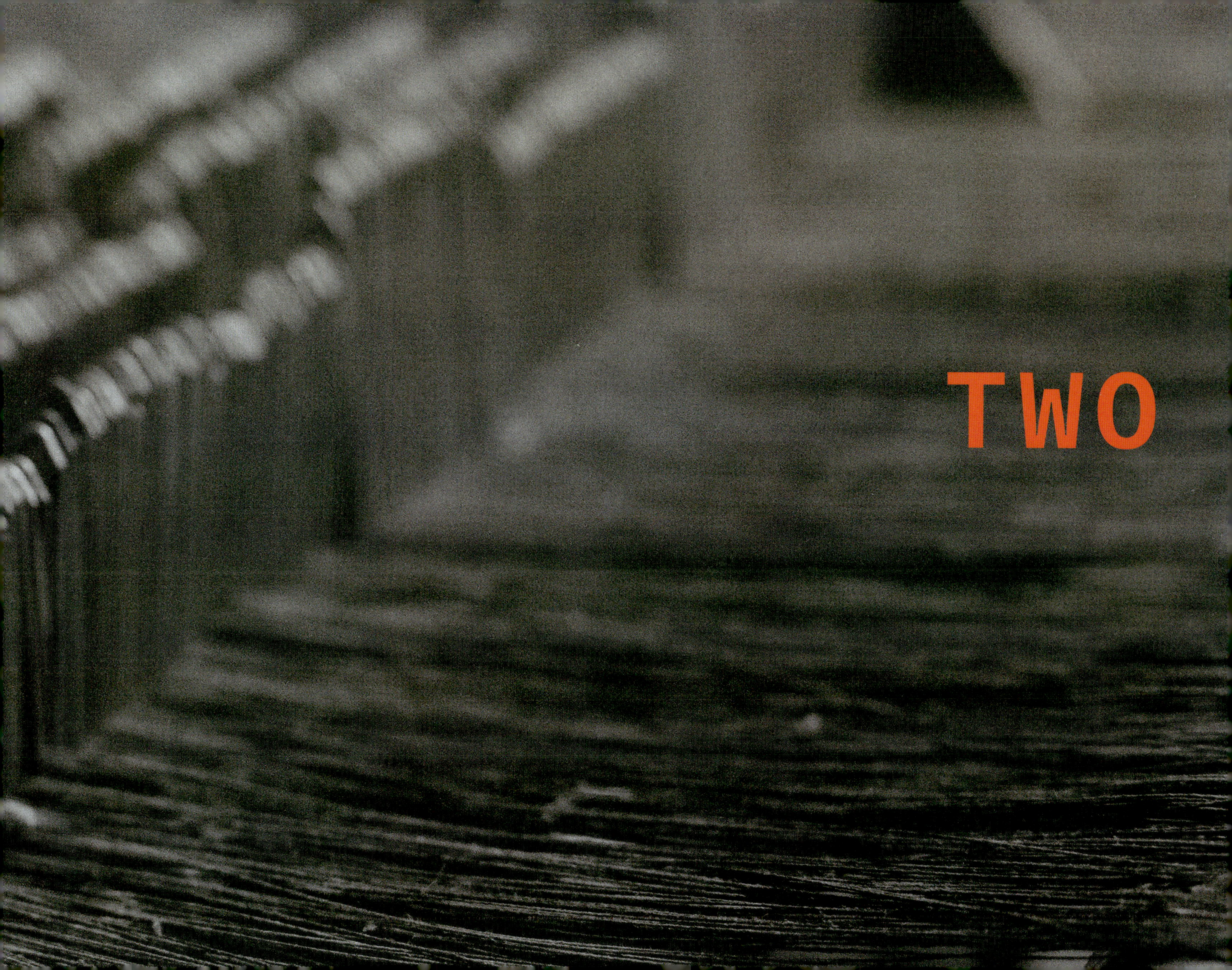

TWO

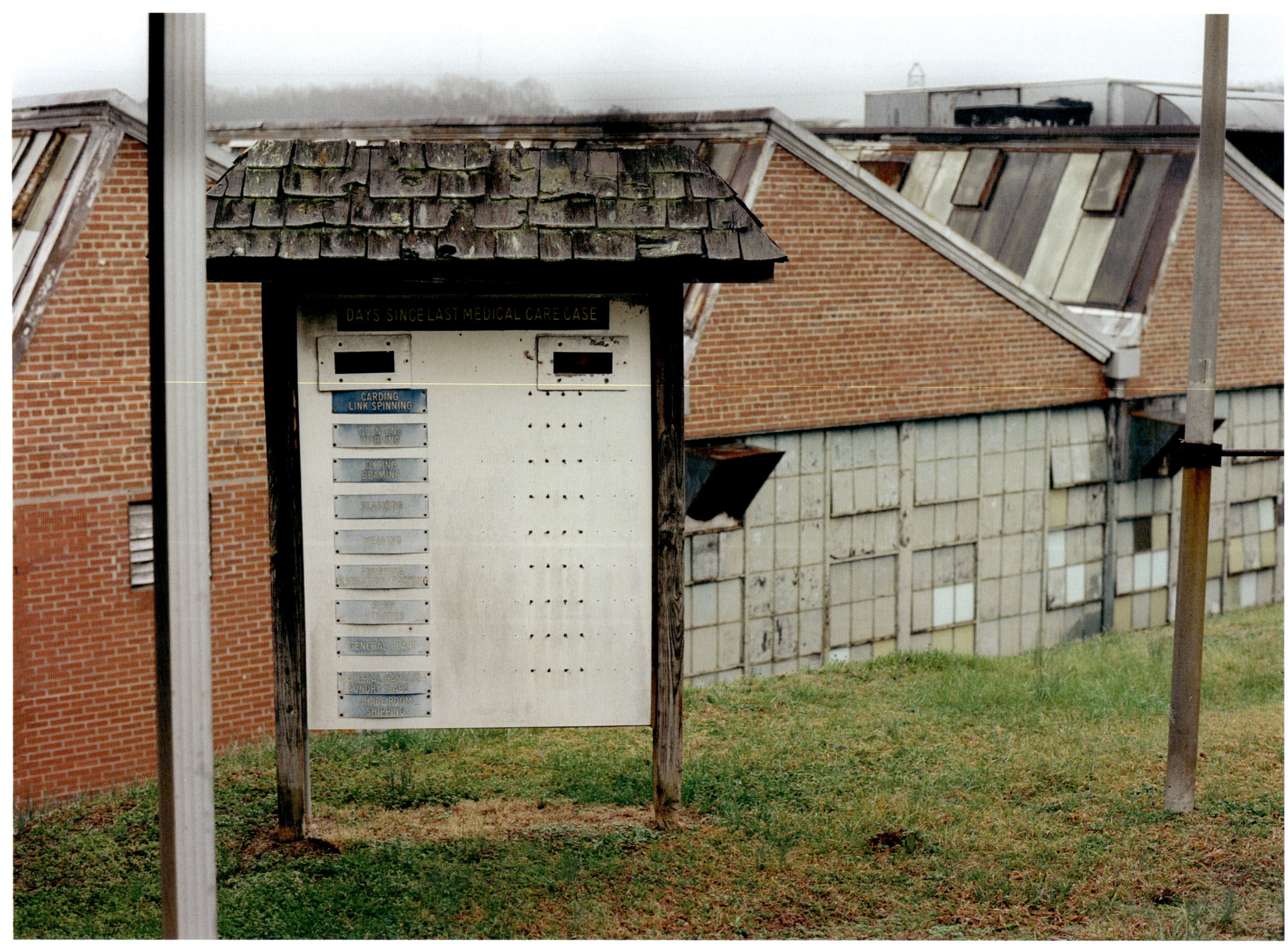

Signage near trucking parking lot overlooking the sawtooth Weave Room

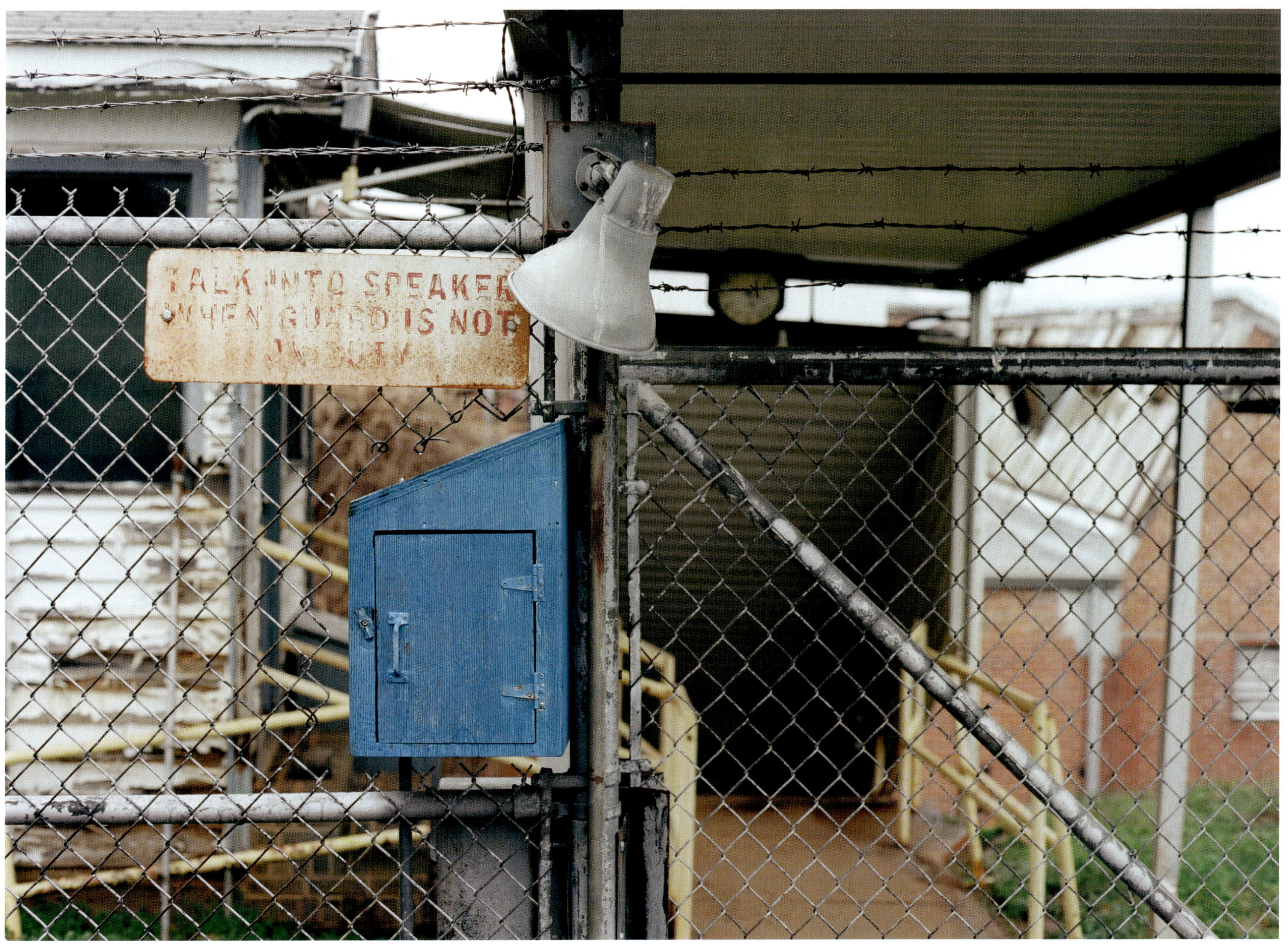
TALK INTO SPEAKER
WHEN GUARD IS NOT
ON DUTY

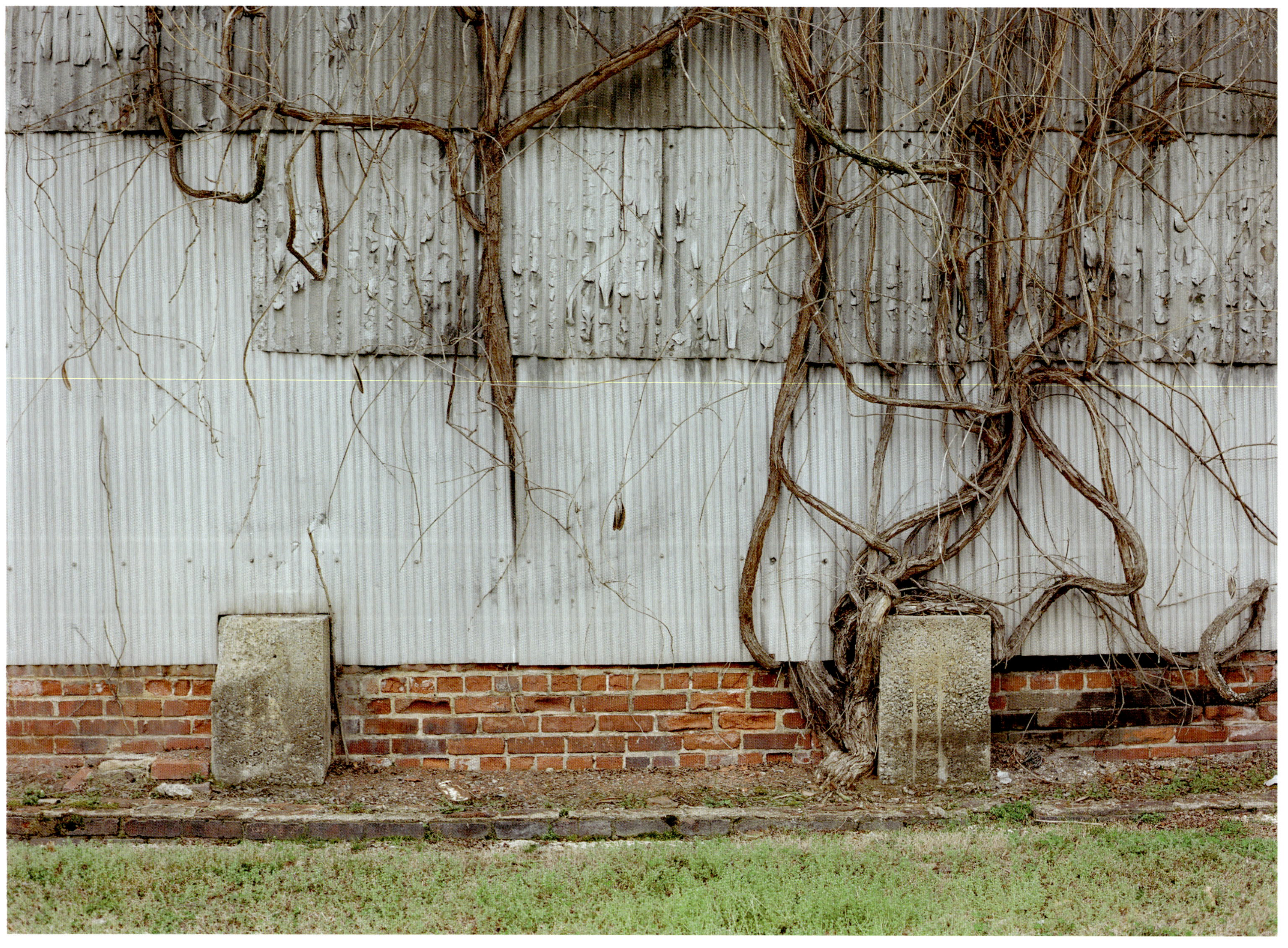

Vines growing alongside the Powerhouse

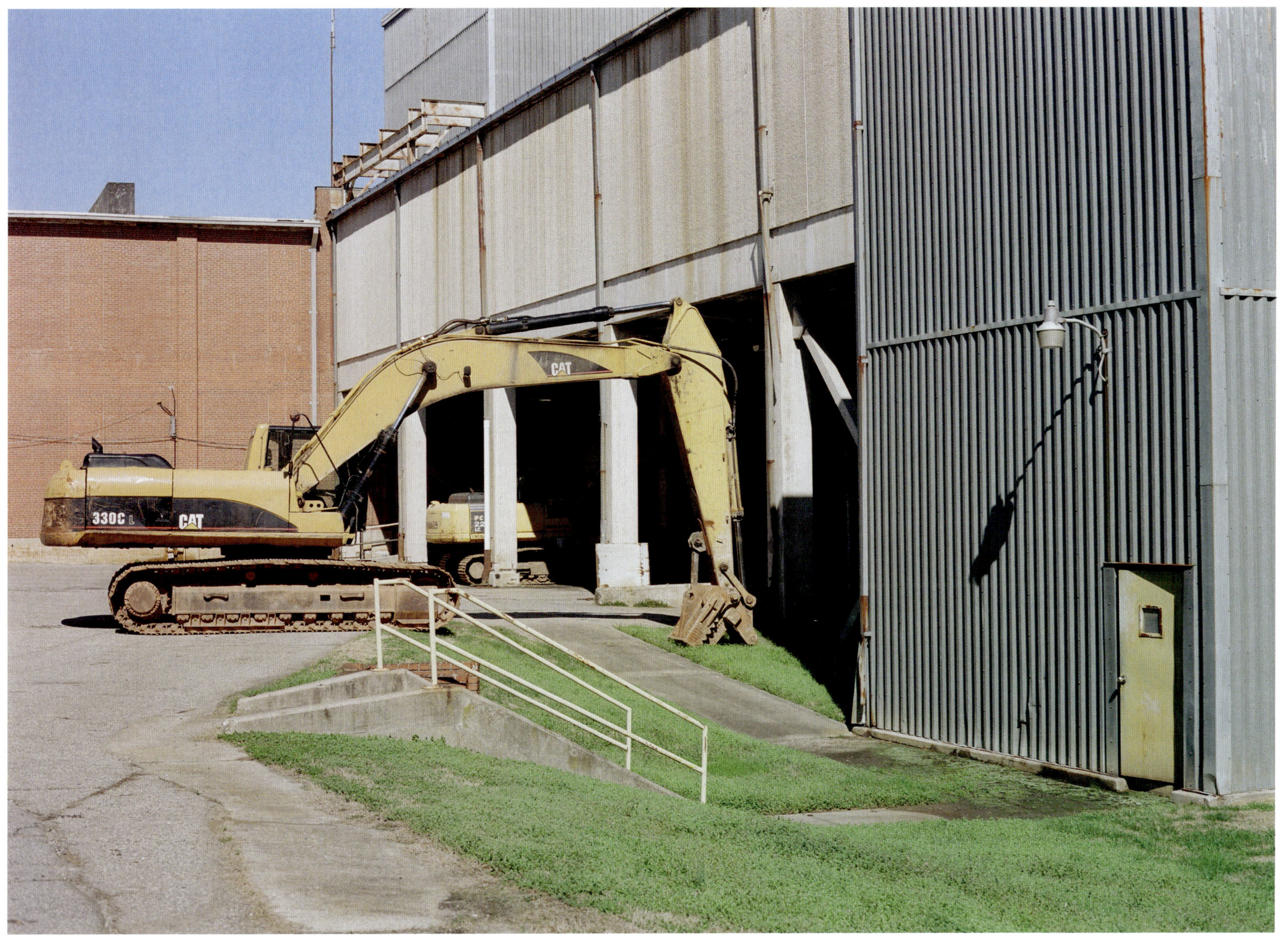

J.W. Demolition vehicles parked at the Weave Room, preparing to begin demolition

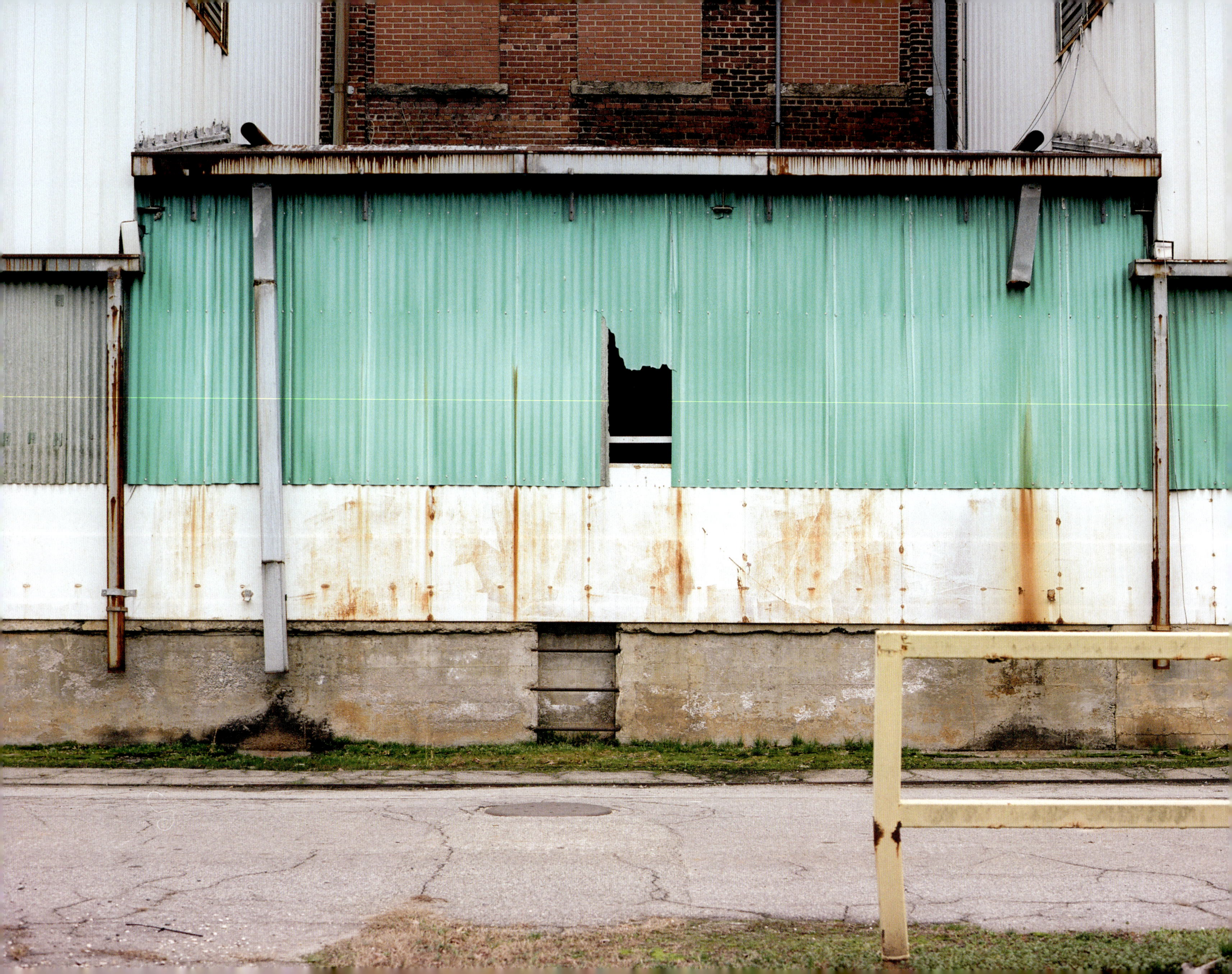

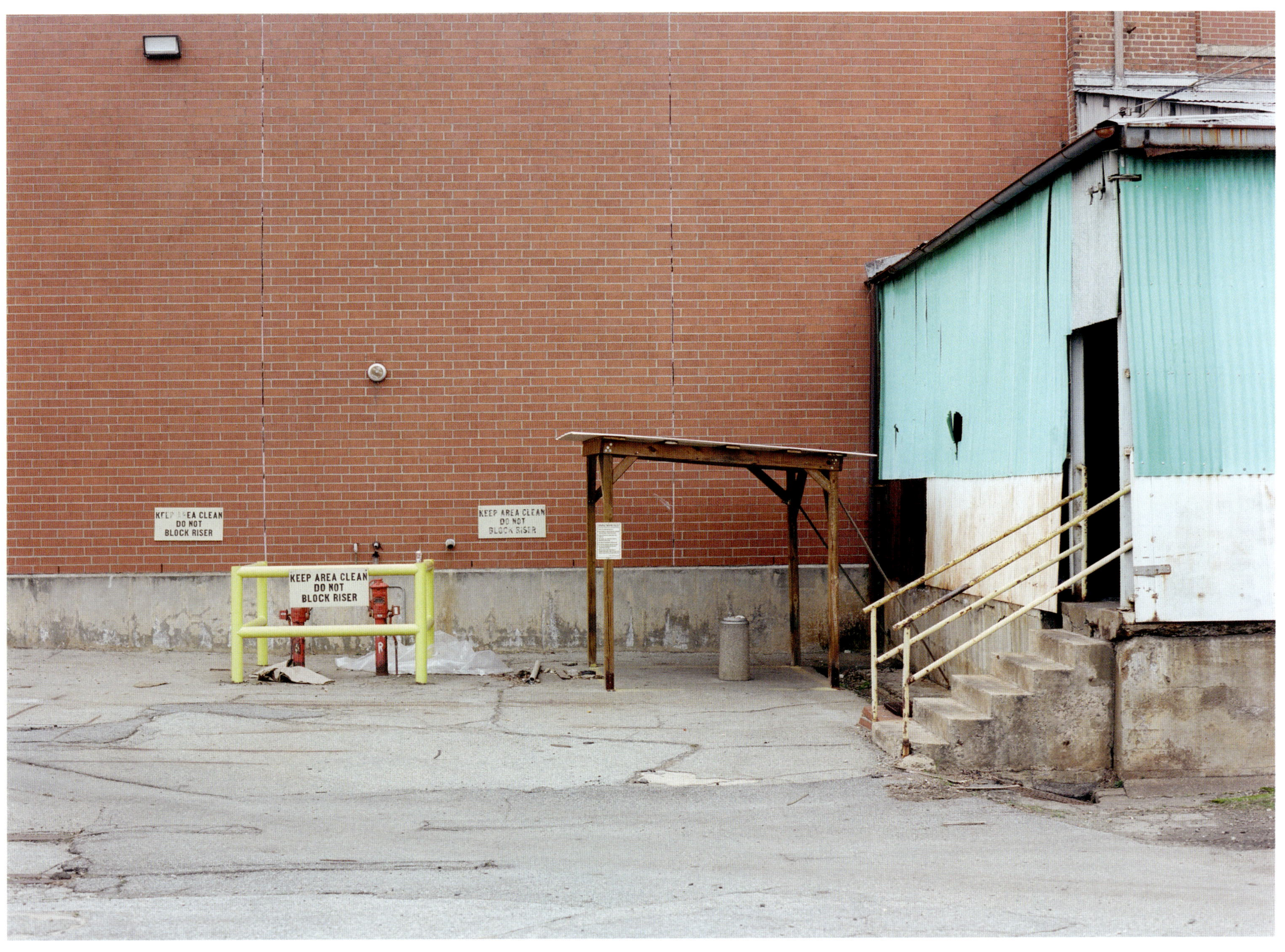

Rear view of the exterior of the Weave Room

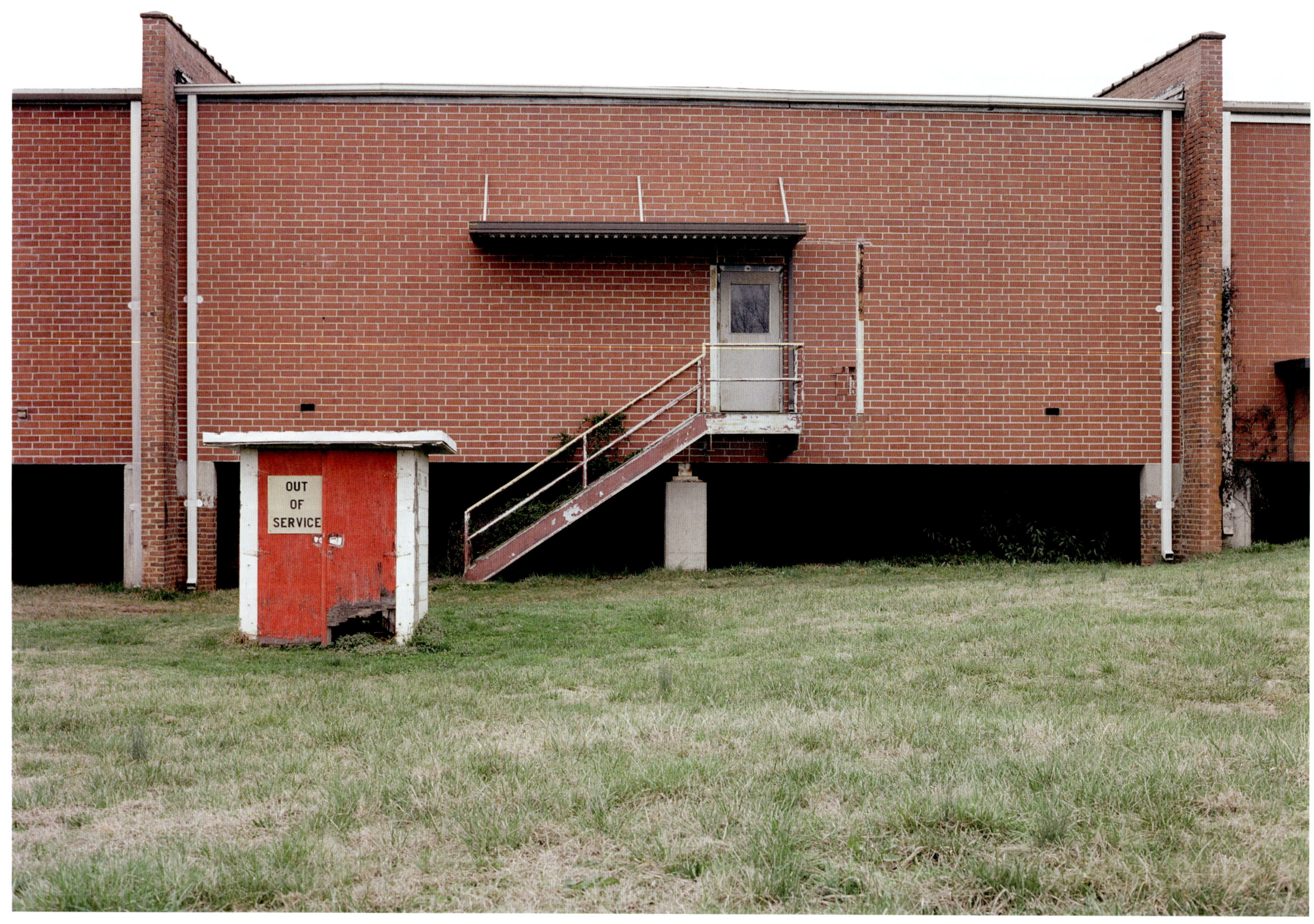

Cotton Warehouses

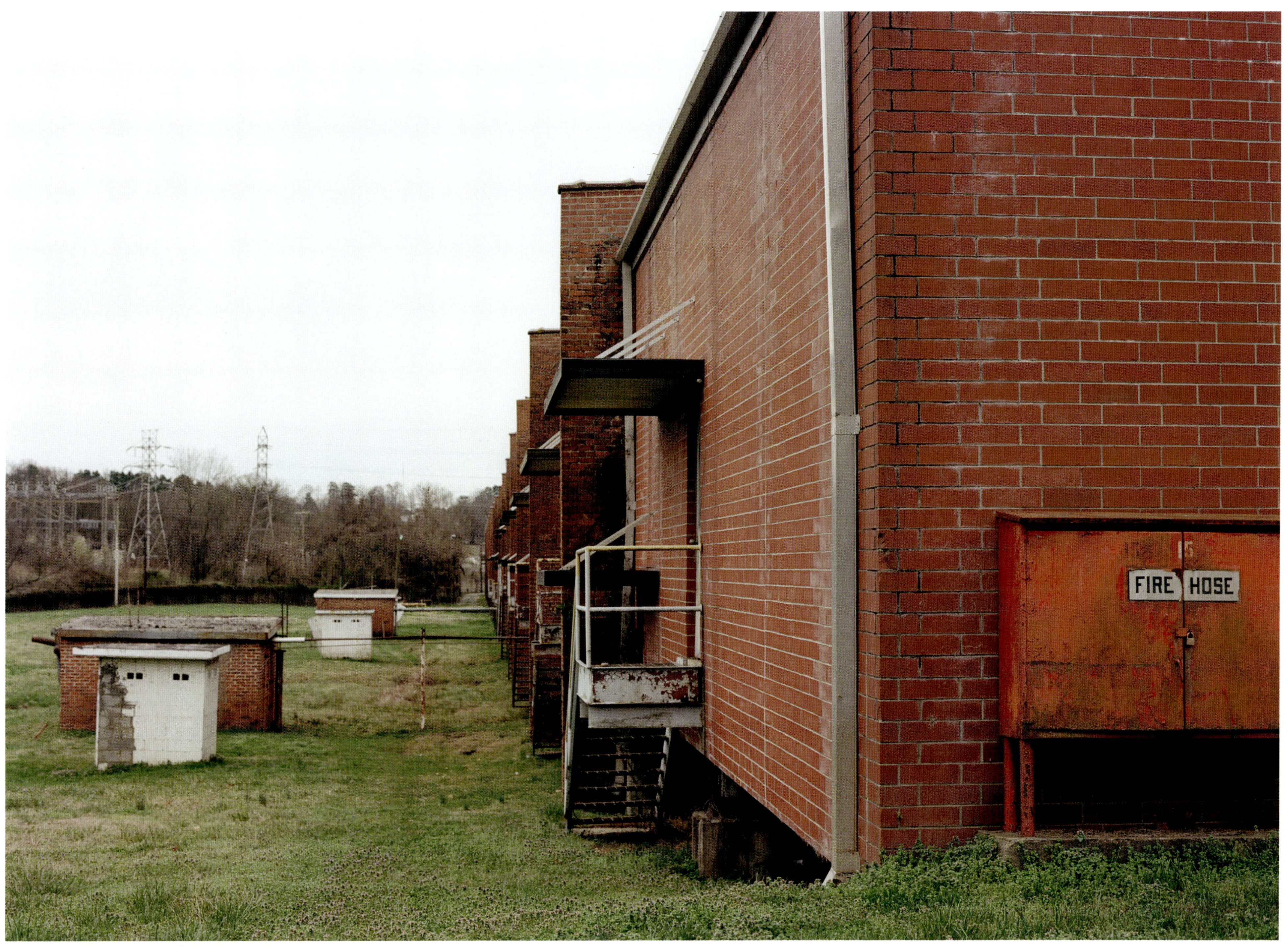
FIRE HOSE

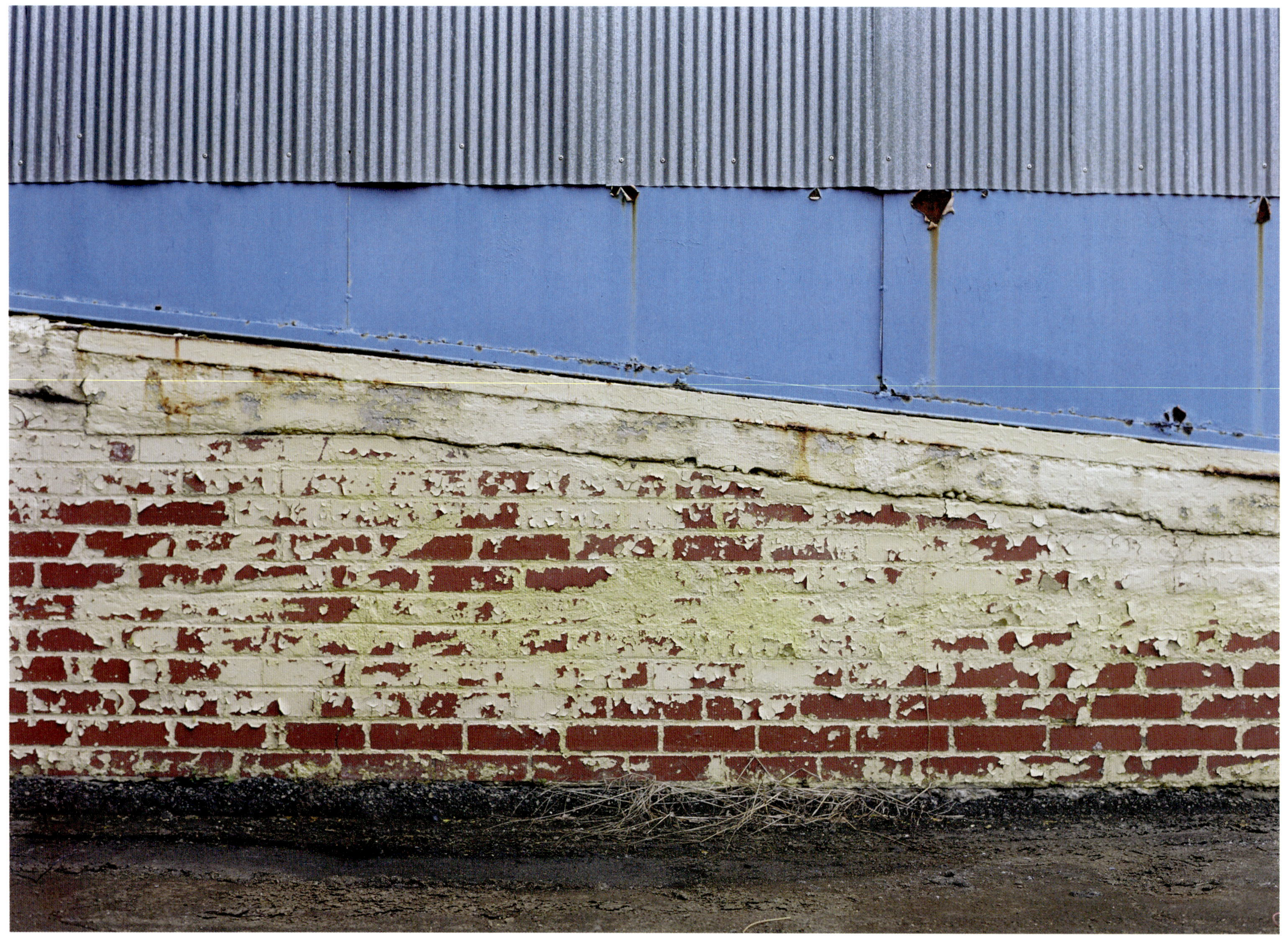

80

Patinated brick on exterior of cotton warehouse loading area // Original entryway into the conveyor room beneath the coal pits of the Powerhouse

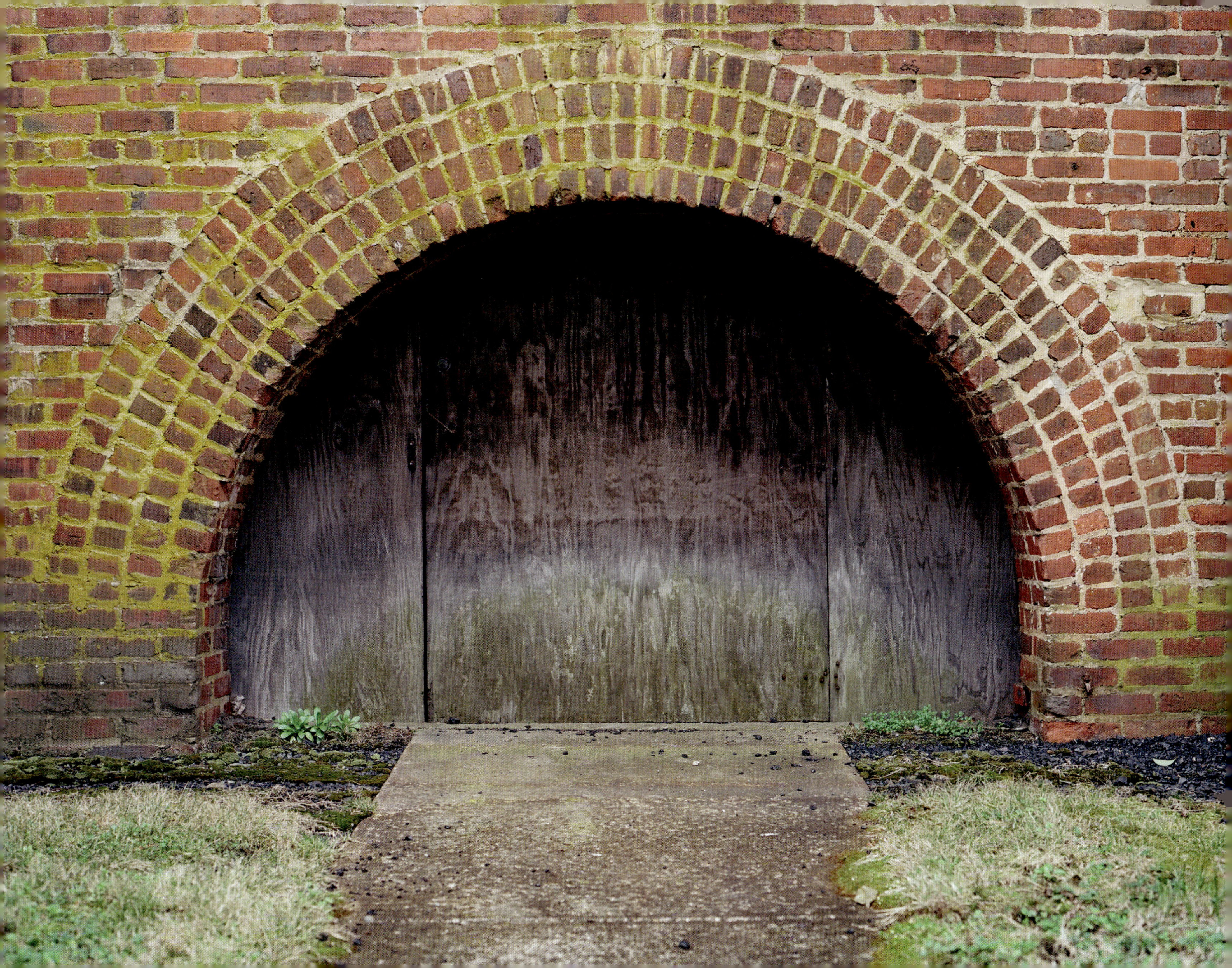

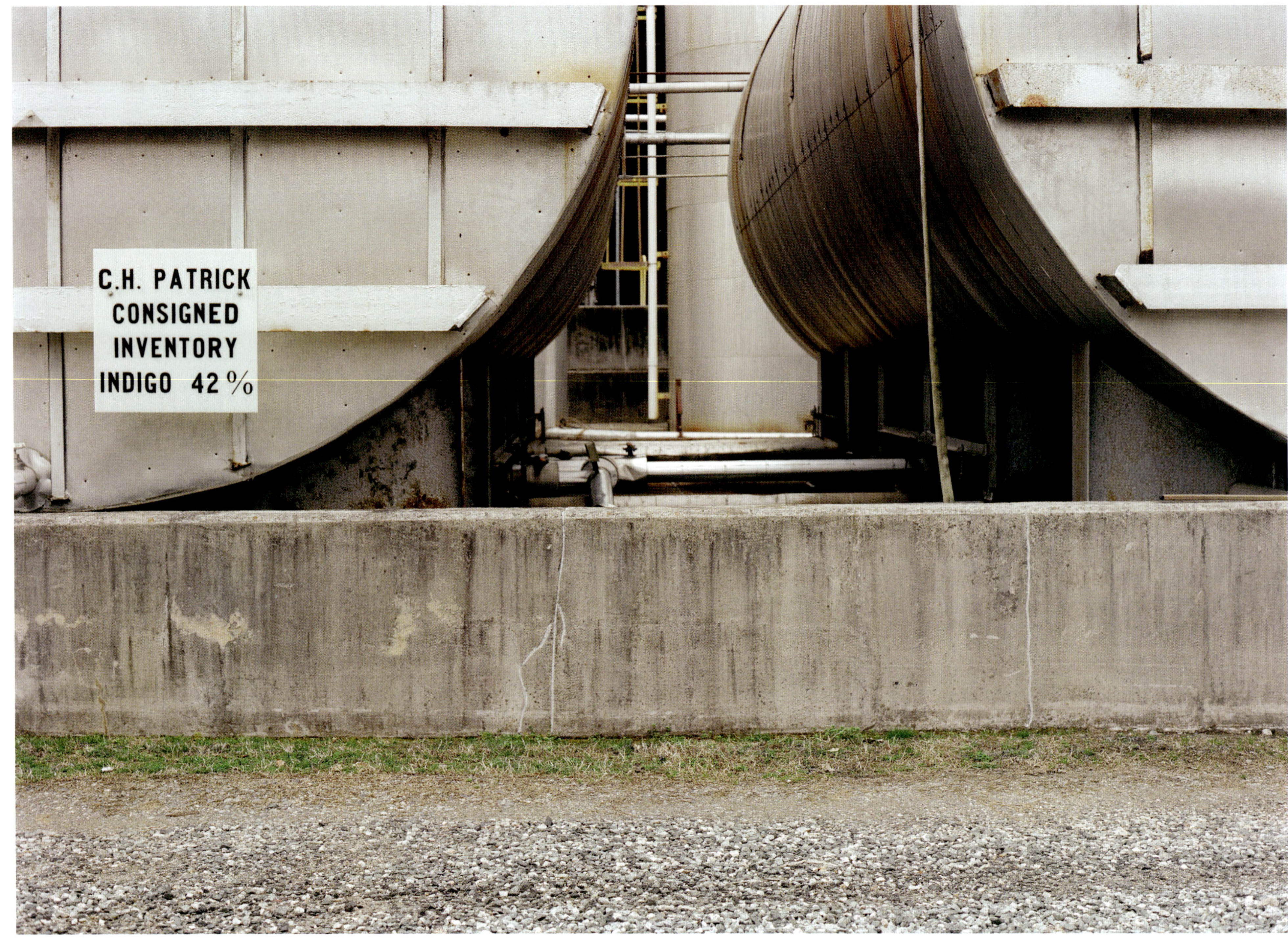

82

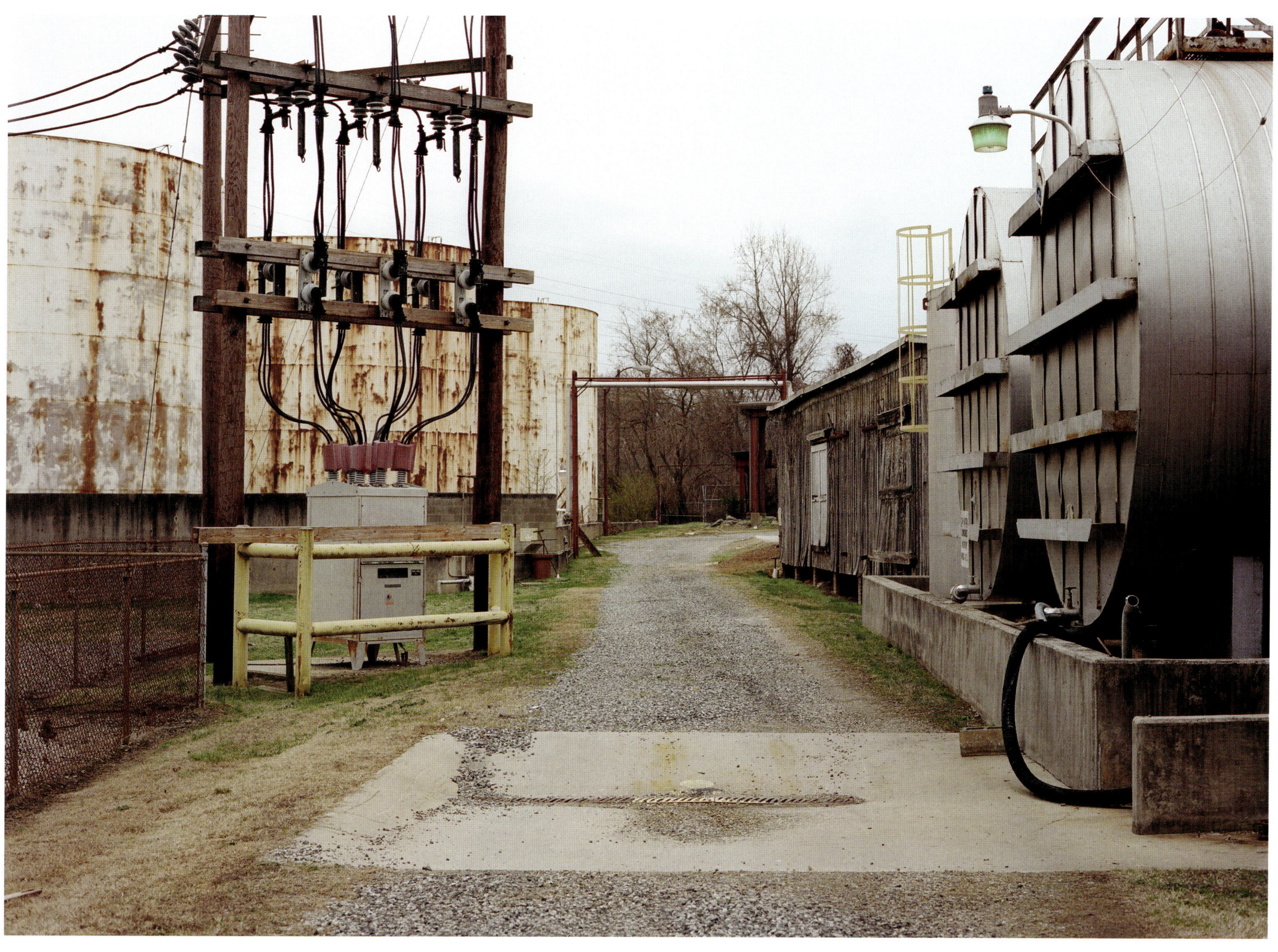

Chemical holding tanks and transformer between the Dye House and Powerhouse

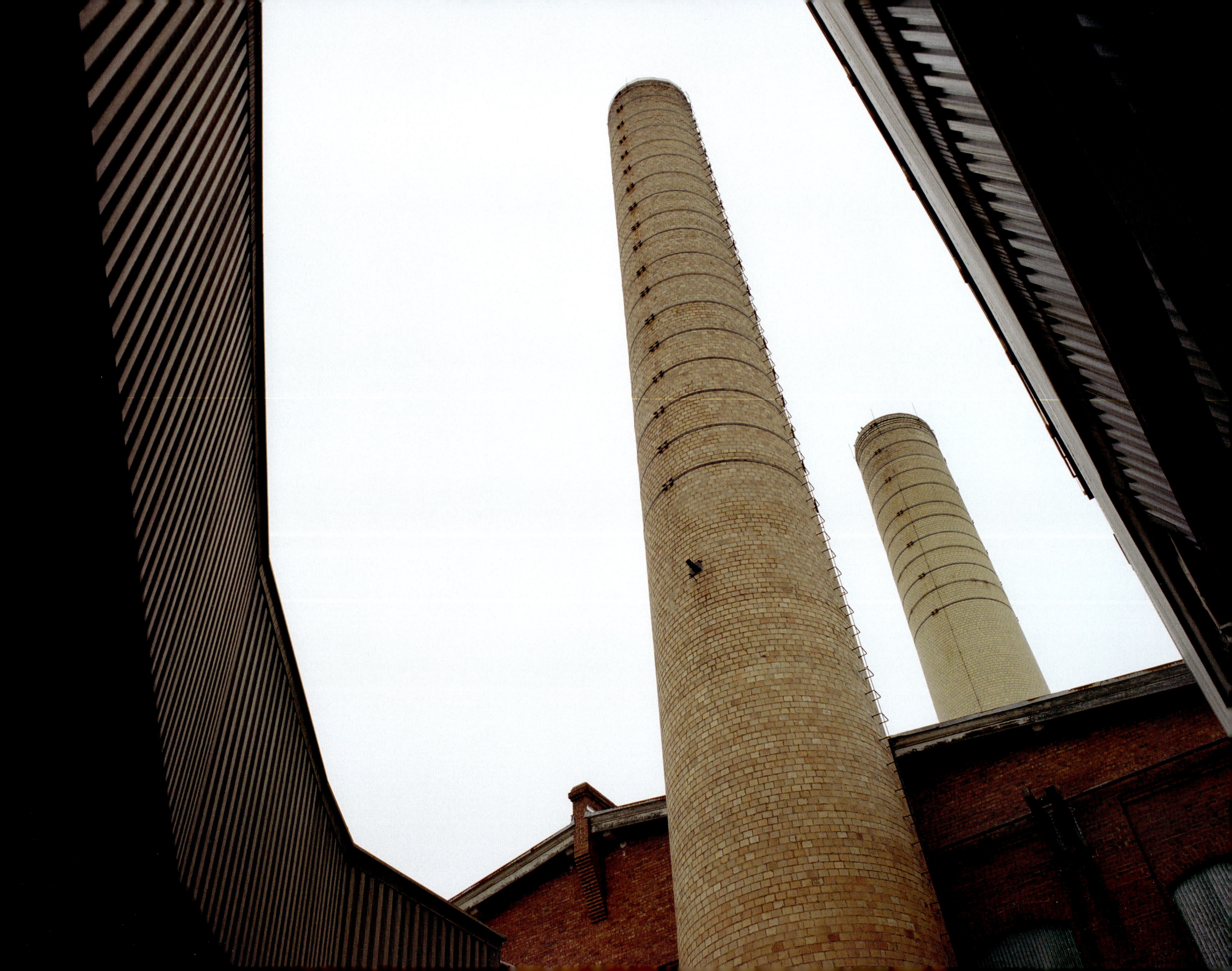

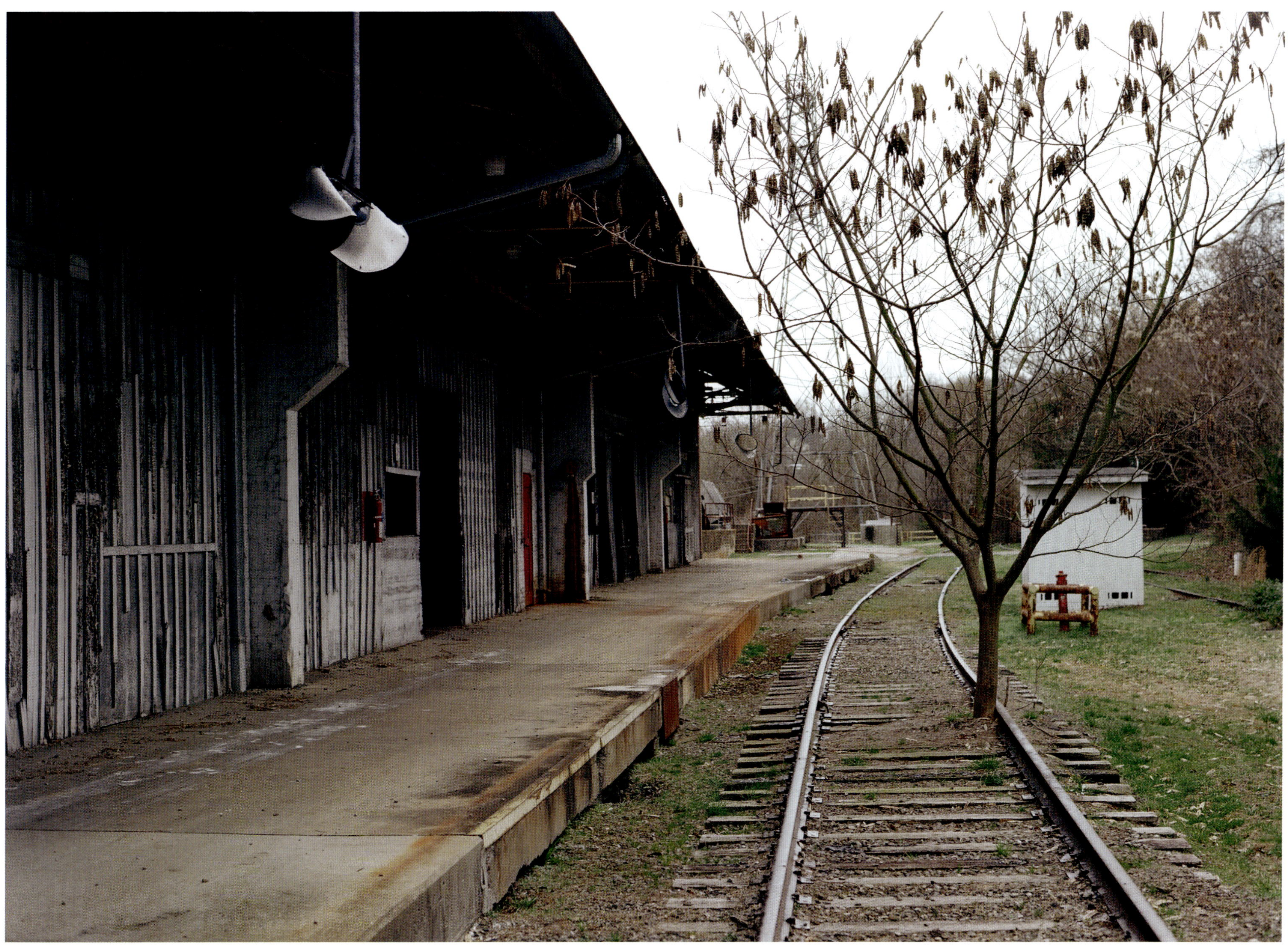

Smokestacks at the Powerhouse // A tree grows from nonoperational tracks in front of Cotton Warehouses

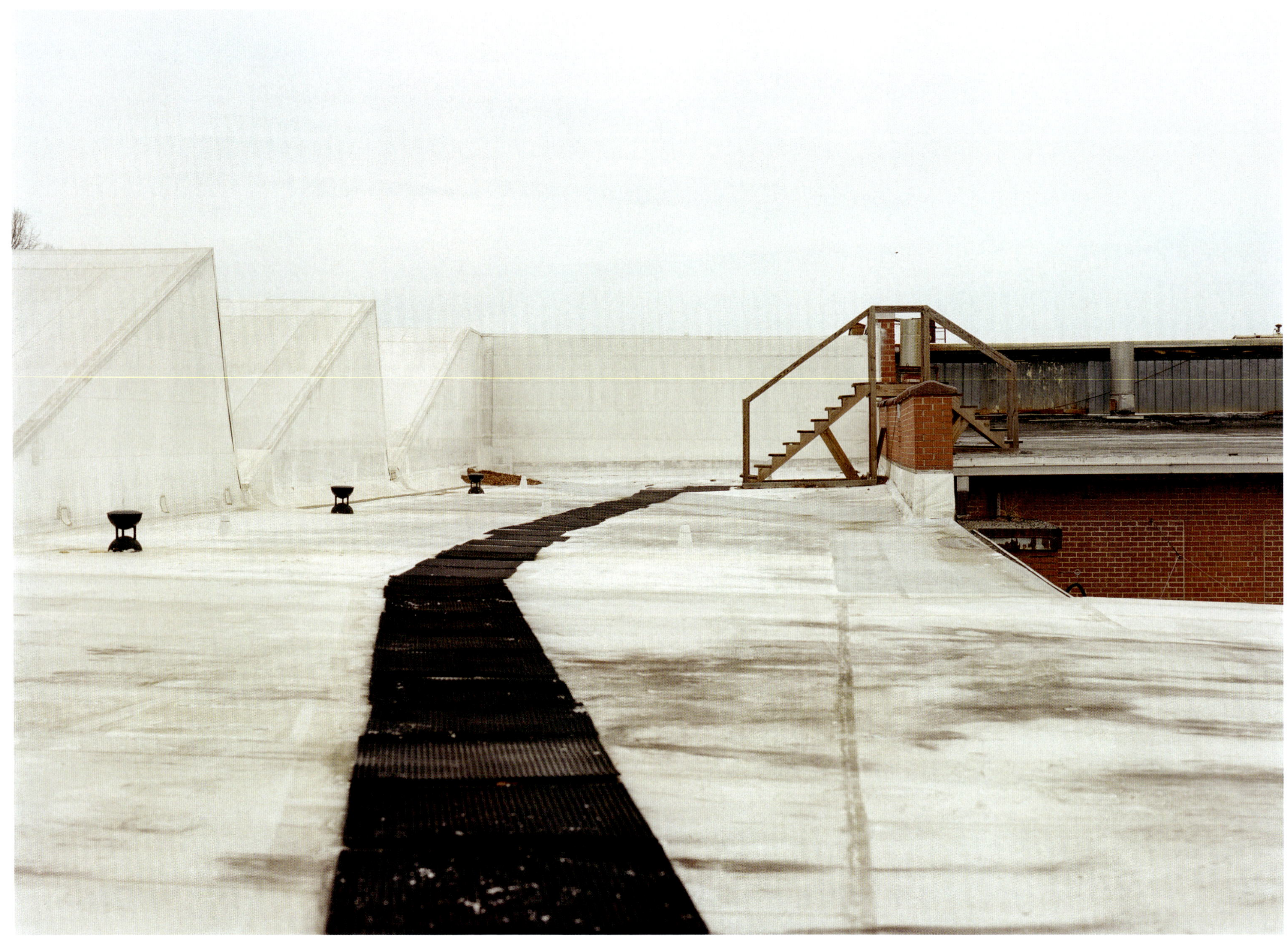

Sawtooth roofline above the Weave Room

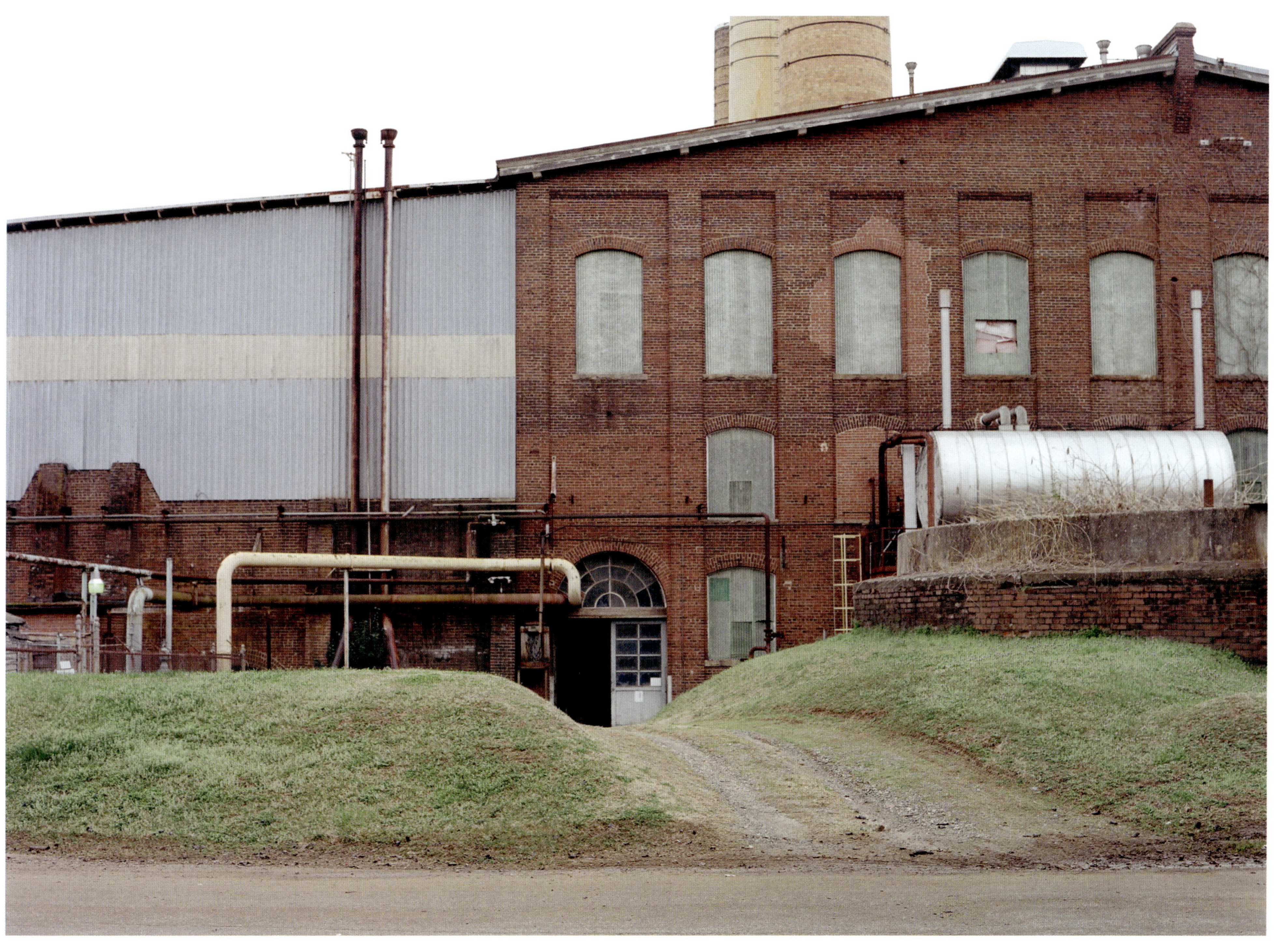

Side entrance to the Powerhouse and Coal Room

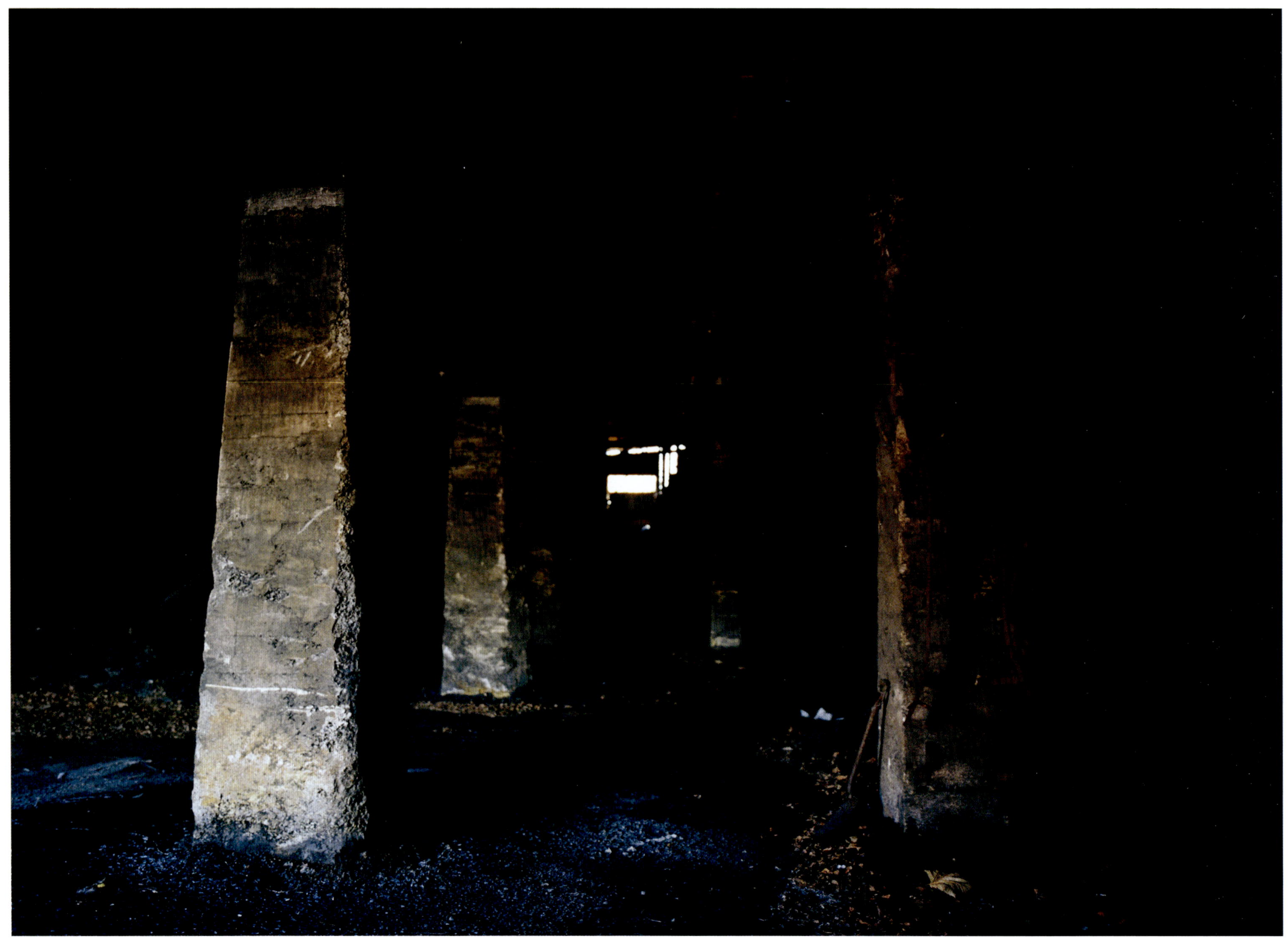

Coal Room under the Powerhouse

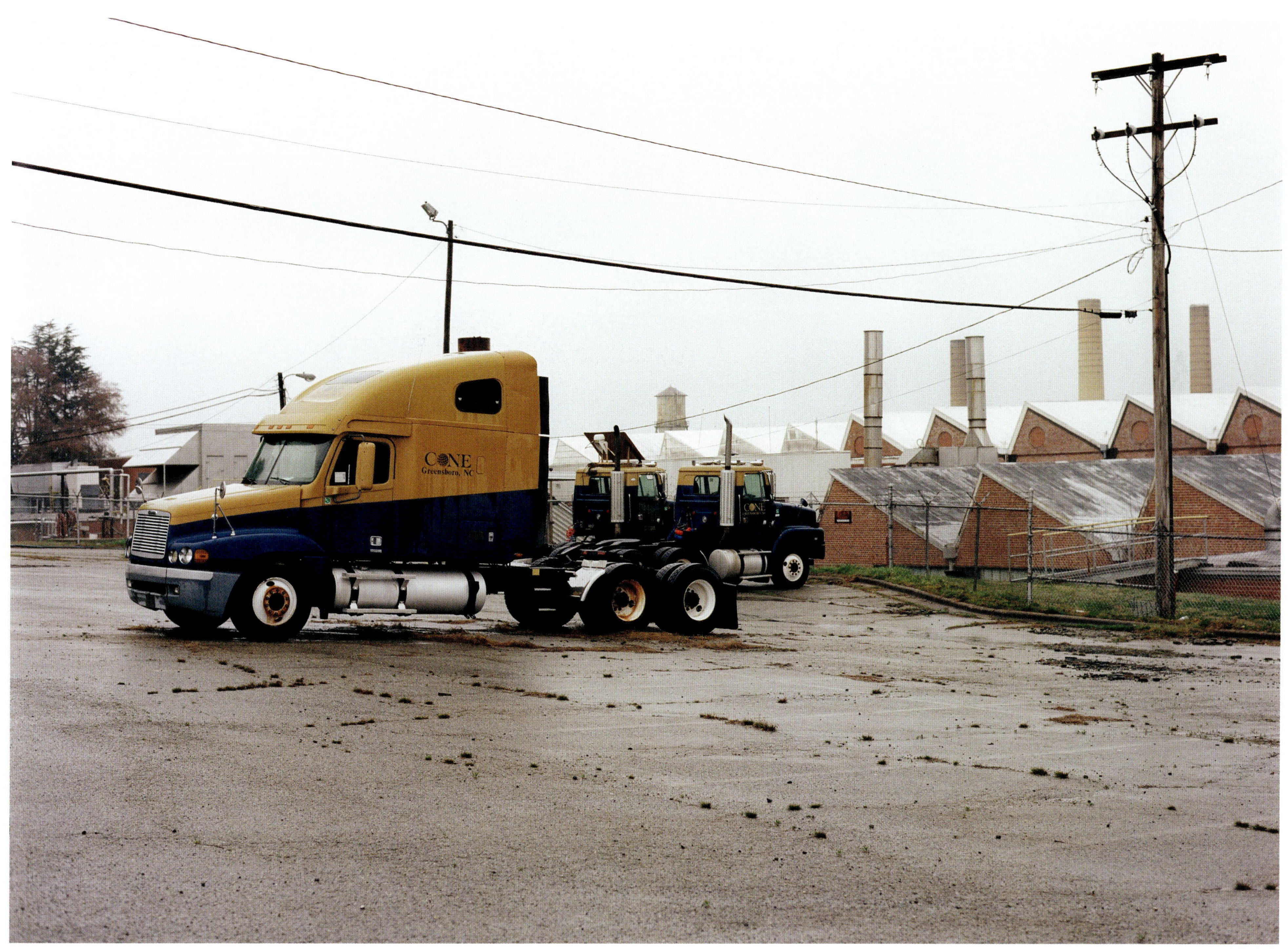

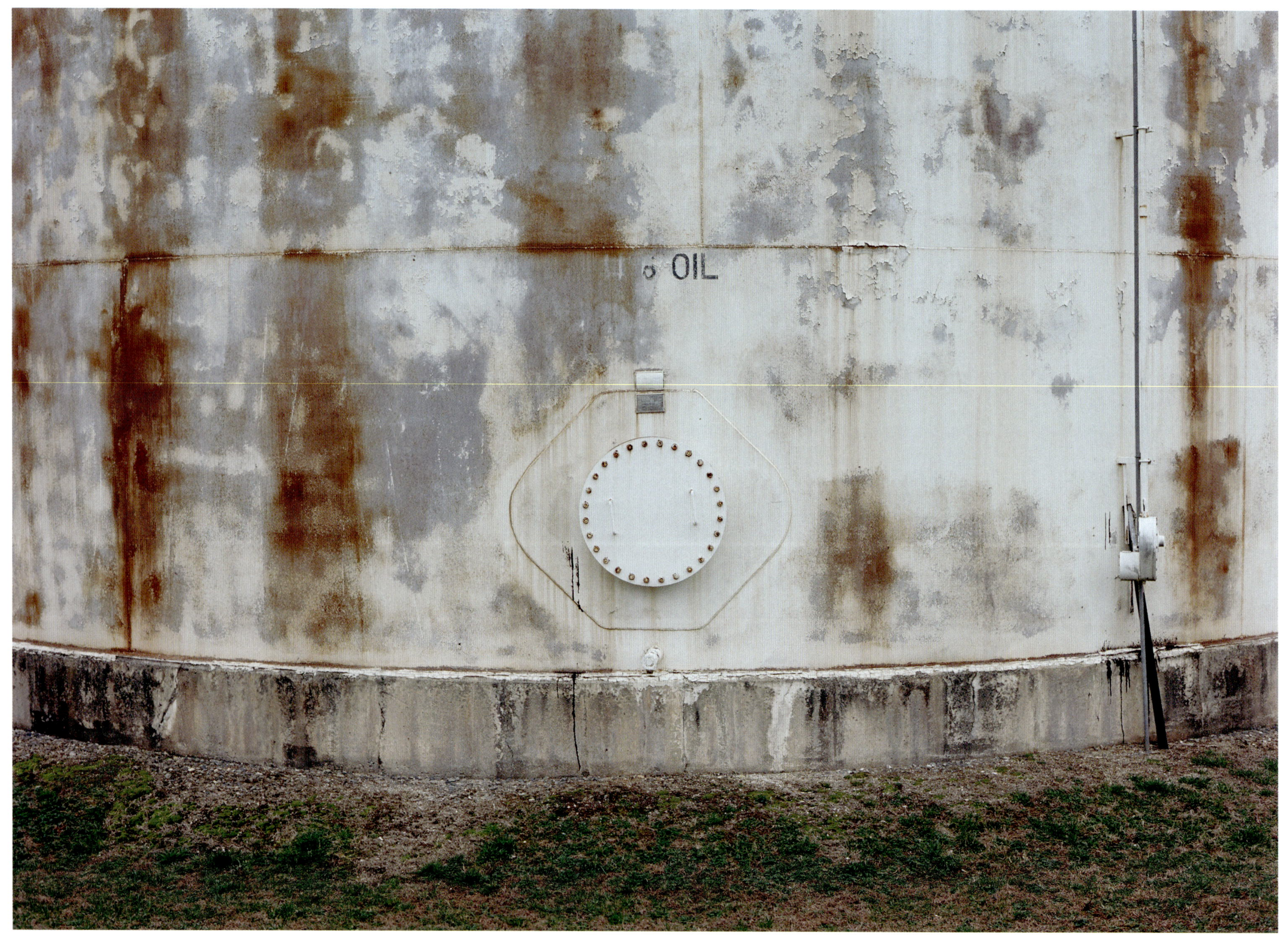

90

Holding tank // Red door at Cotton Warehouse

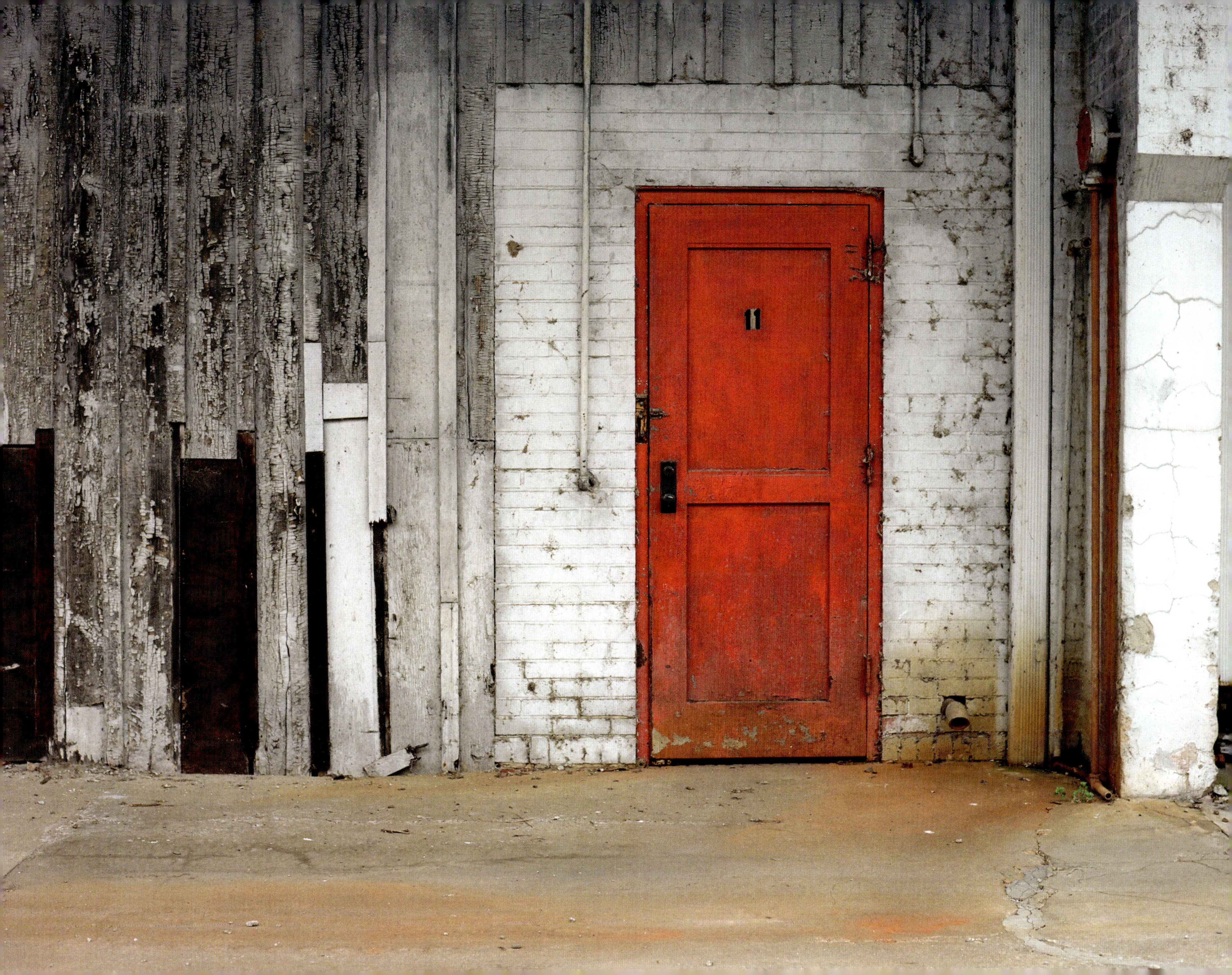

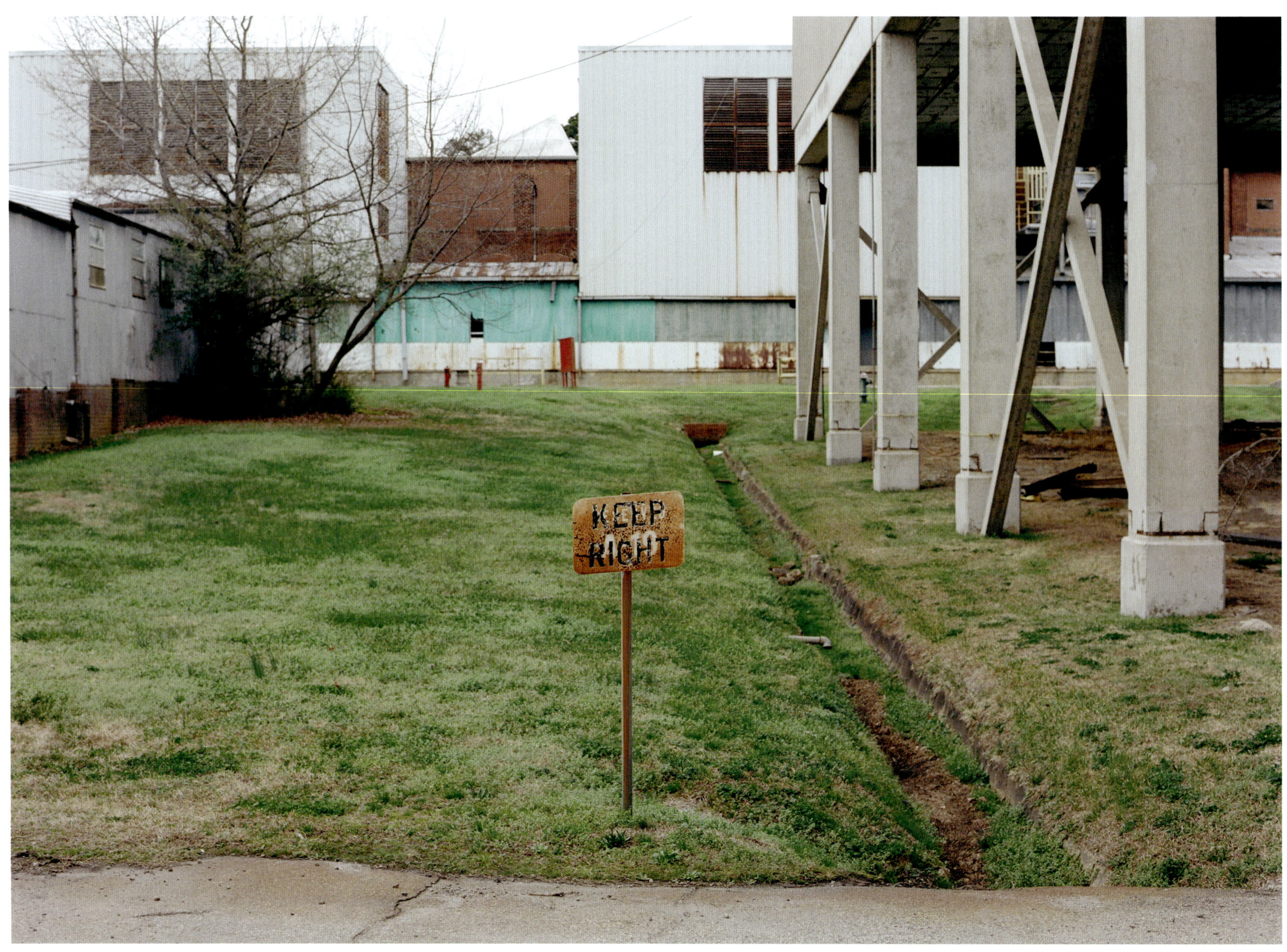

Old trucking signage behind the Weave Room

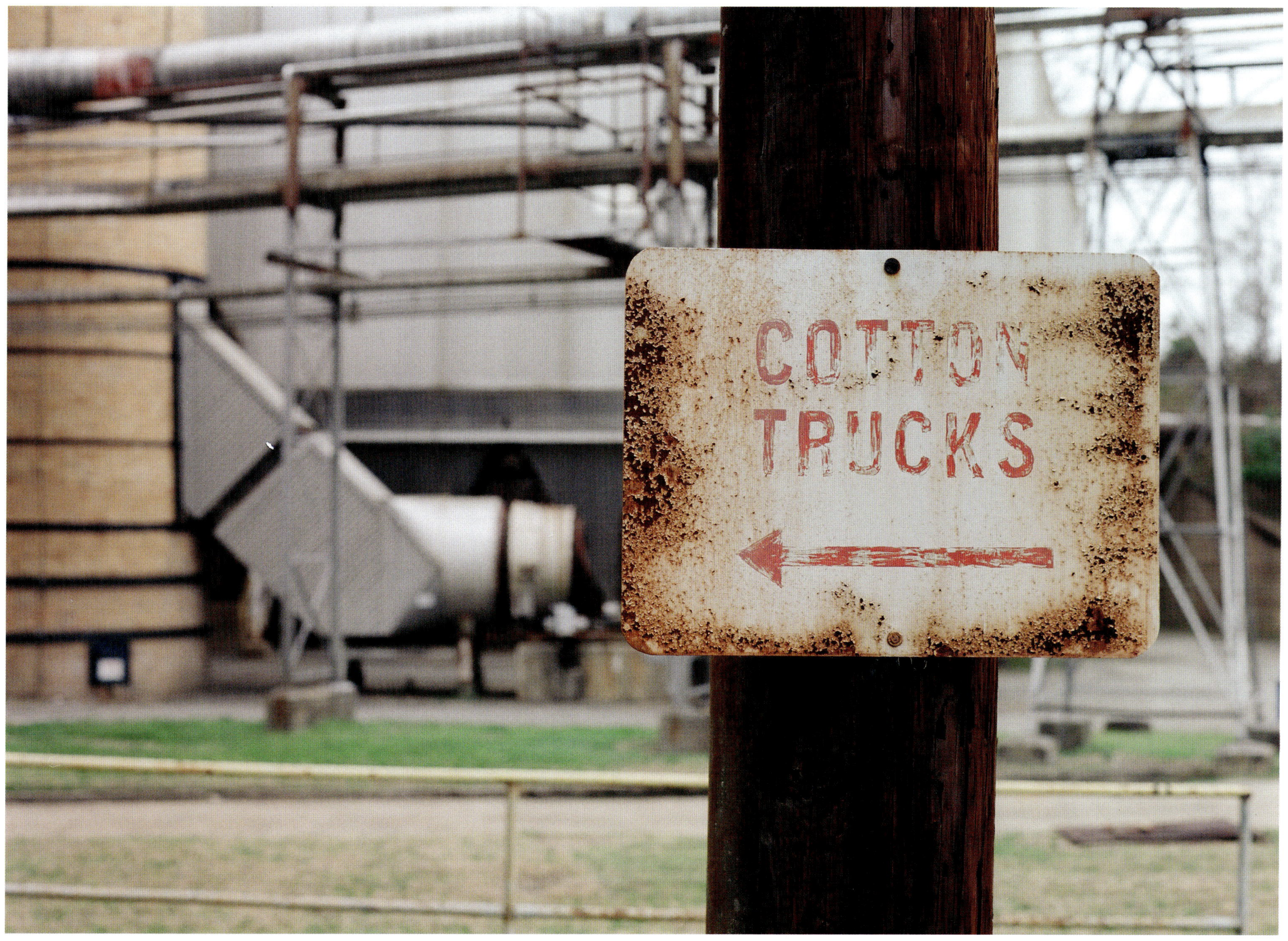

COTTON
TRUCKS

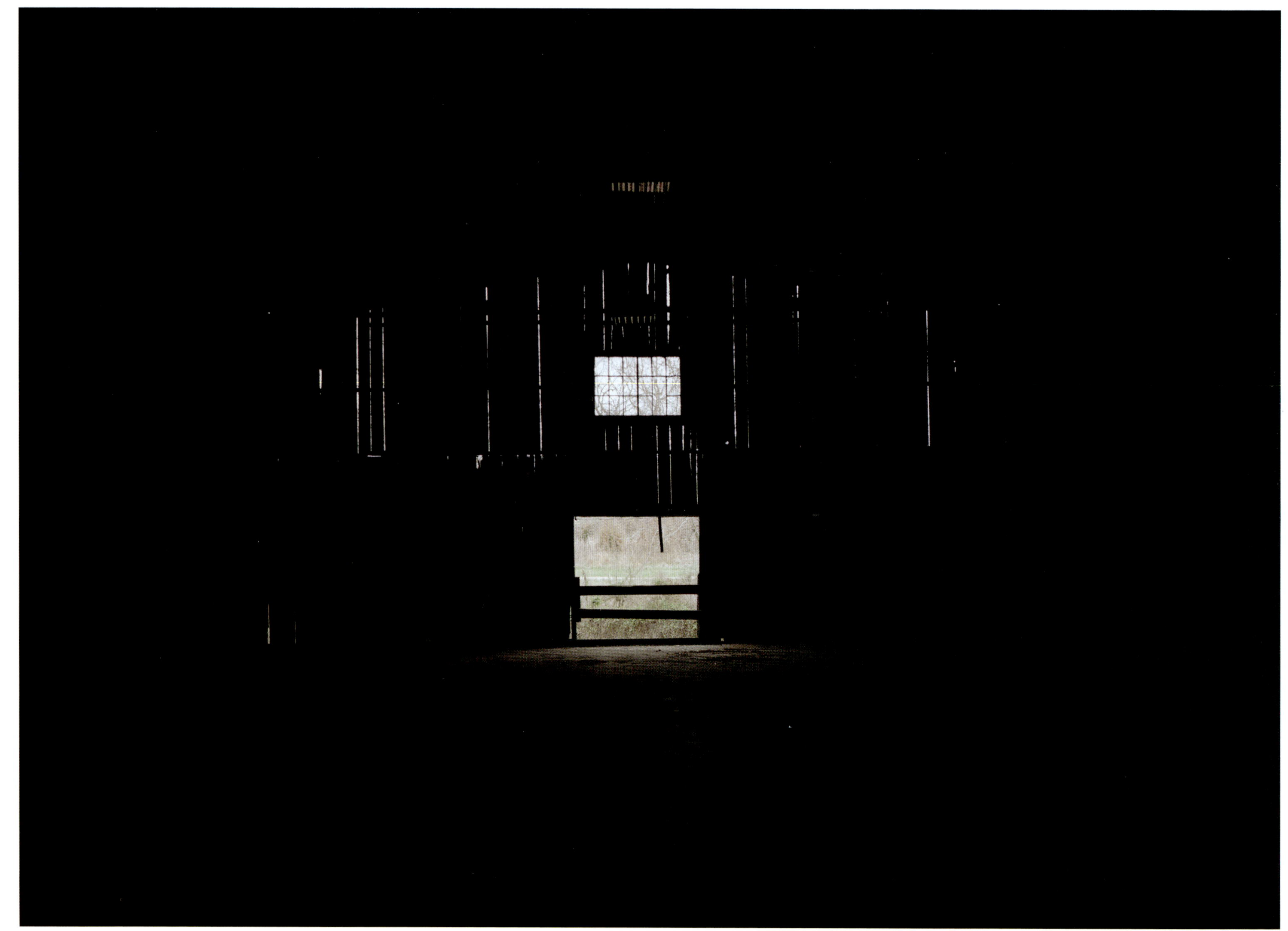

Empty Cotton Warehouse

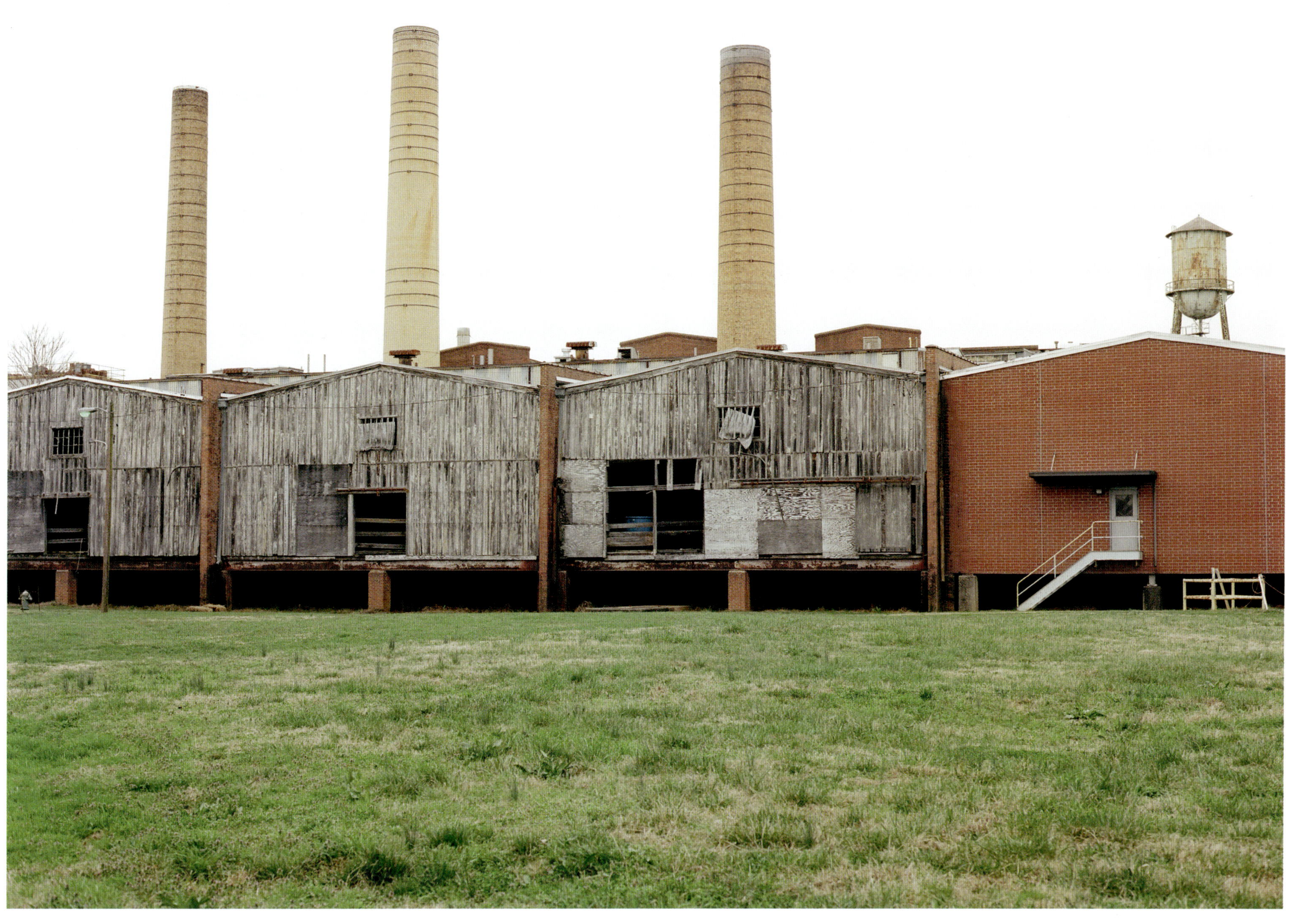

Cotton Warehouses in front of the smokestacks, water tower, and Powerhouse

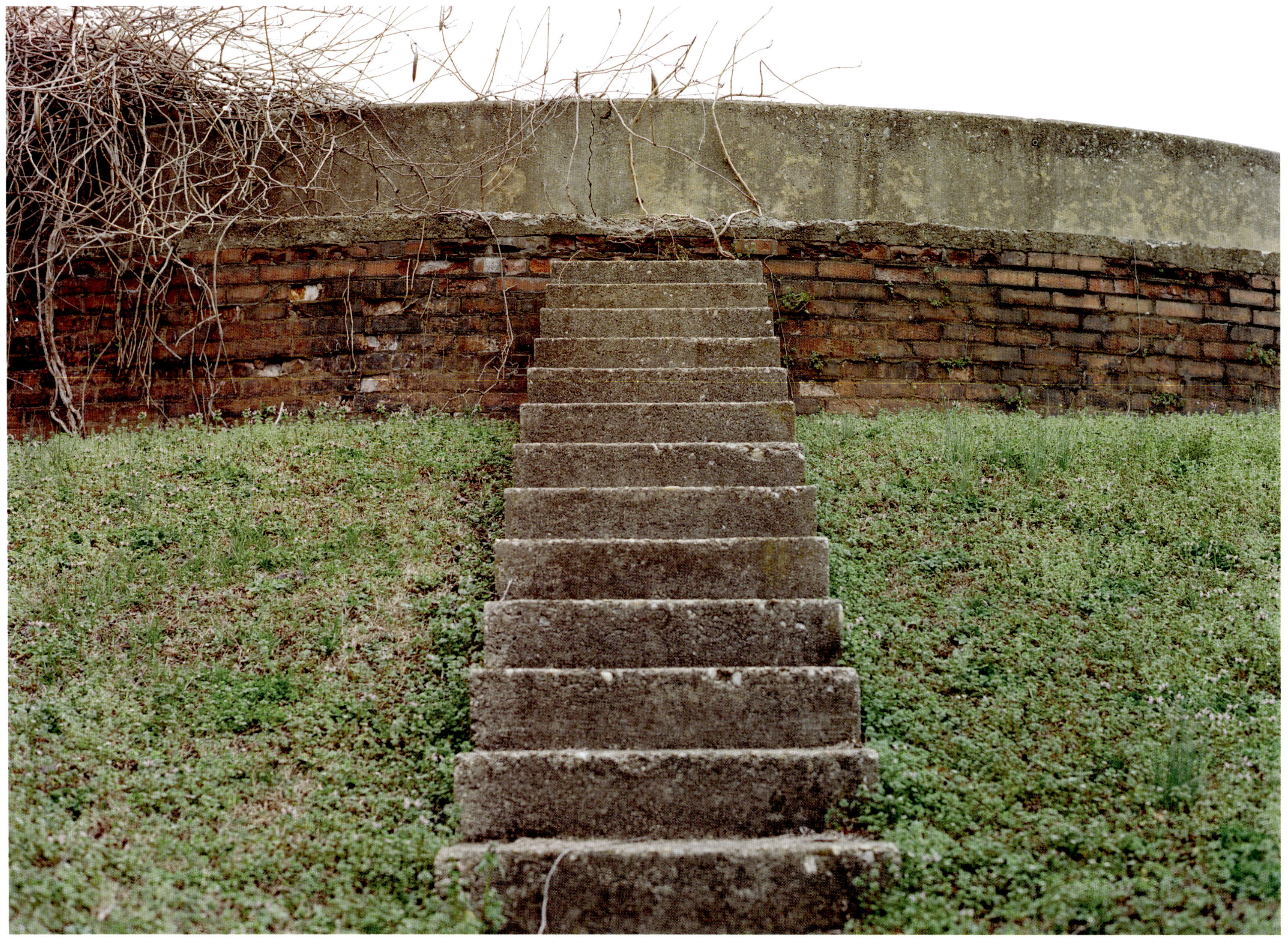

Steps leading up to a water reservoir

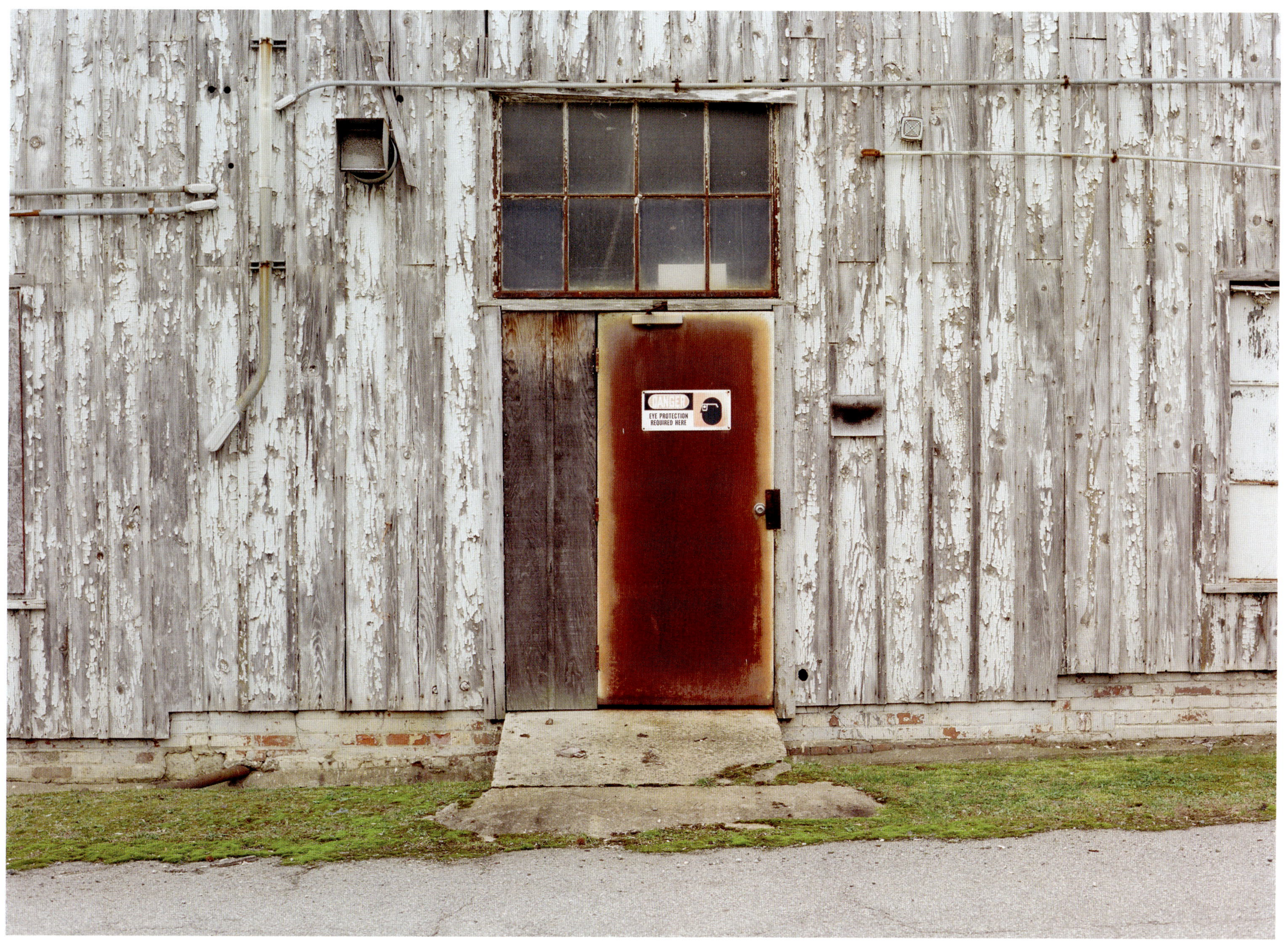

Maintenance Building behind the Weave Room

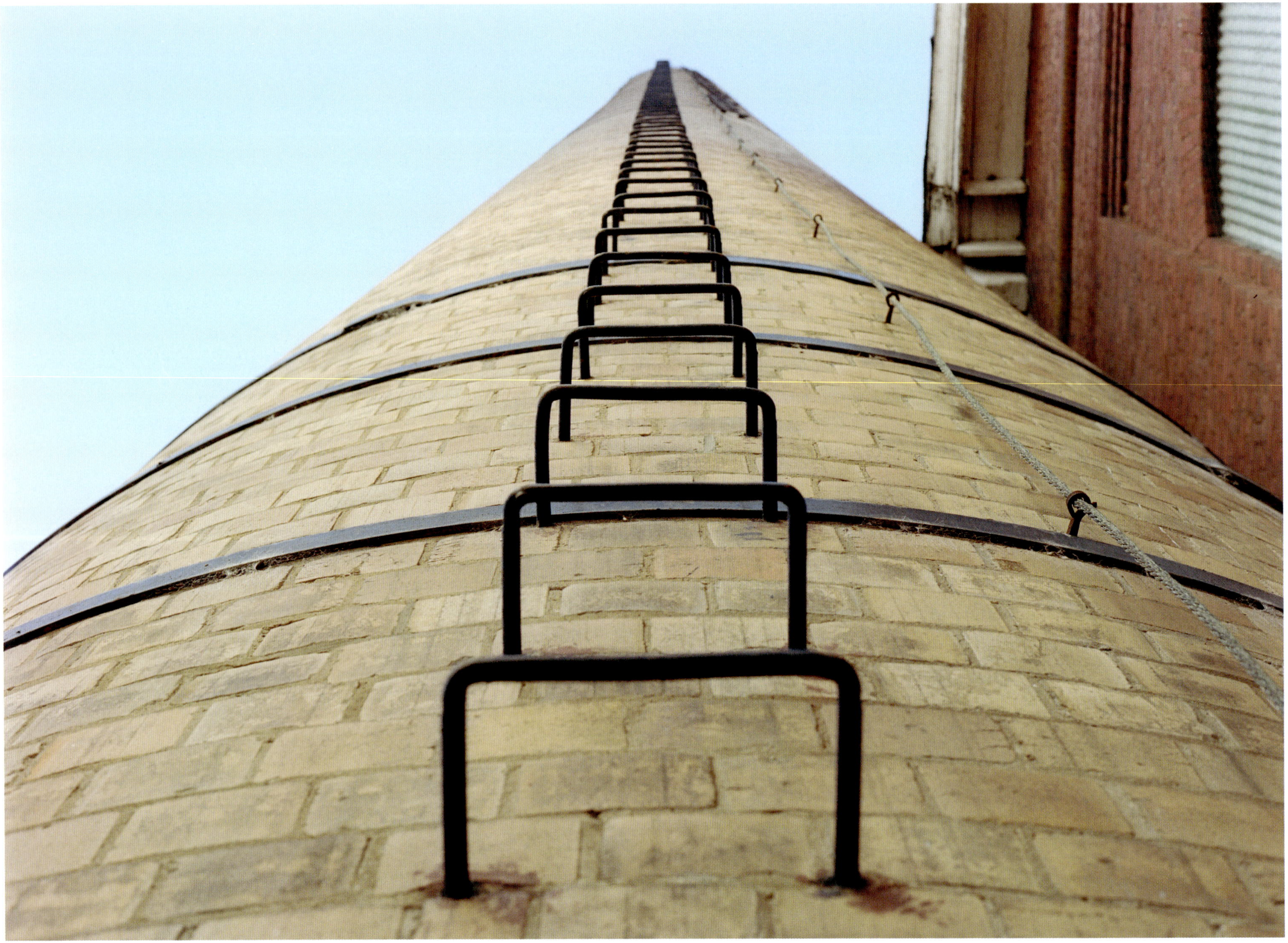

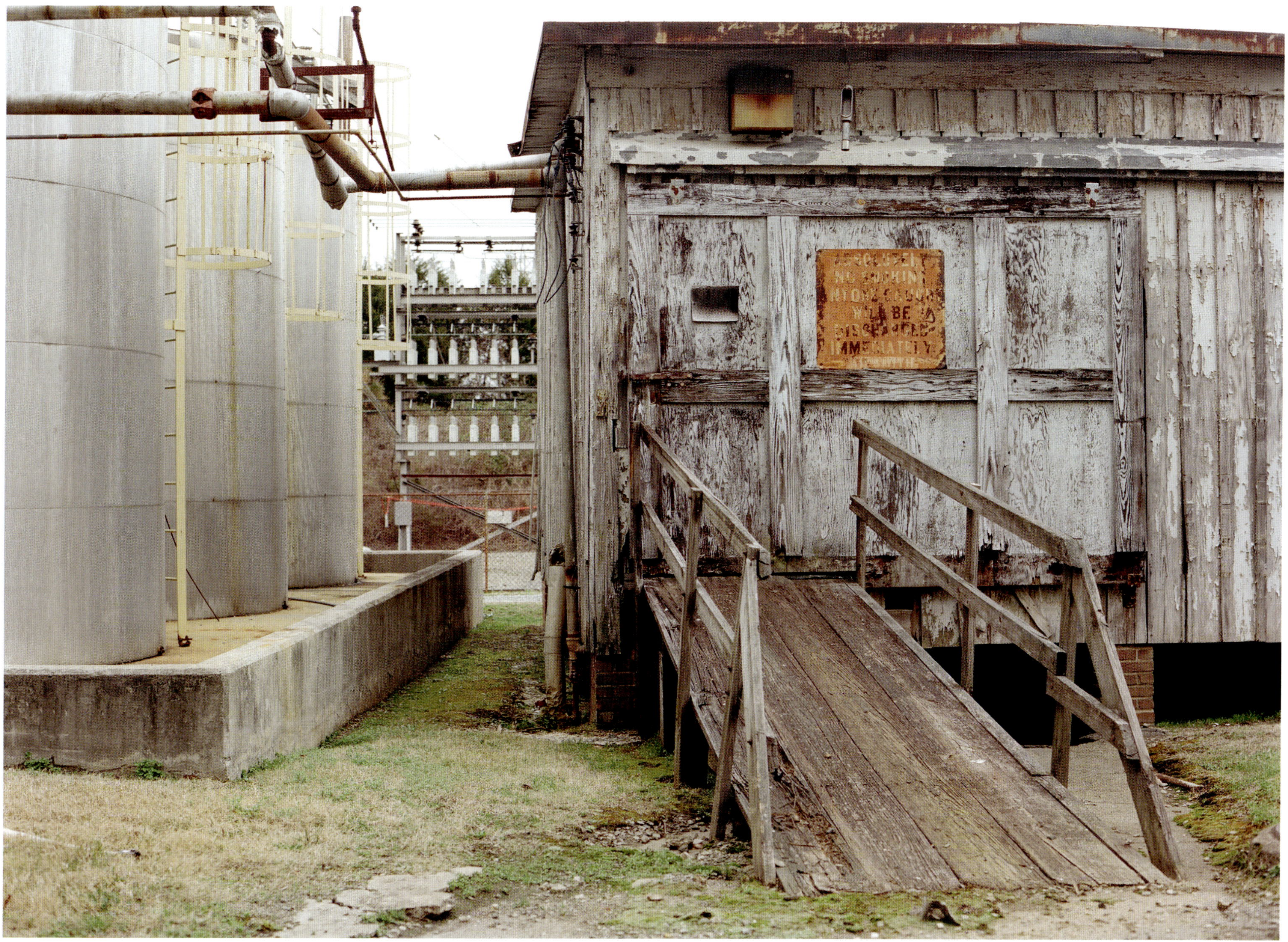

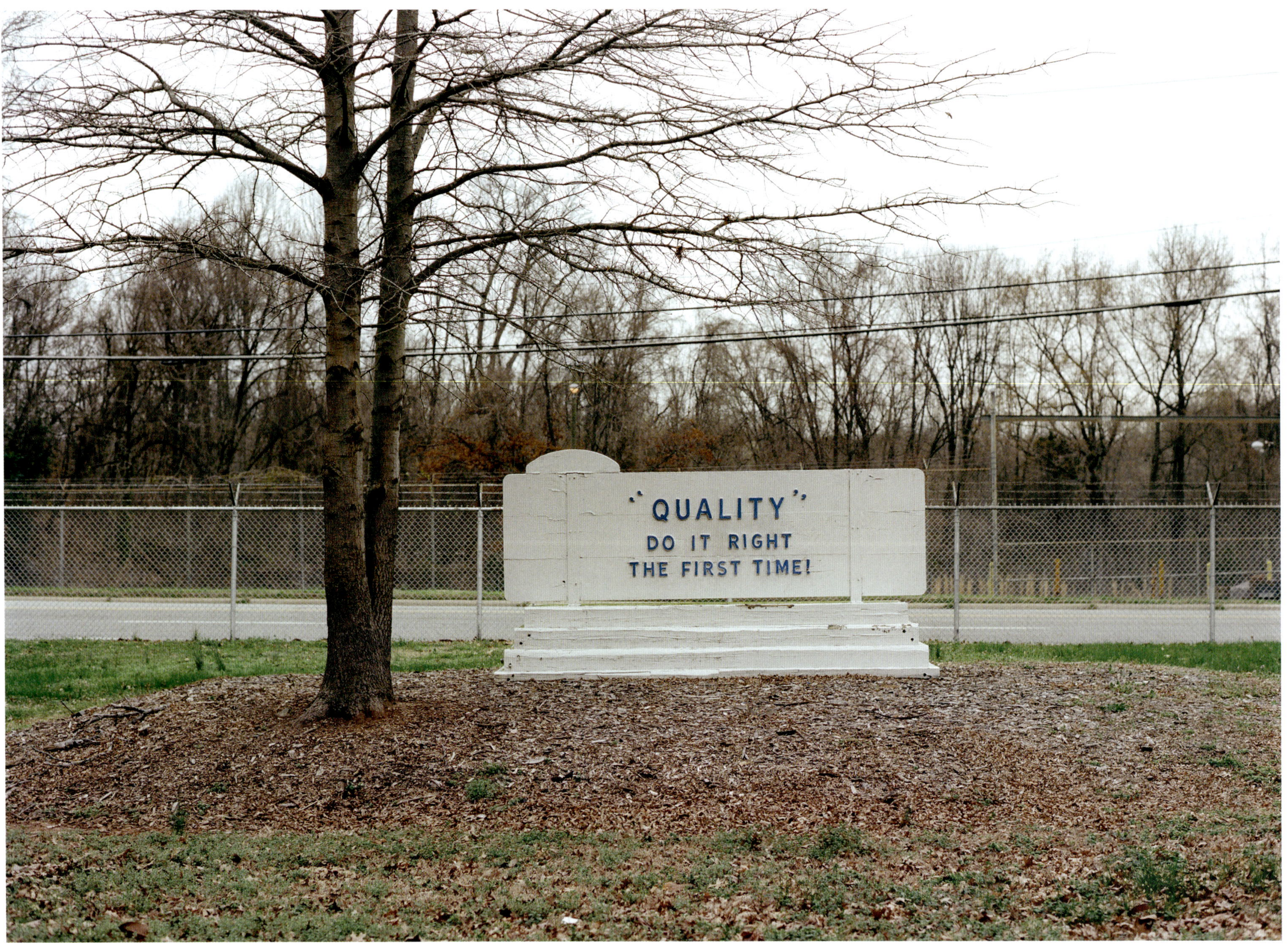

Signage on the 16th Street side of the property

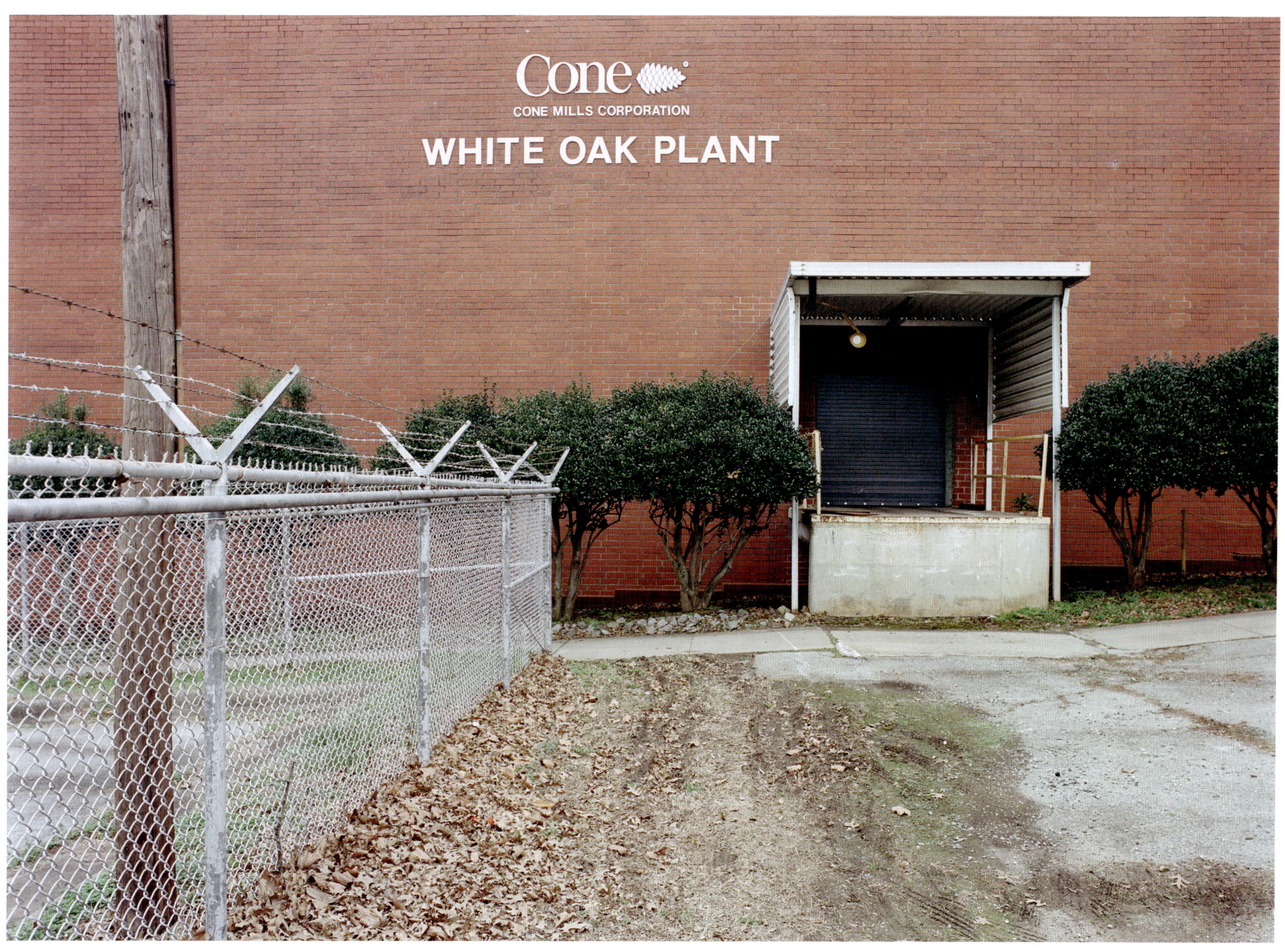

Trucking dock alongside 16th Street Annex Building

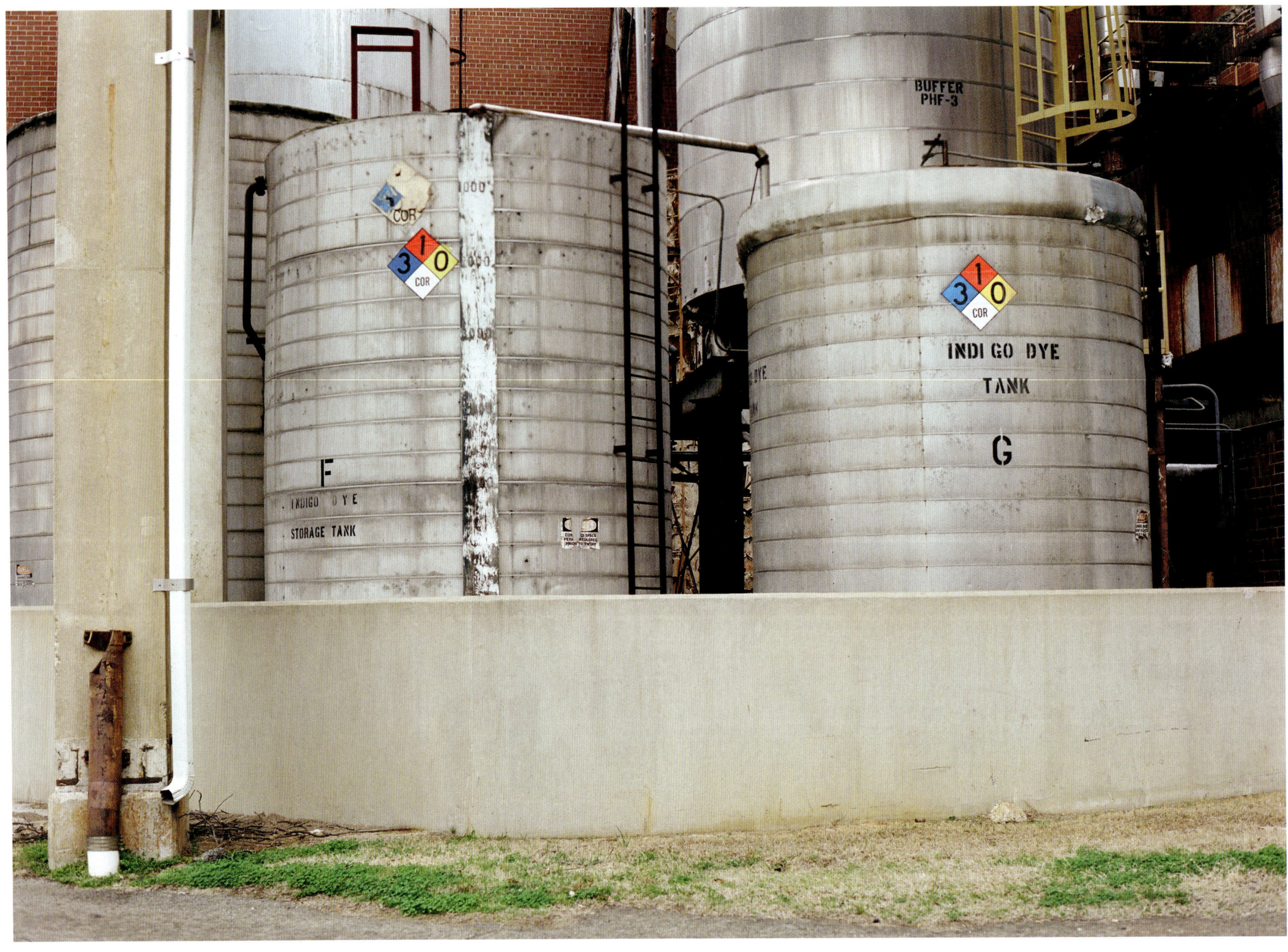

Chemical holding tanks at the Dye House   //   Doorway and signage at the Dye House

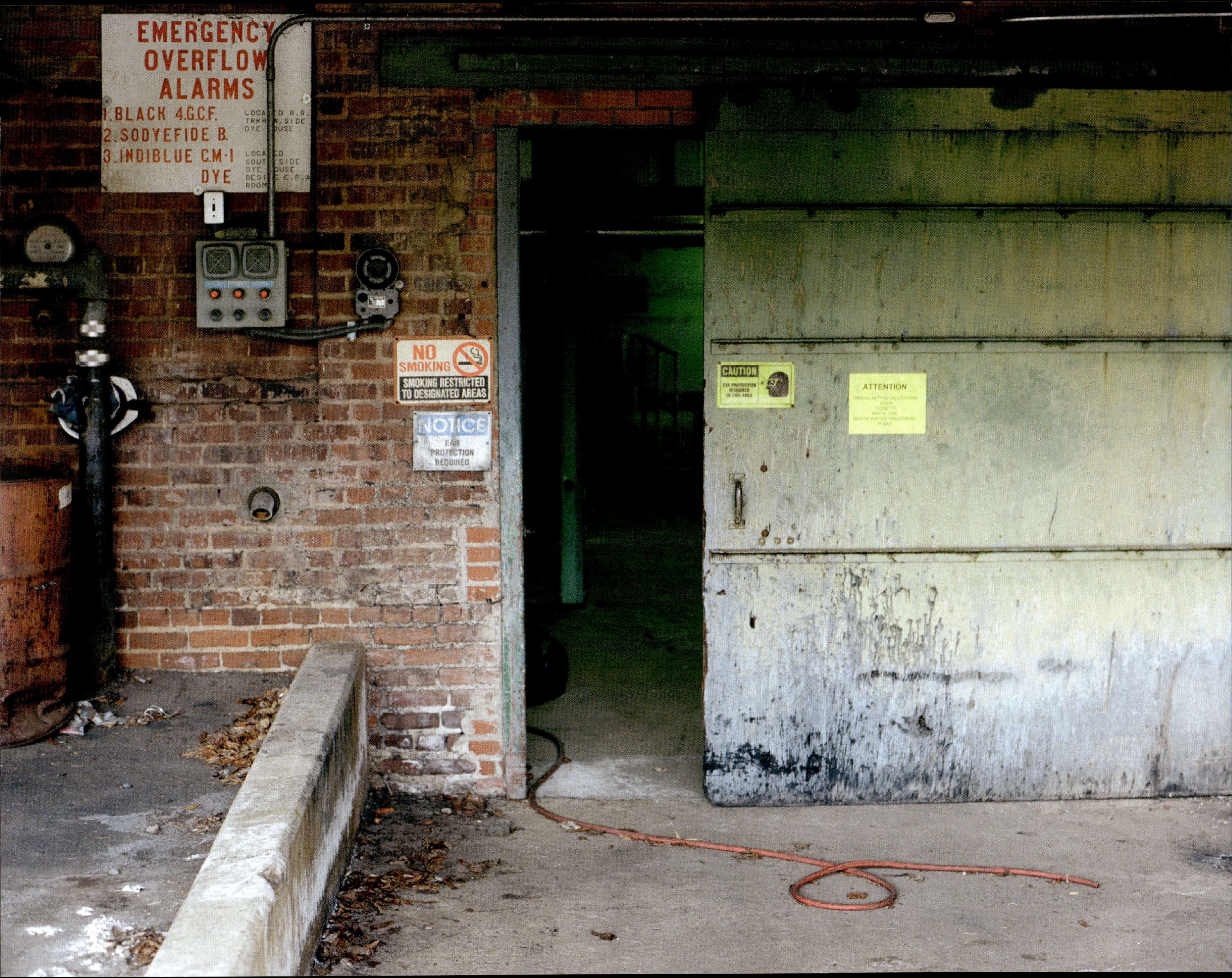

EMERGENCY
OVERFLOW
ALARMS
1.BLACK 4.G.C.F.   LOCATED R.R.
                          TRK R.N.SIDE
2.SODYEFIDE B.   DYE HOUSE
3.INDIBLUE C.M·1   LOCATED
                          SOUTH SIDE
          DYE   DYE HOUSE
                          BESIDE E.P.A.
                          ROOM
NO SMOKING
SMOKING RESTRICTED
TO DESIGNATED AREAS
NOTICE
EAR
PROTECTION
REQUIRED
CAUTION
EYE PROTECTION
REQUIRED
IN THIS AREA
ATTENTION
DRAINS IN THIS UNLOADING
AREA
FLOW TO
WHITE OAK
WASTE WATER TREATMENT
PLANT

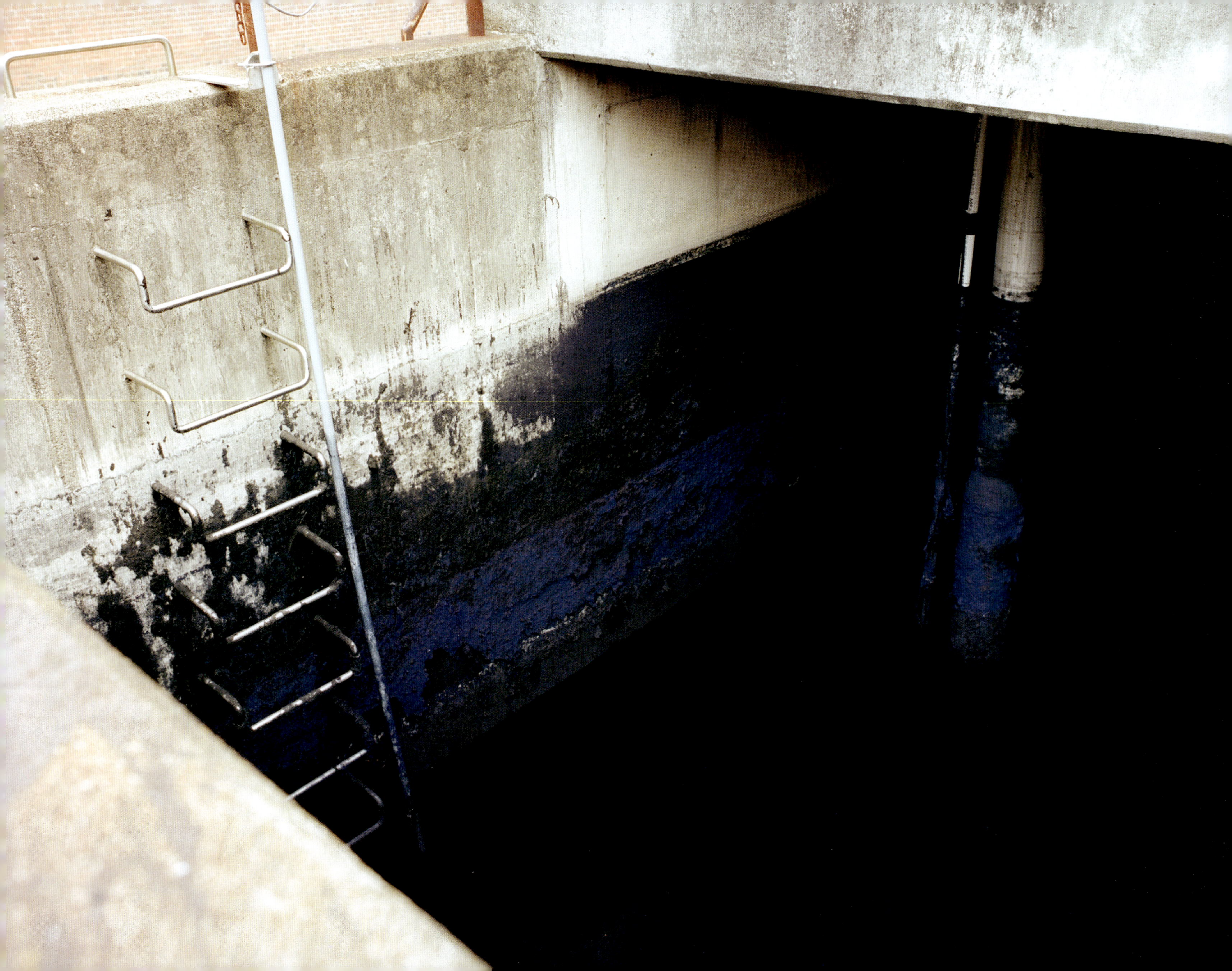

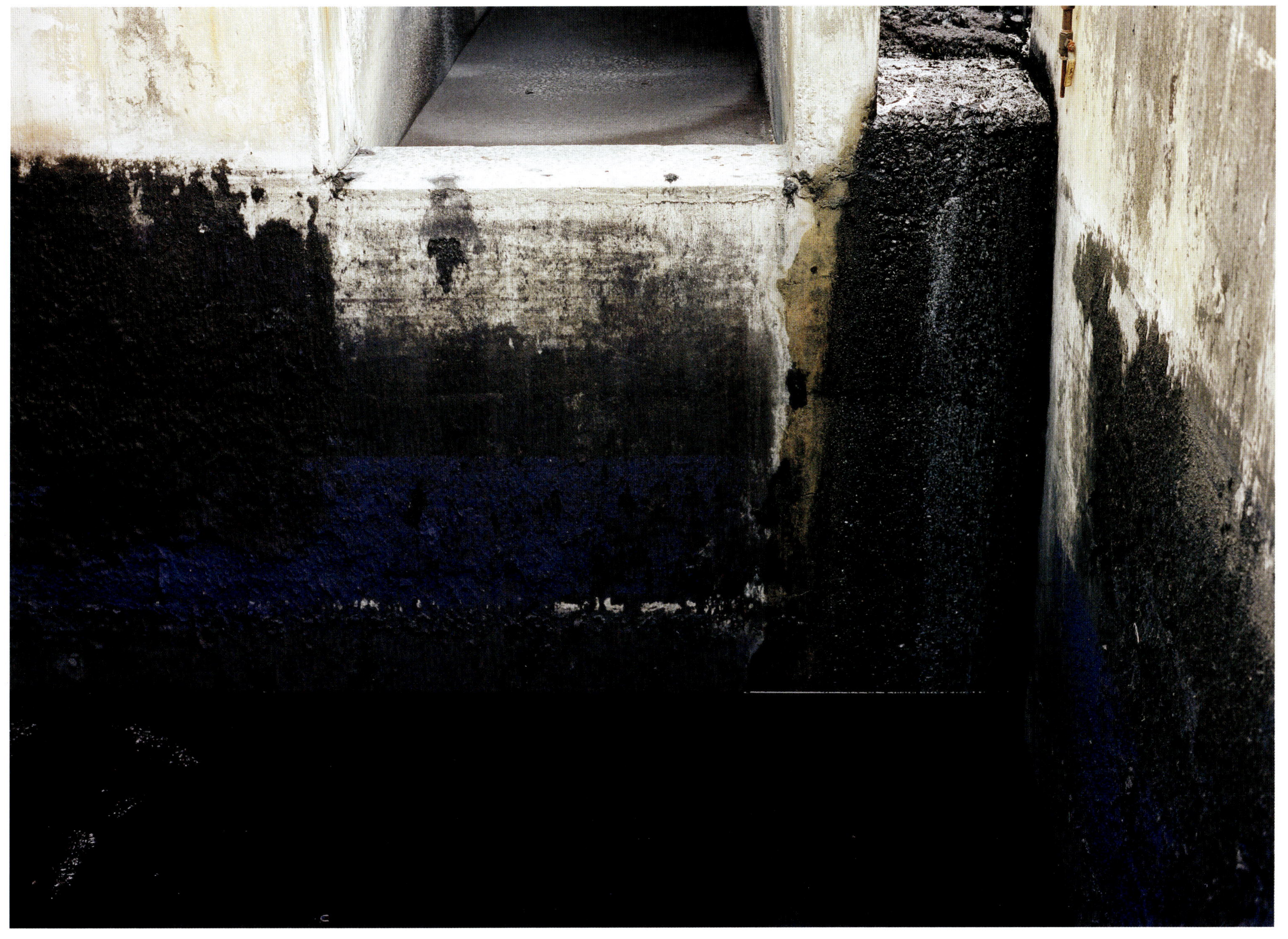

Indigo dye pits

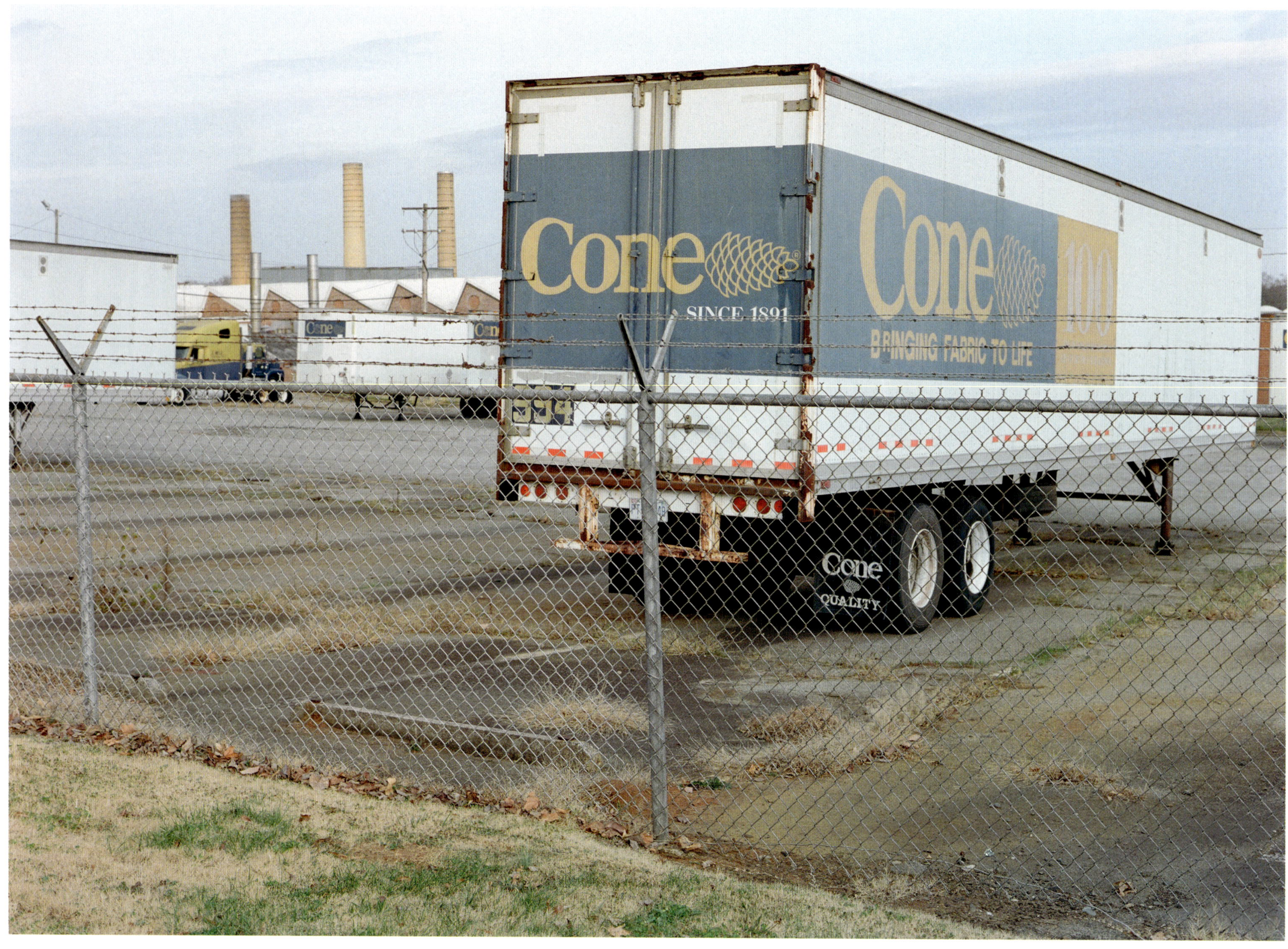
Cone®
SINCE 1891
Cone® 100
BRINGING FABRIC TO LIFE
Cone
QUALITY

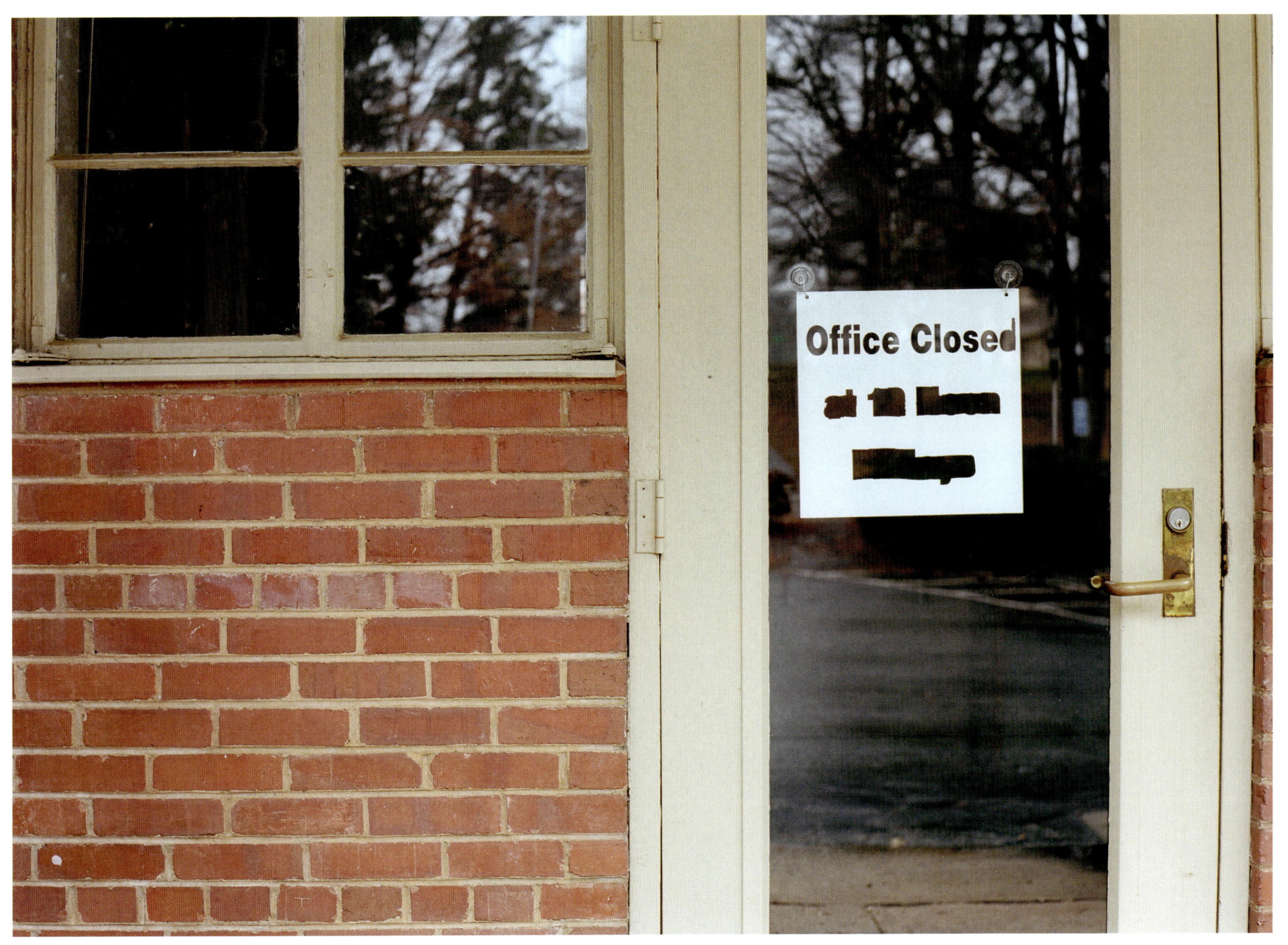

Front door at Main Office

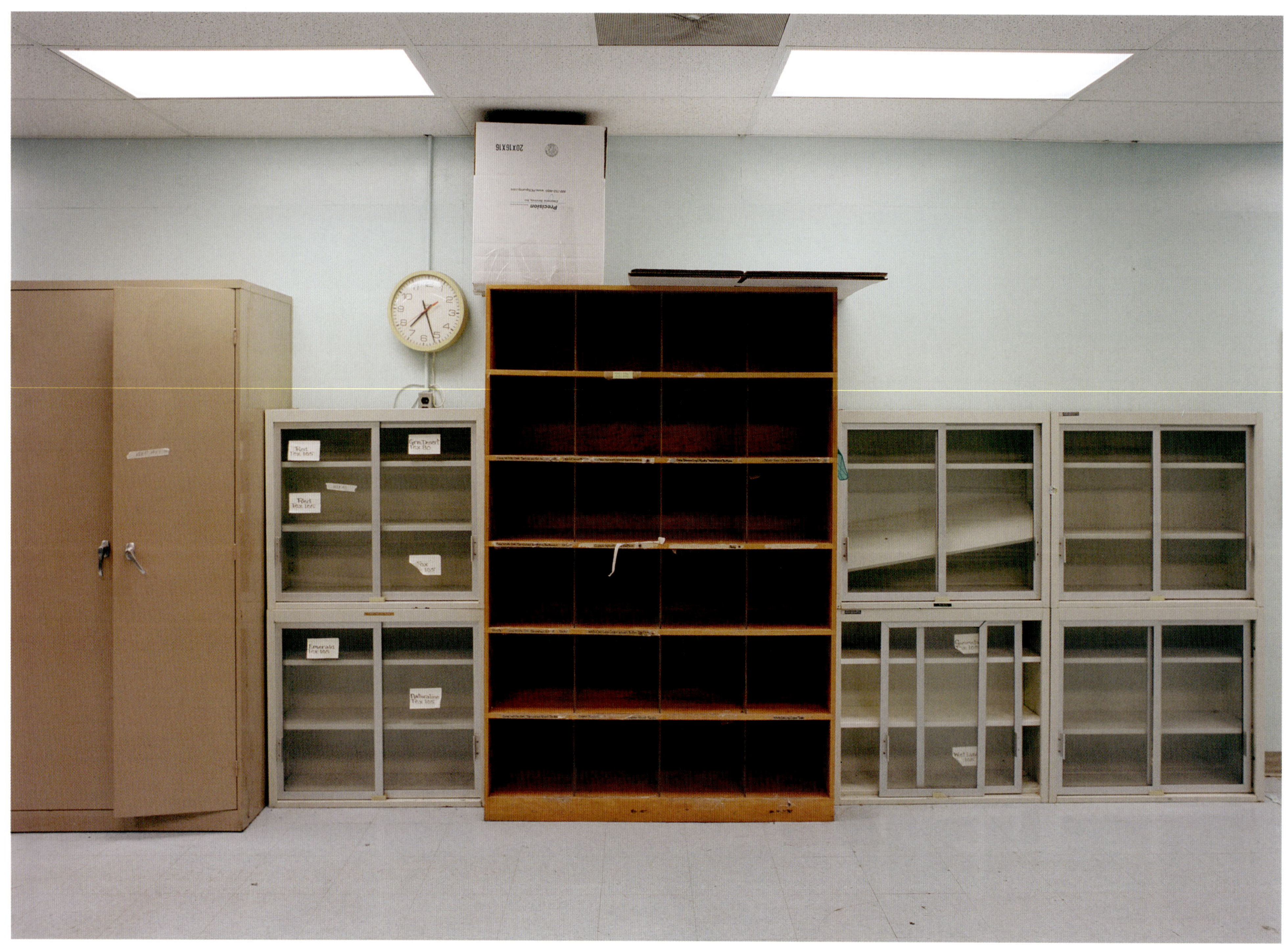

Empty cabinets in the Garment Lab

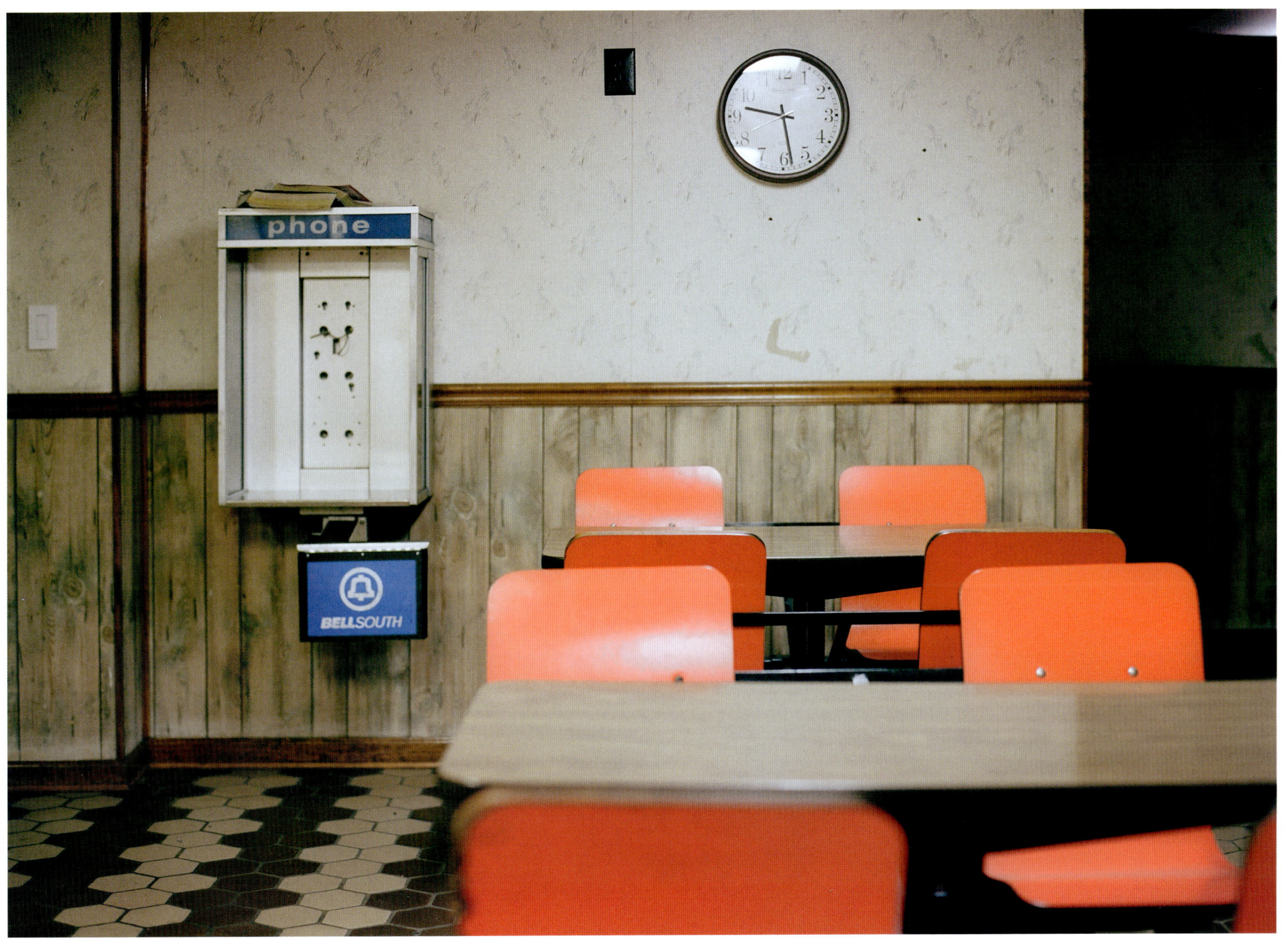

Clock still ticking in the "Canteen," and inoperable phone

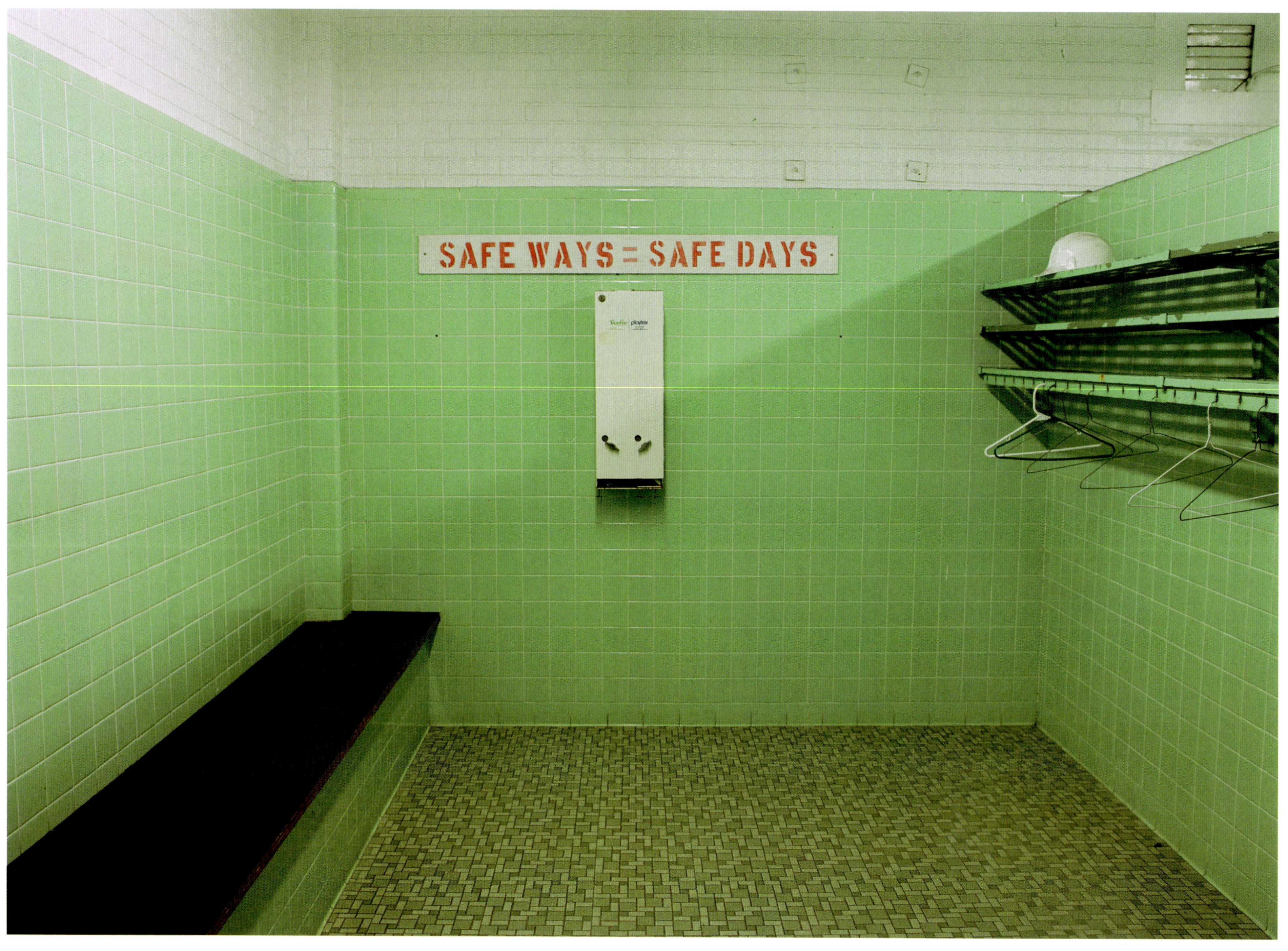

Women's locker room

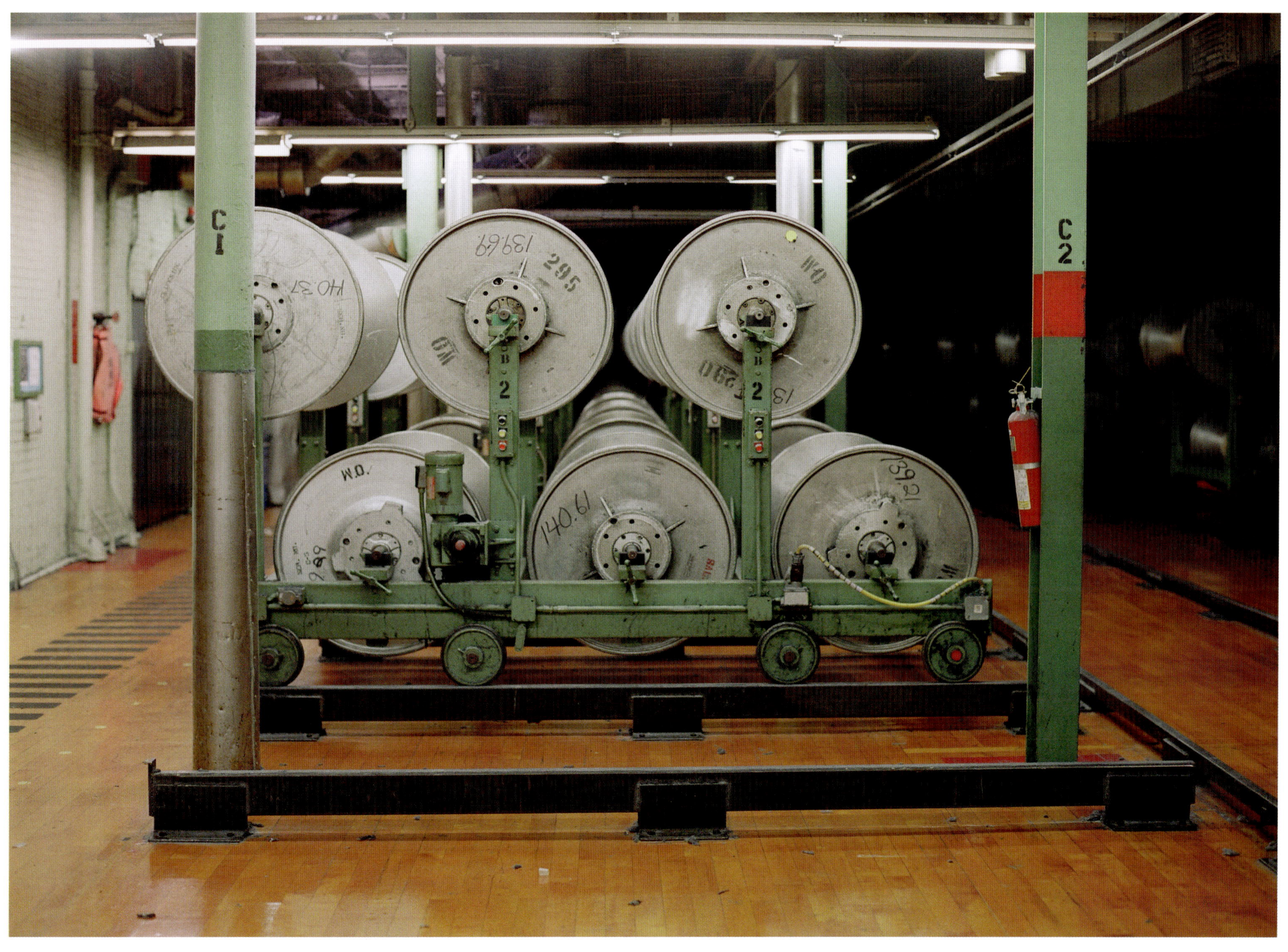

Section beams on cart in the Slashing Department

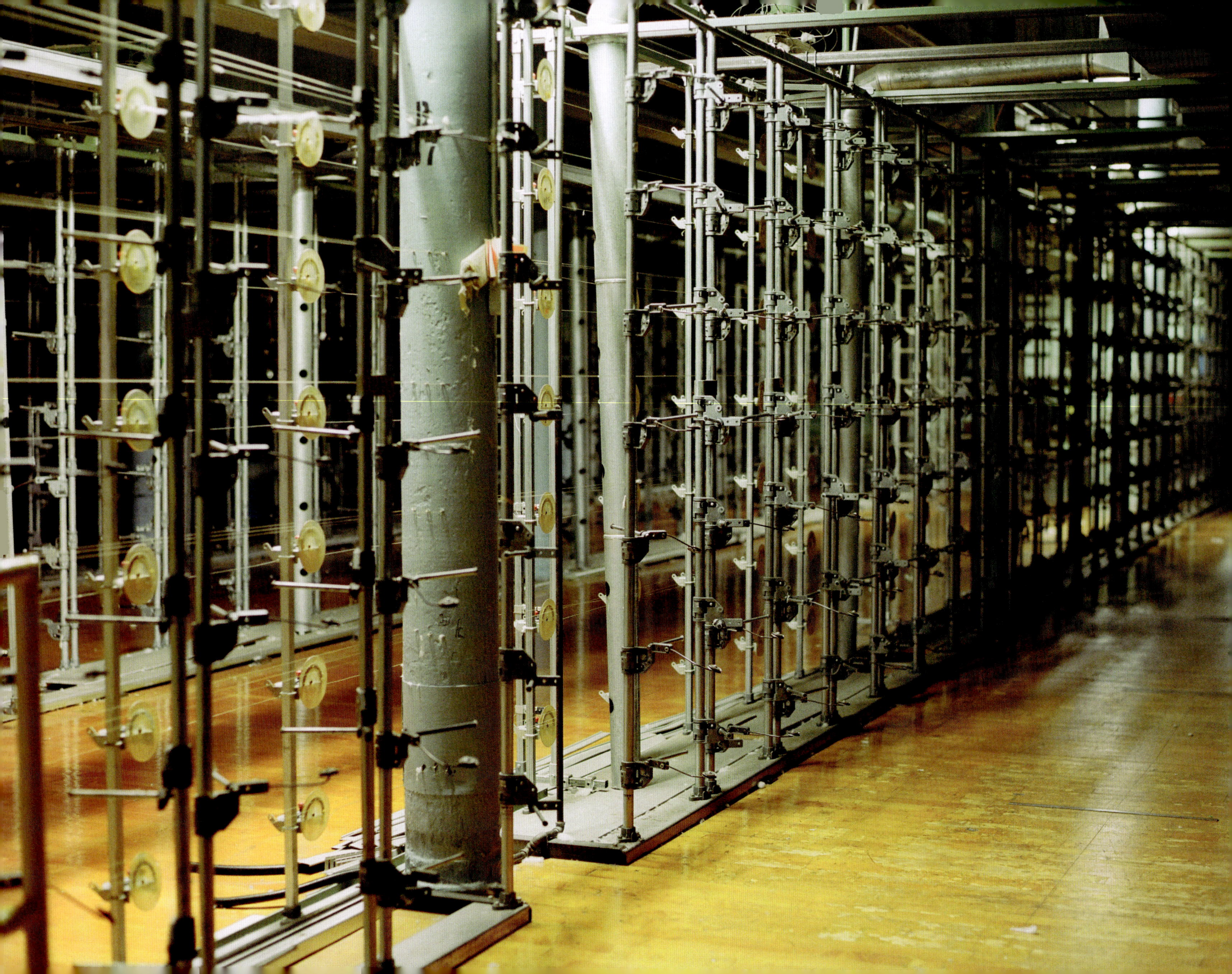

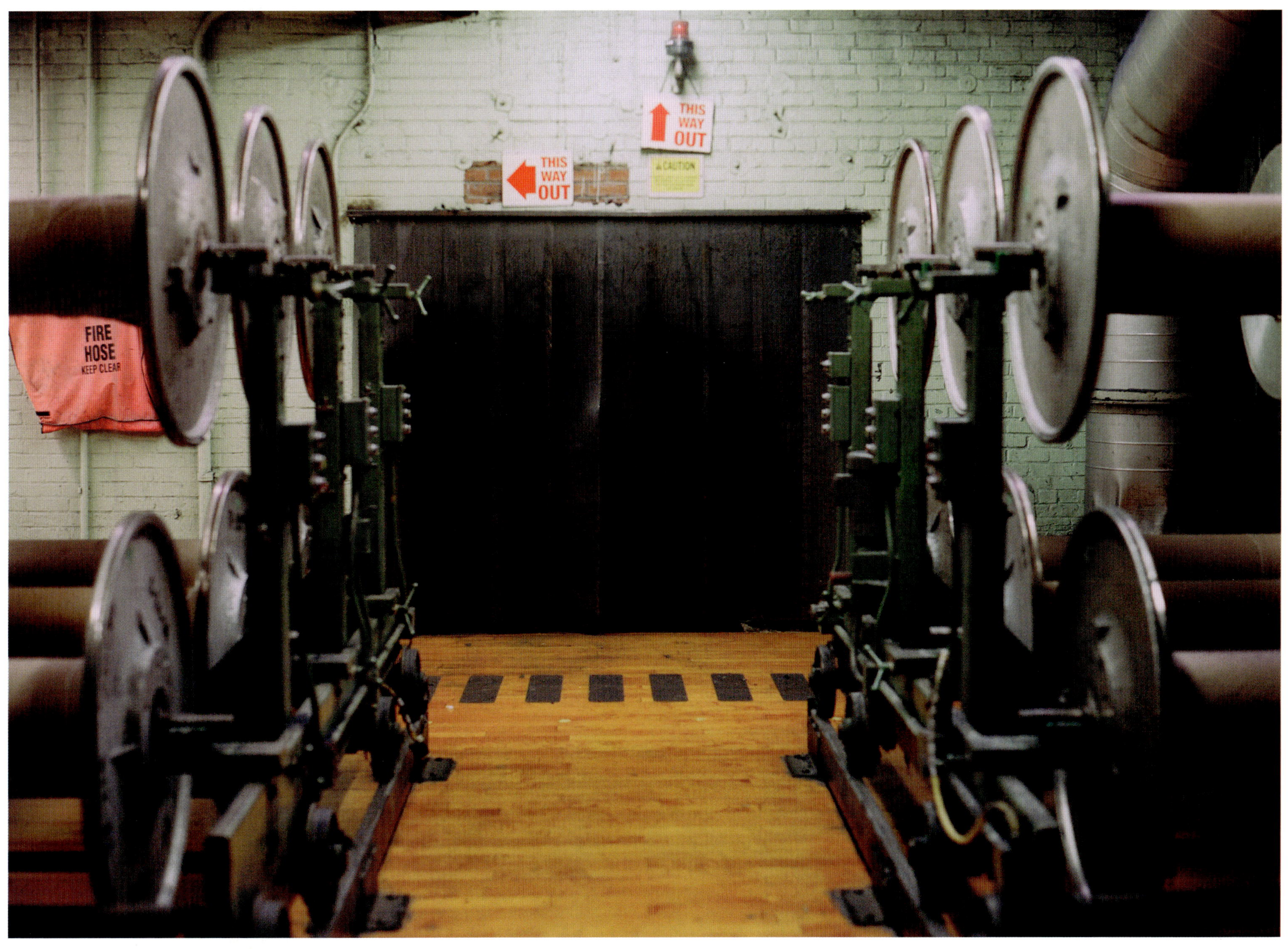

Empty yarn creels in the Warping Department   //   Empty section beams in the Slashing Department

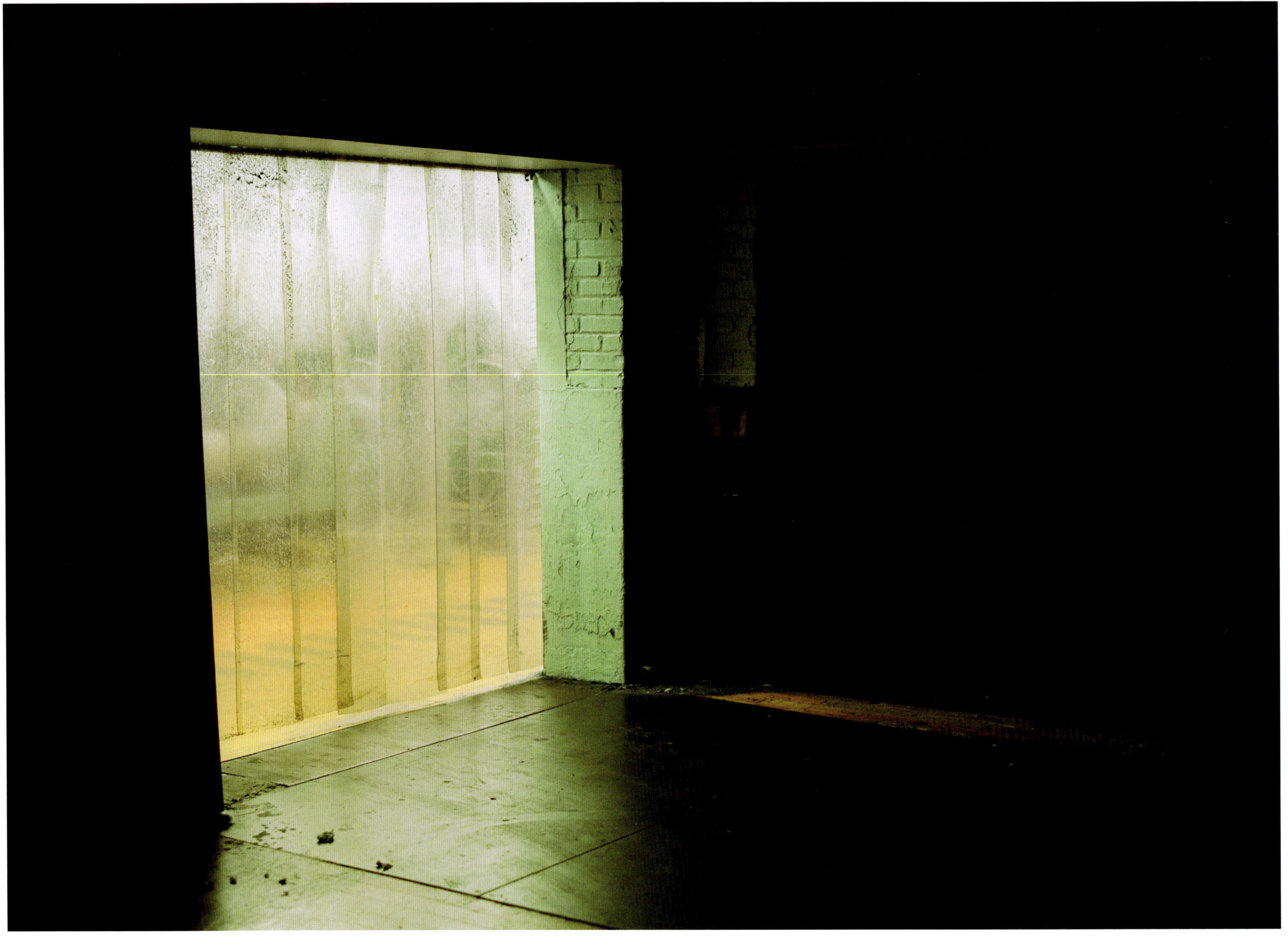

Entryway to main Weave Room from the 16th Street Annex Building

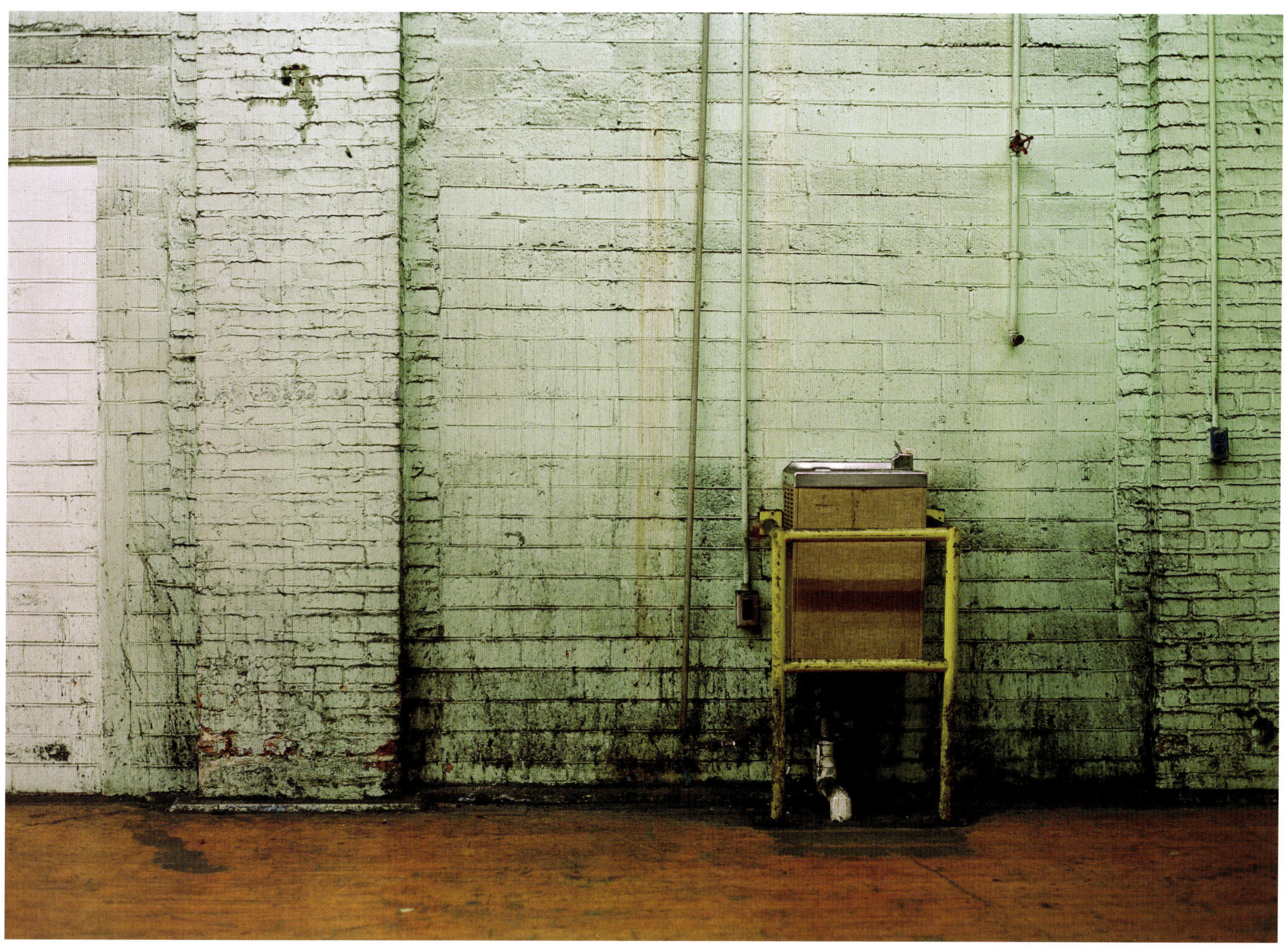

Drinking fountain in the 16th Street Annex Building

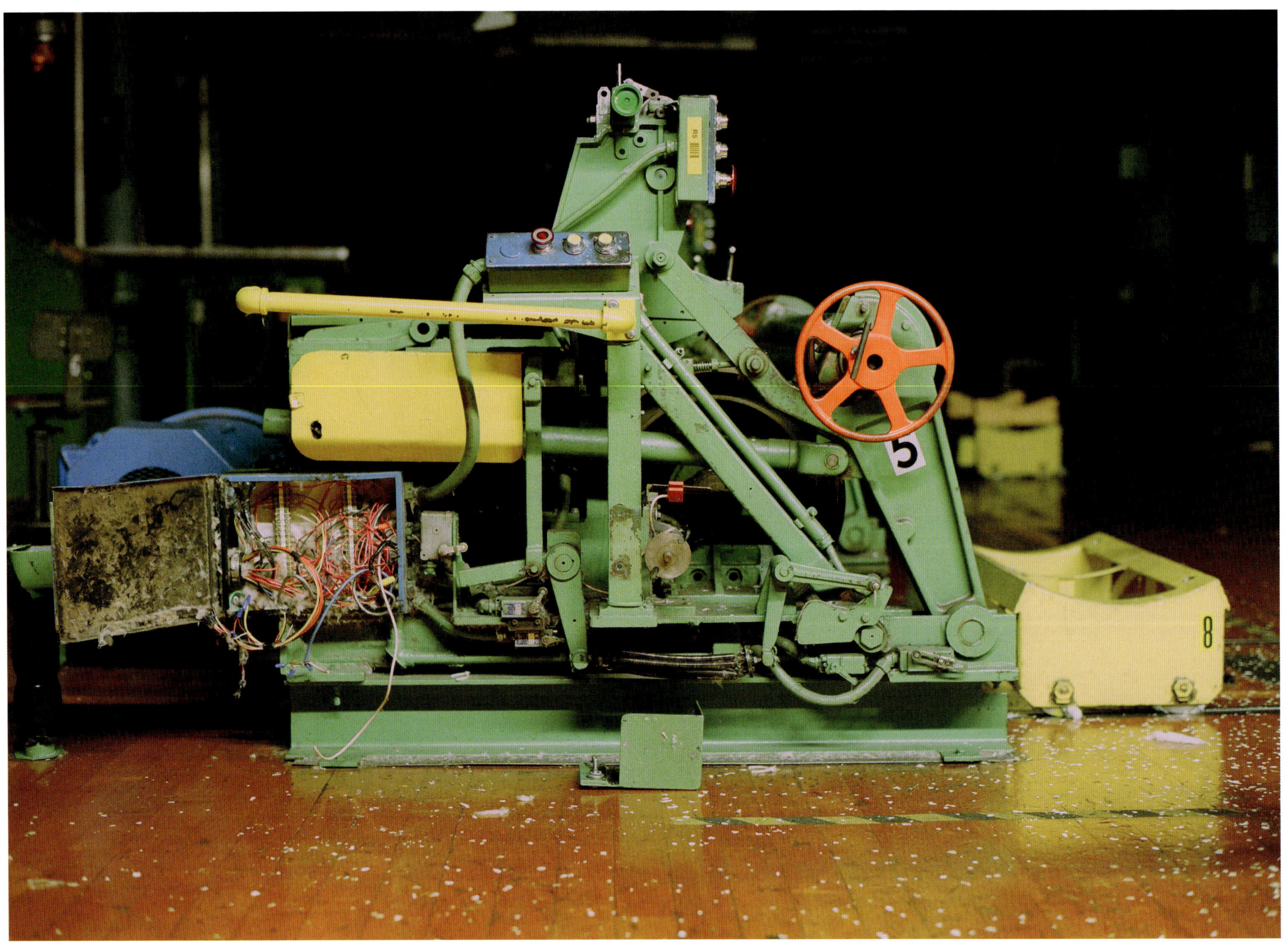

Ball warper in the Warping Department

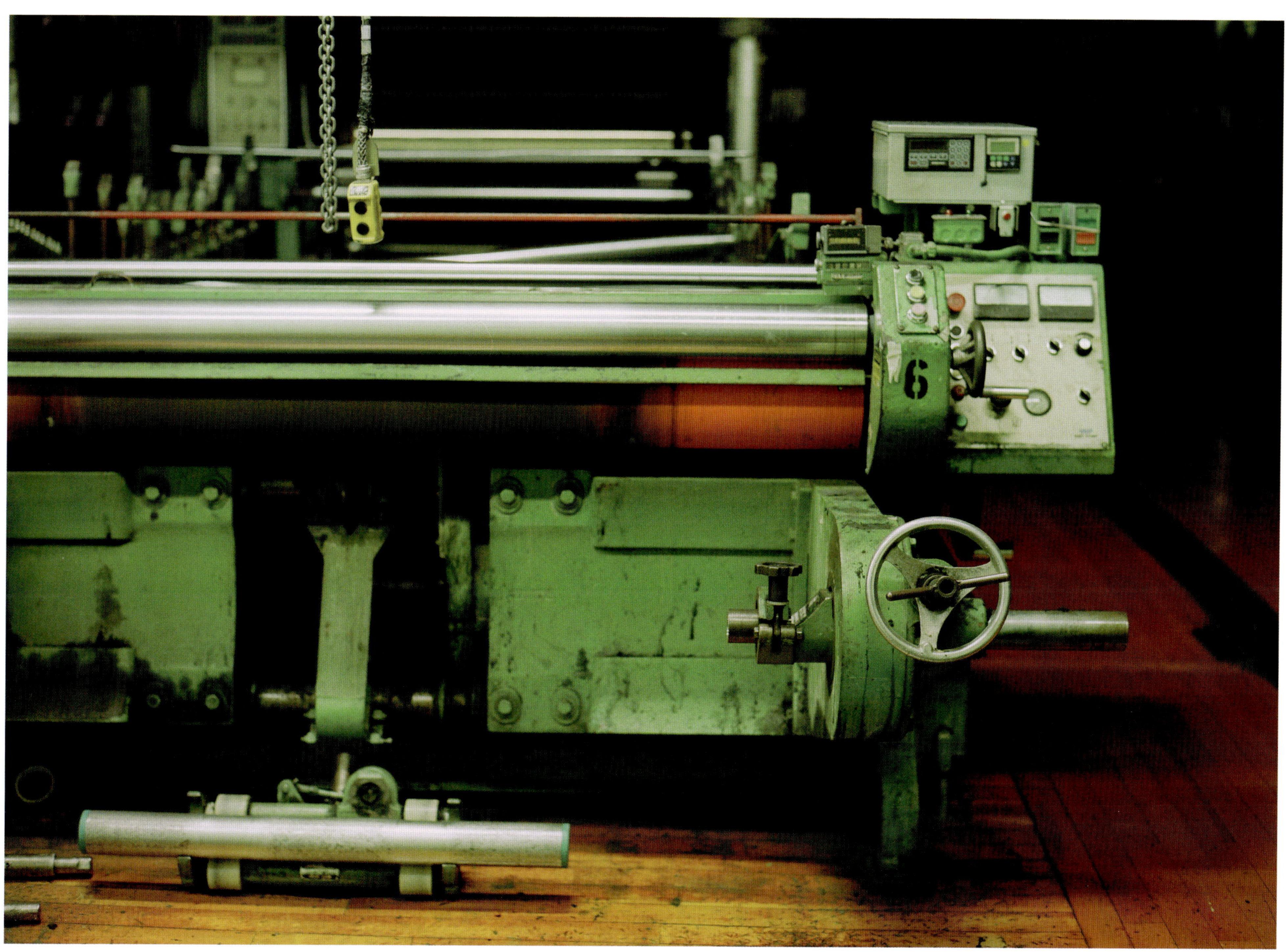

Slasher in the Slashing Department

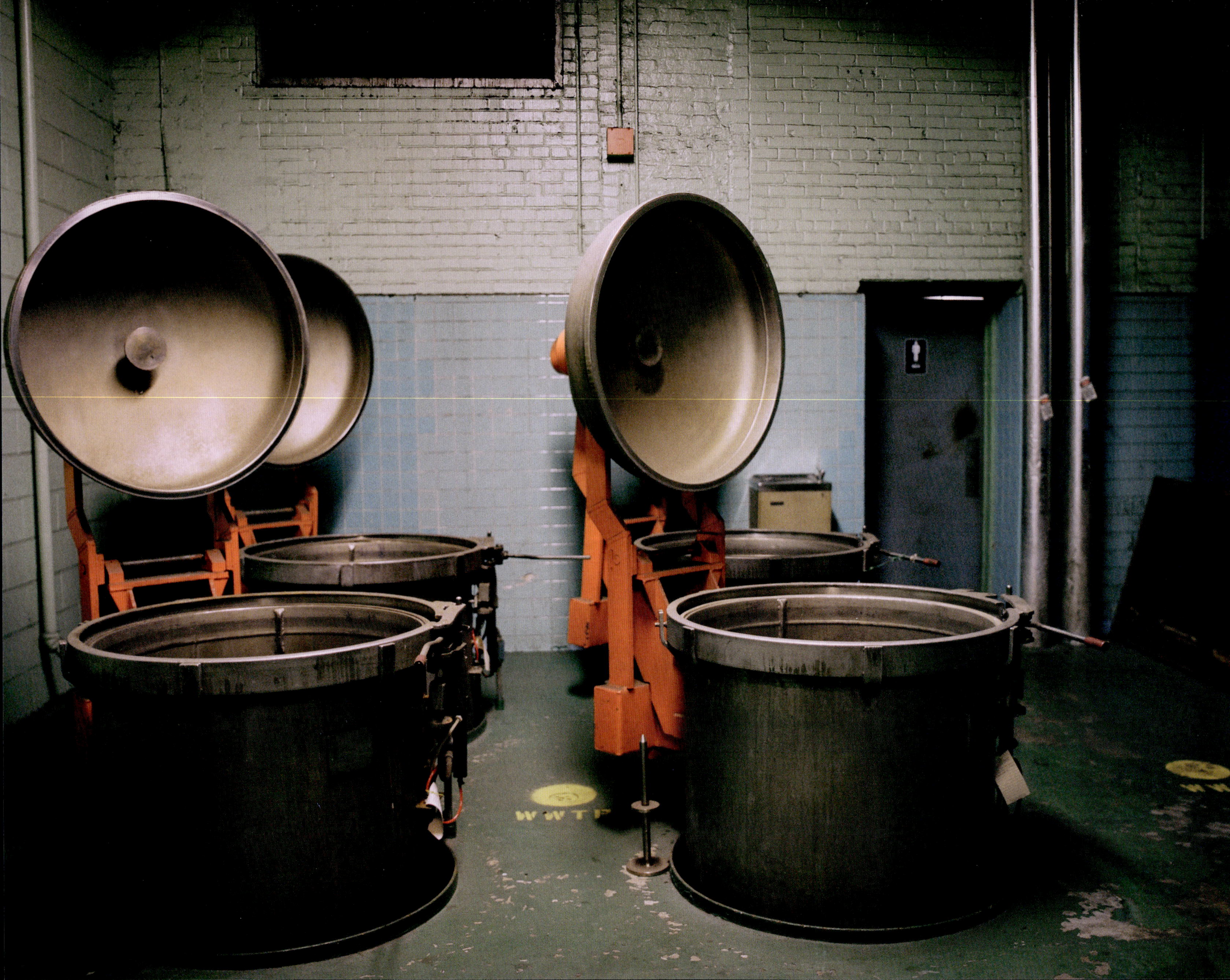

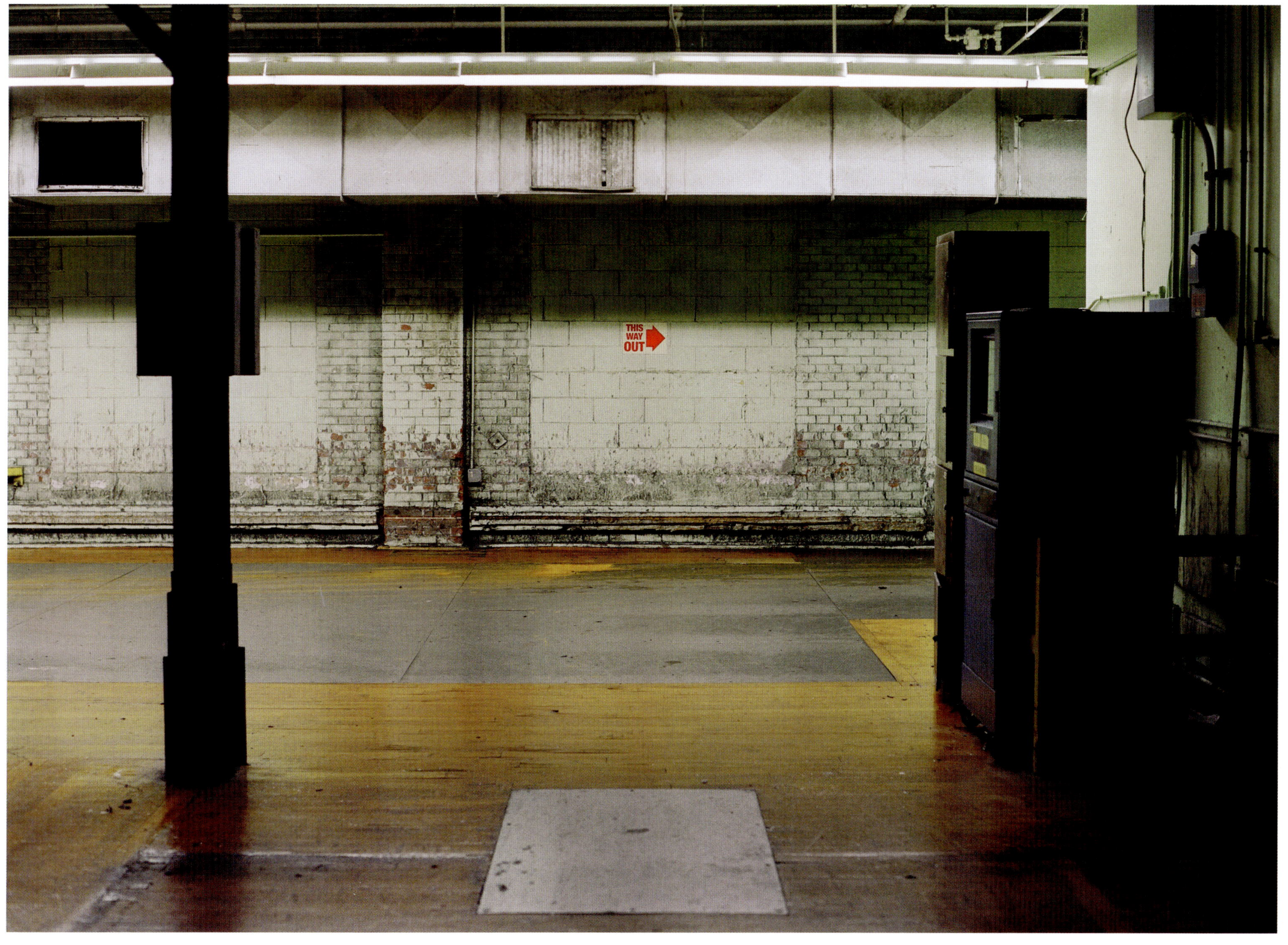

Dye vats for selvage yarn dyeing near the Weave Room   //   "This Way Out" sign and electrical pannels in the Tying-In Room

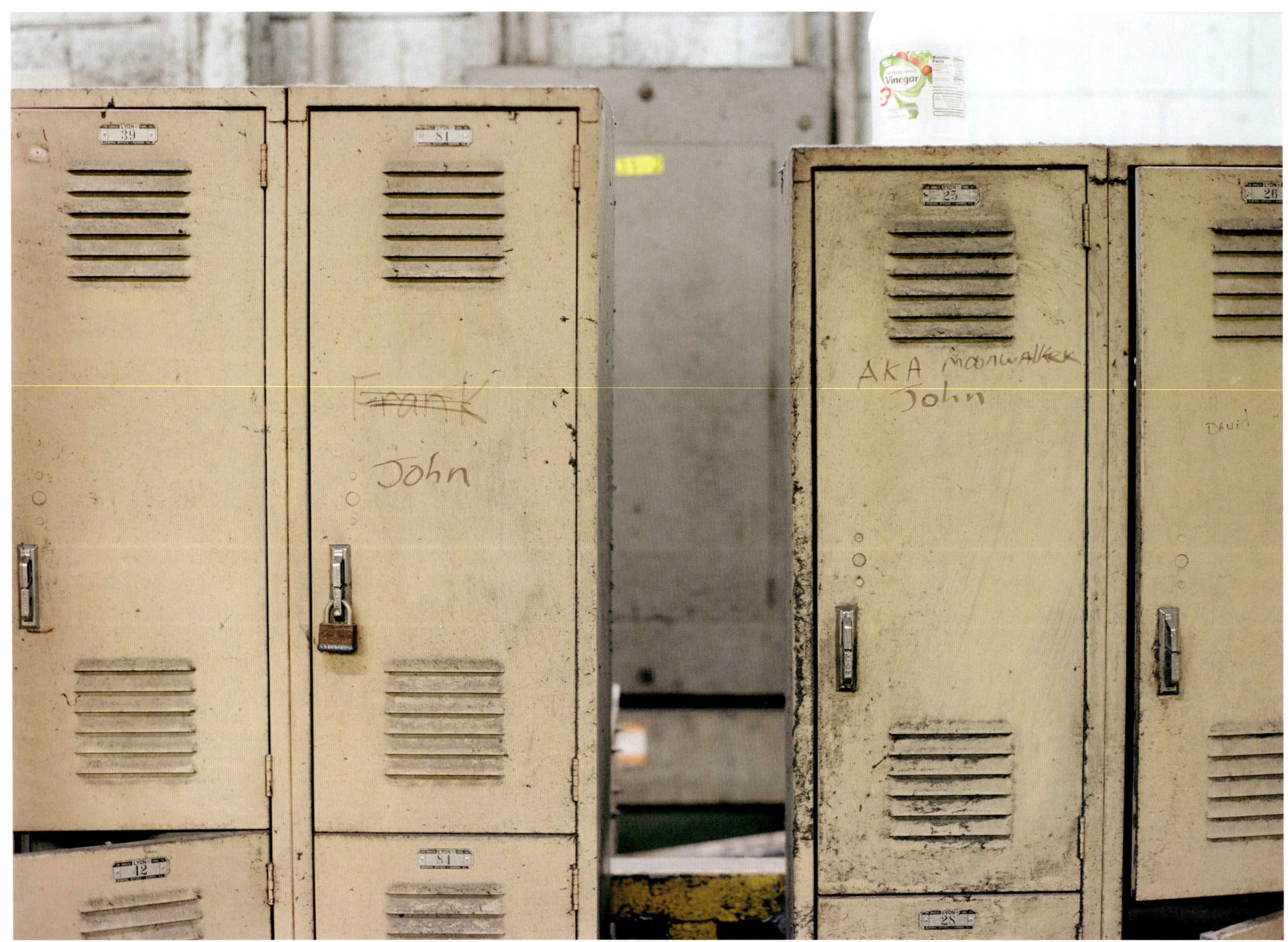

Employee lockers in the Slashing Department

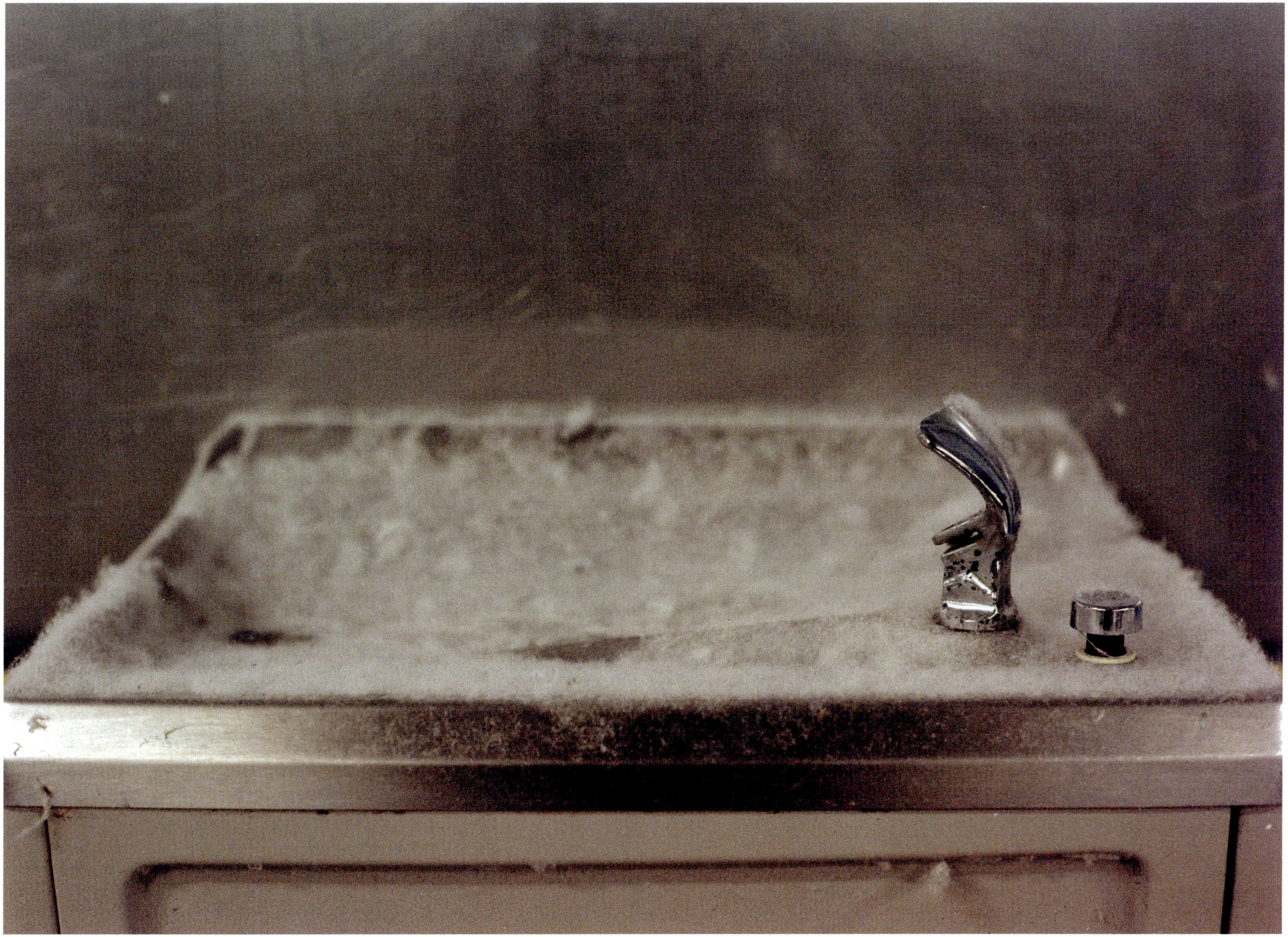

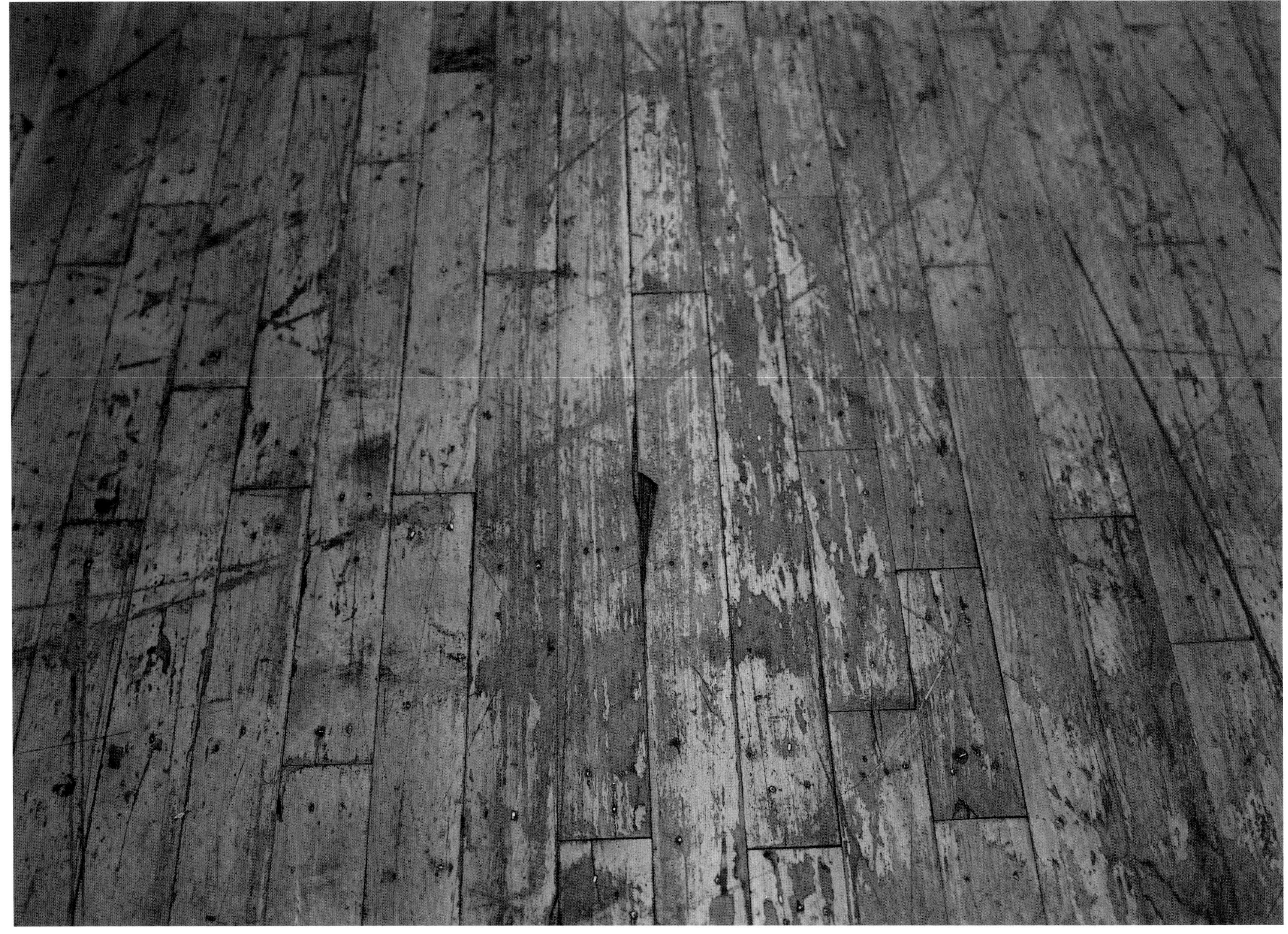

Maple flooring in the Spinning Department   //   Signage in the Rebeaming Department

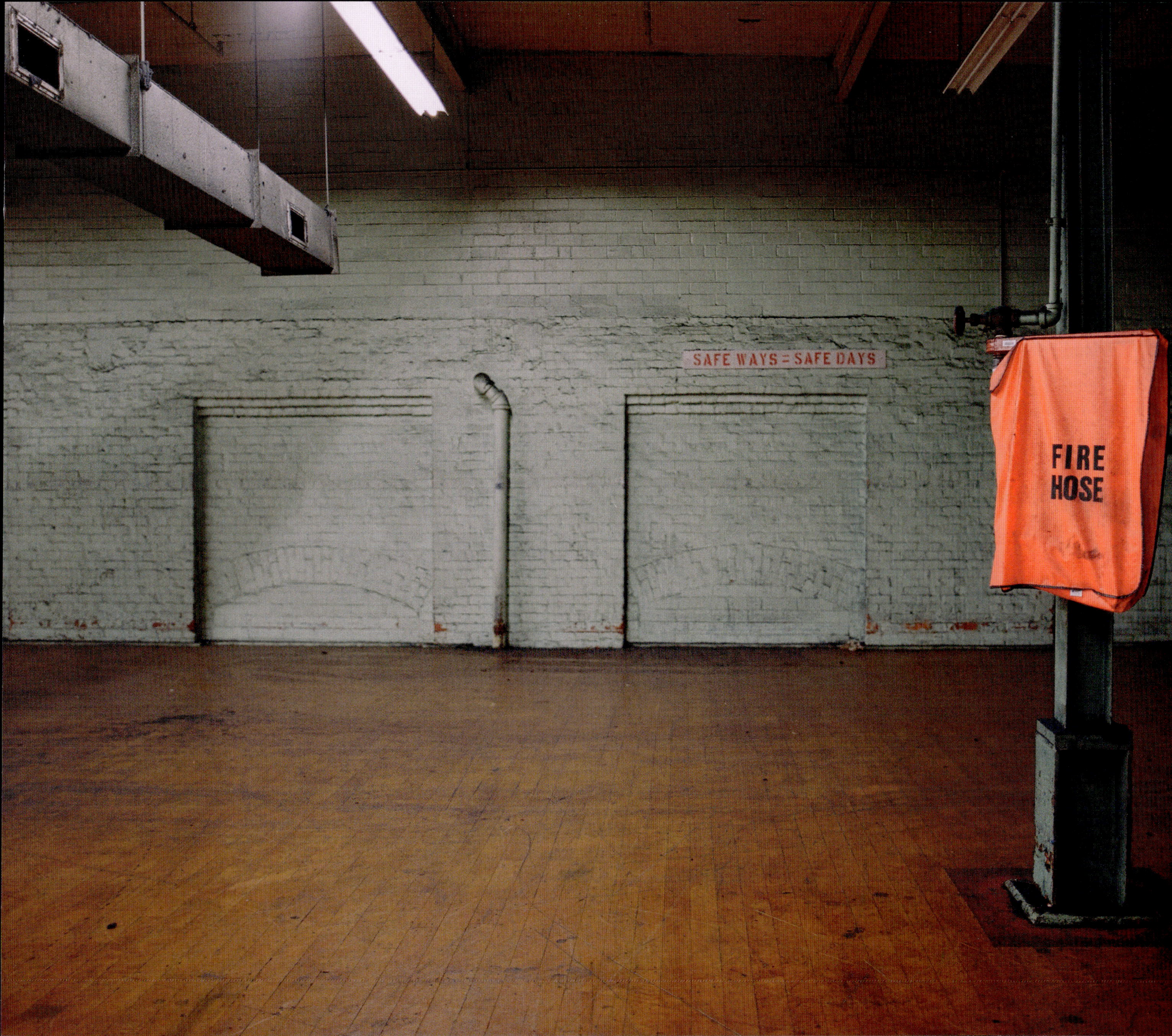

SAFE WAYS = SAFE DAYS
FIRE HOSE

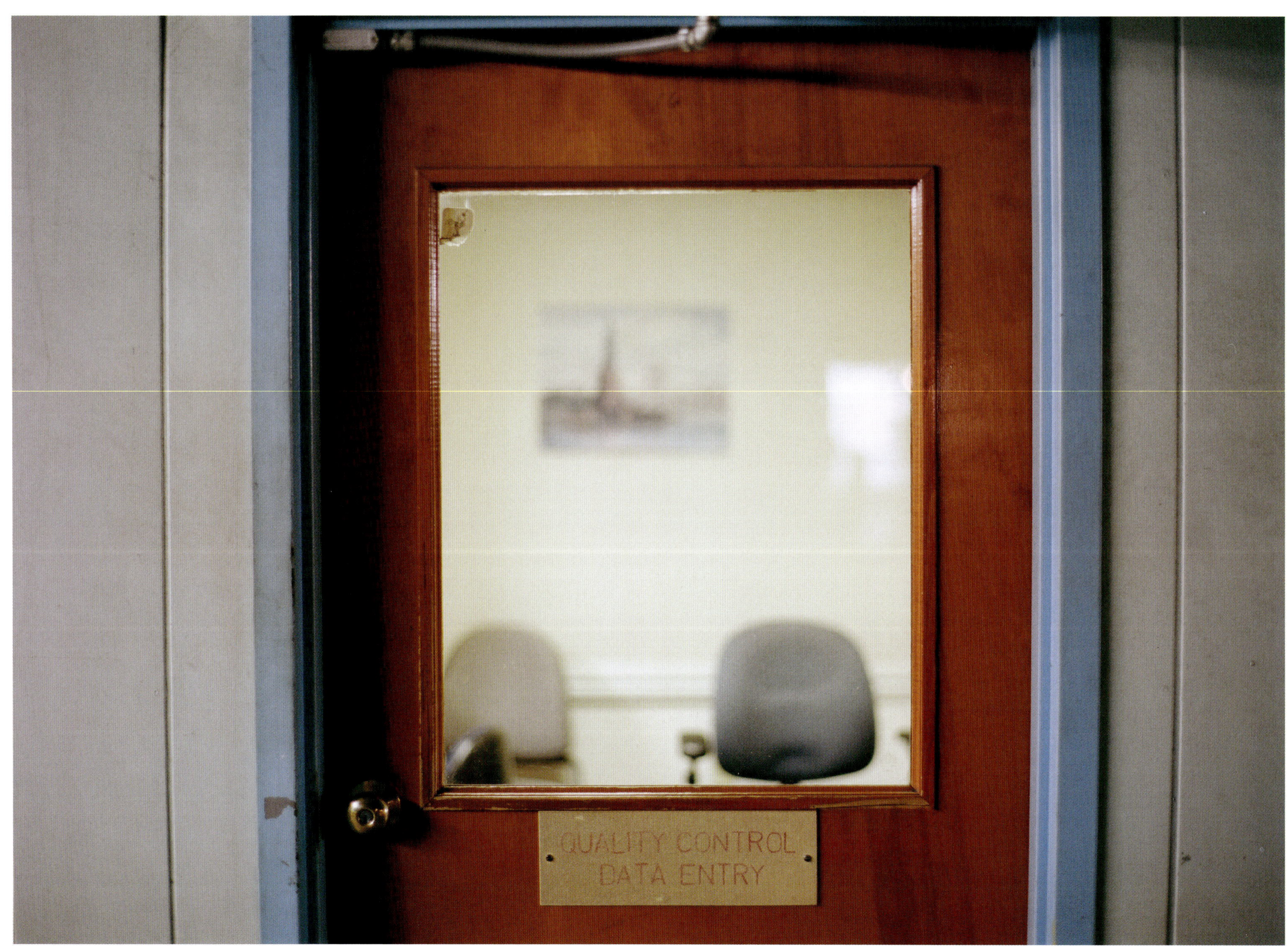

Quality Control Room in the Finishing Department

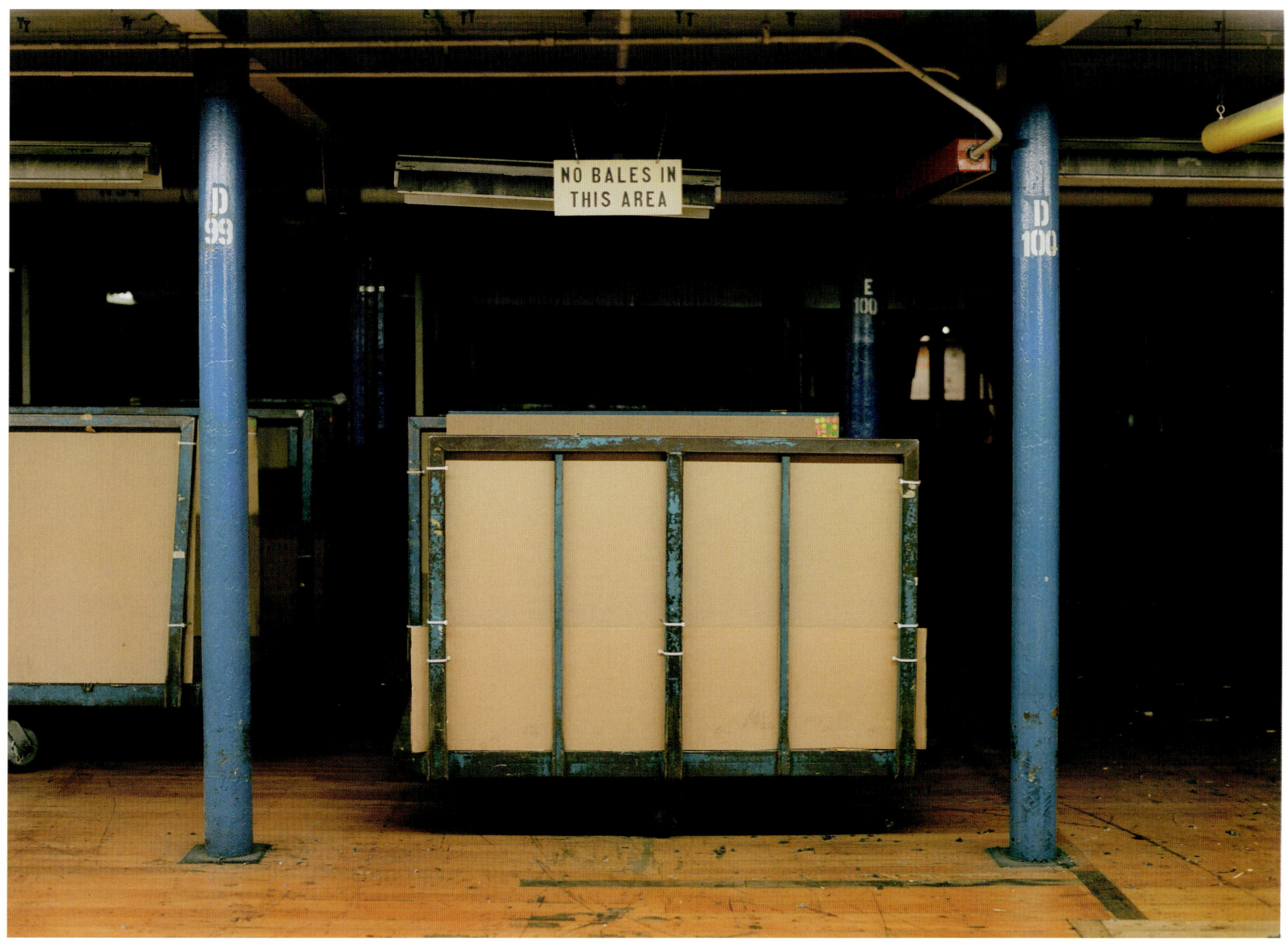

Textile carts on the Finishing Floor

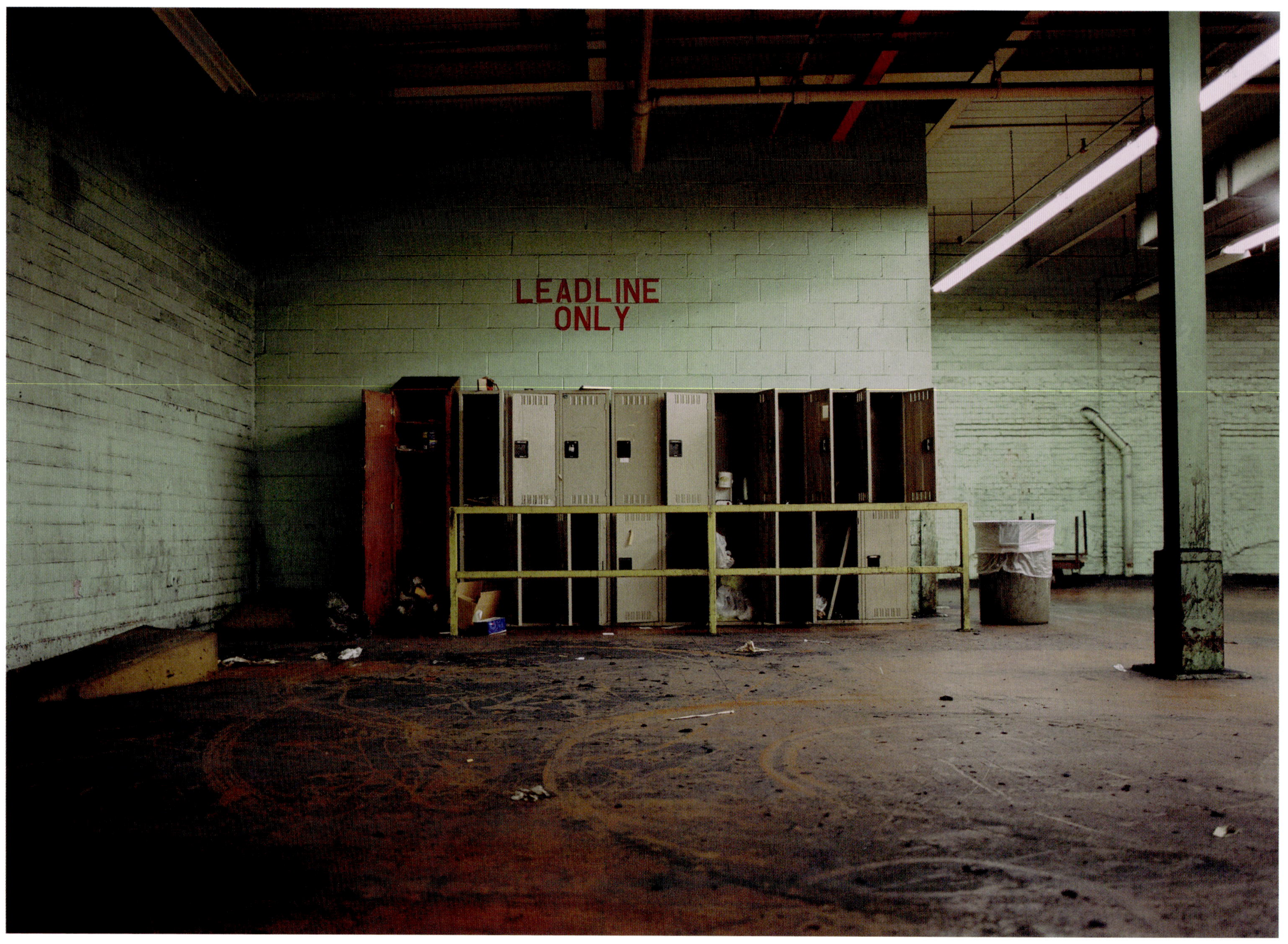

Employee lockers in the Rebeaming Department

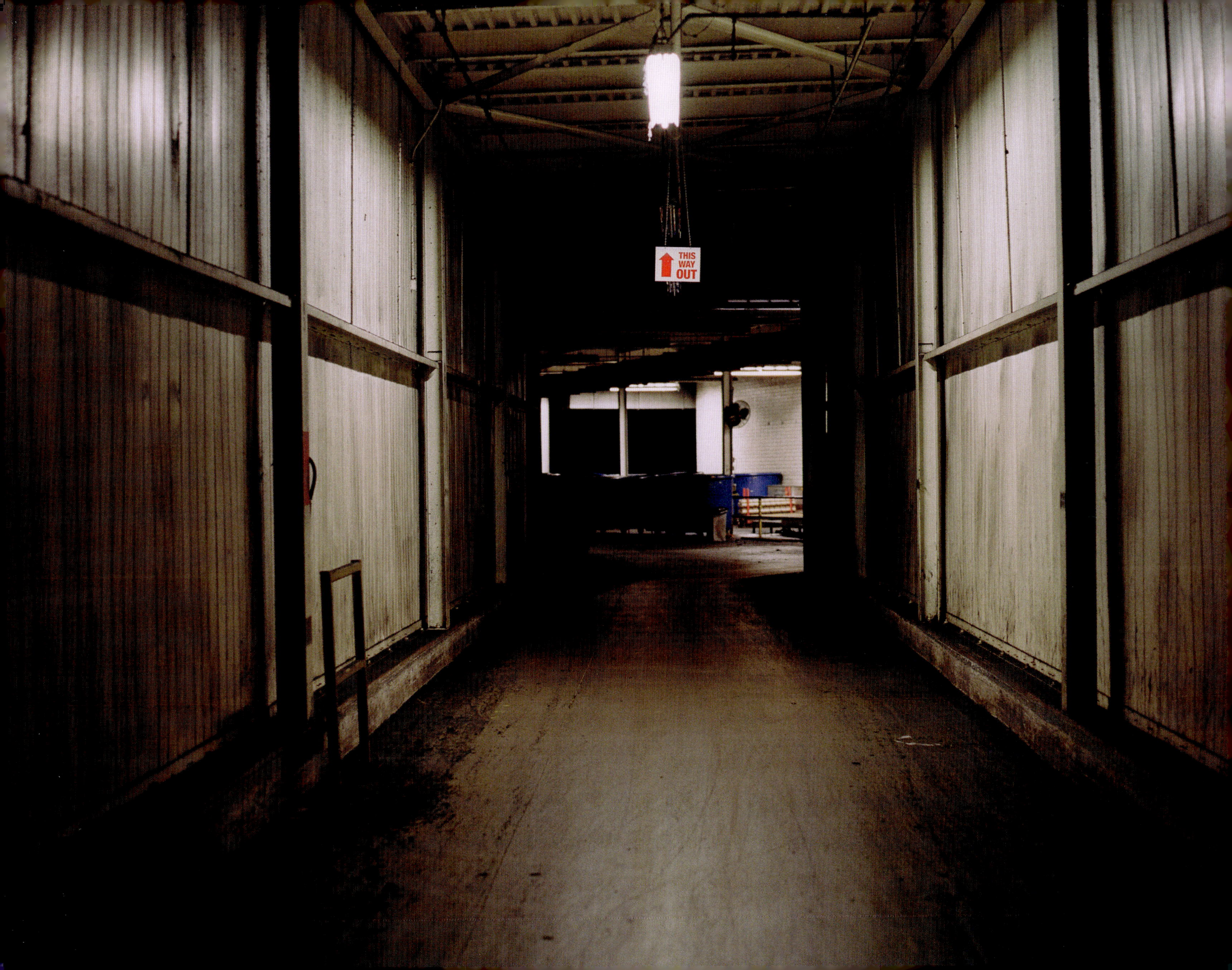
THIS
WAY
OUT

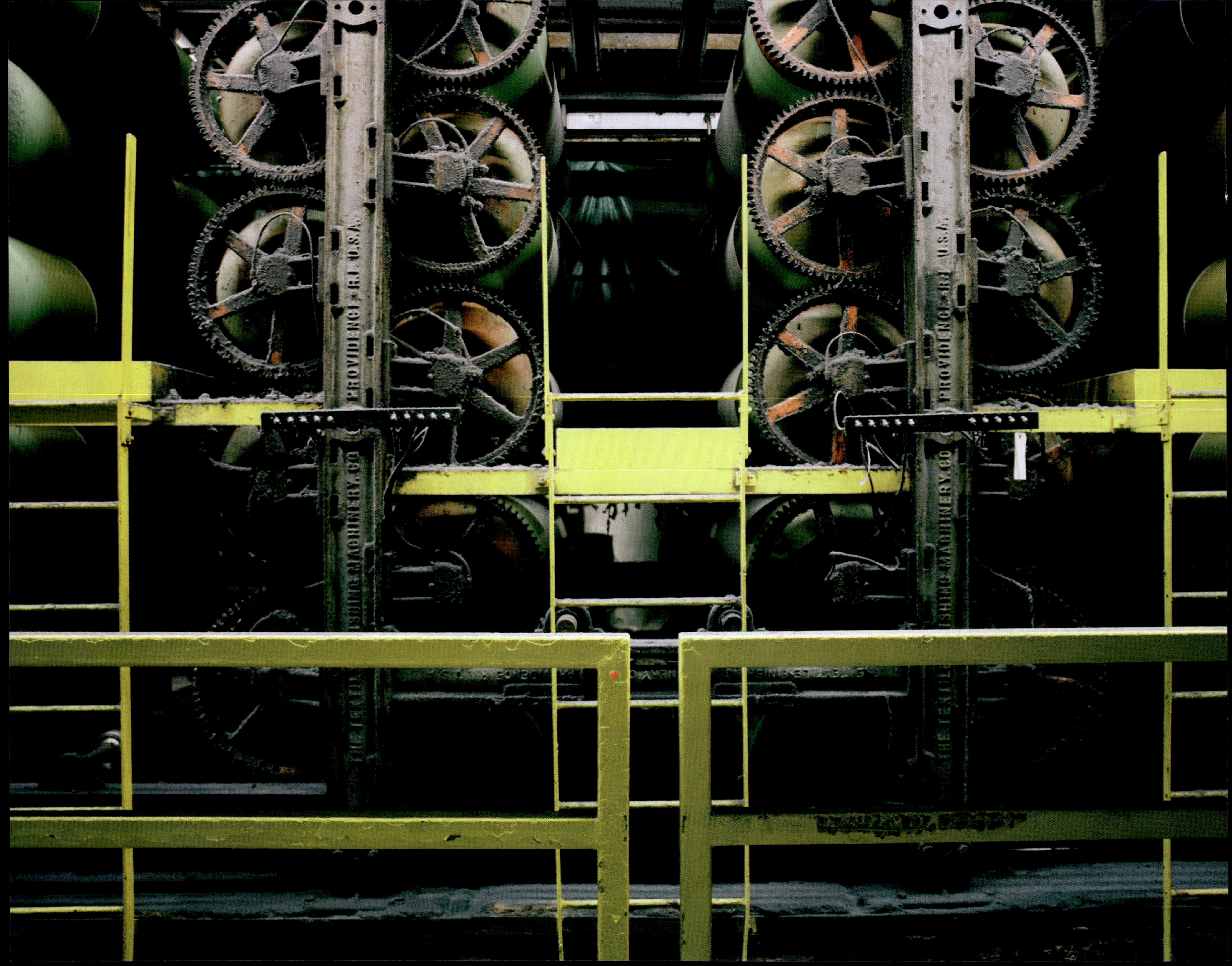

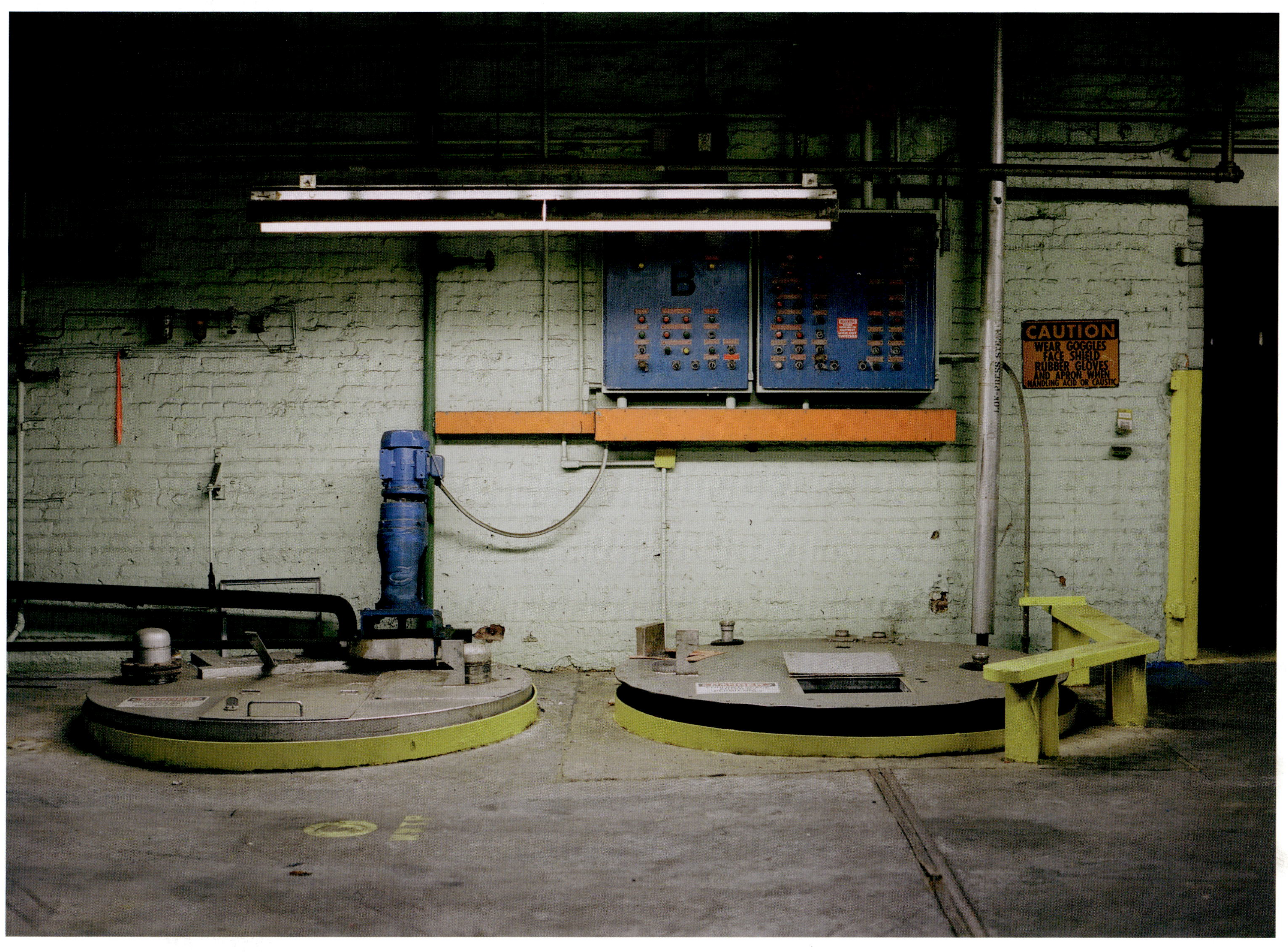

Side view of tenter frame in the Finishing Department   //   Vats and control panel in the Finishing Department

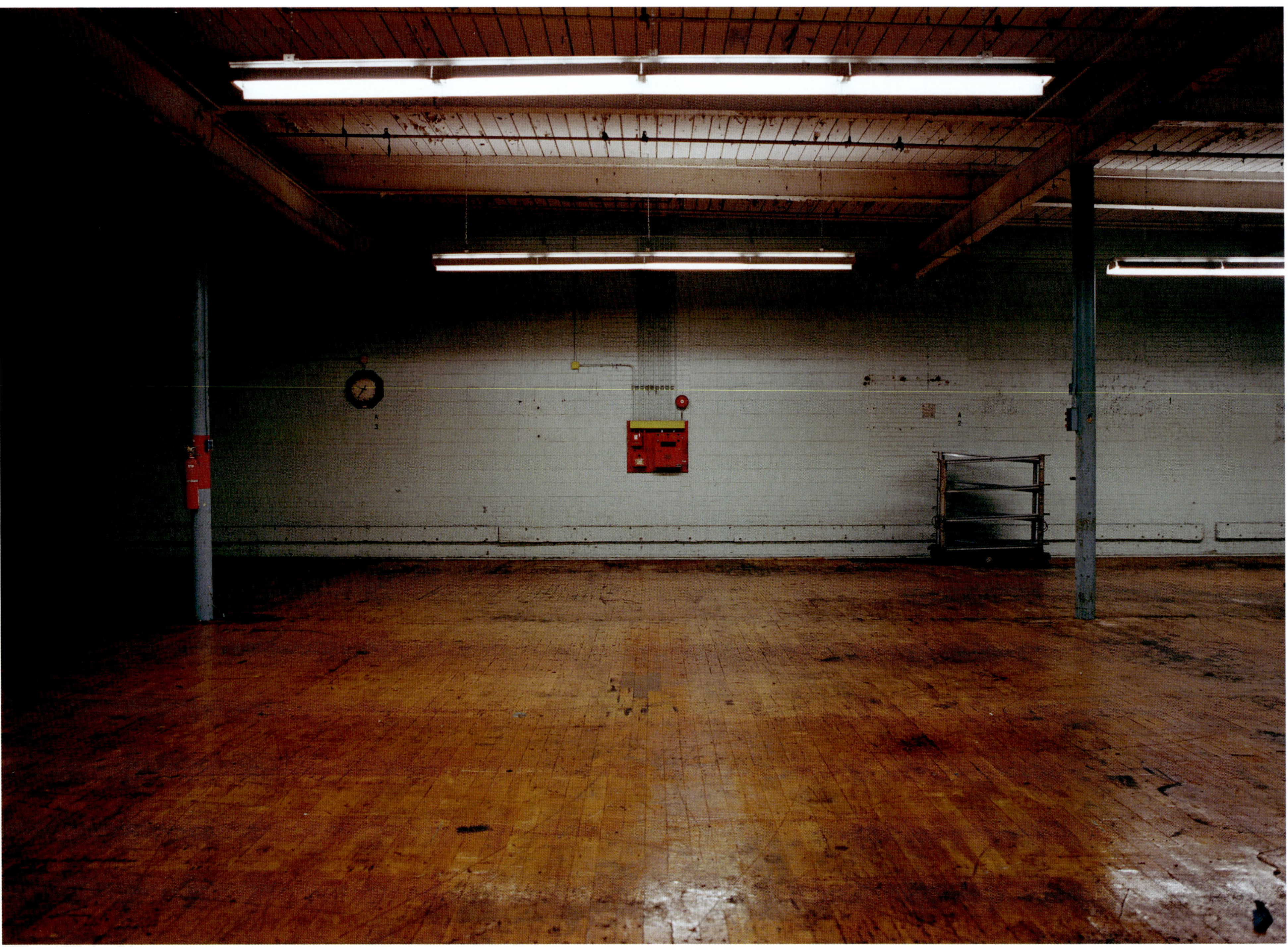

Upper floor of the 16th Street Annex Building

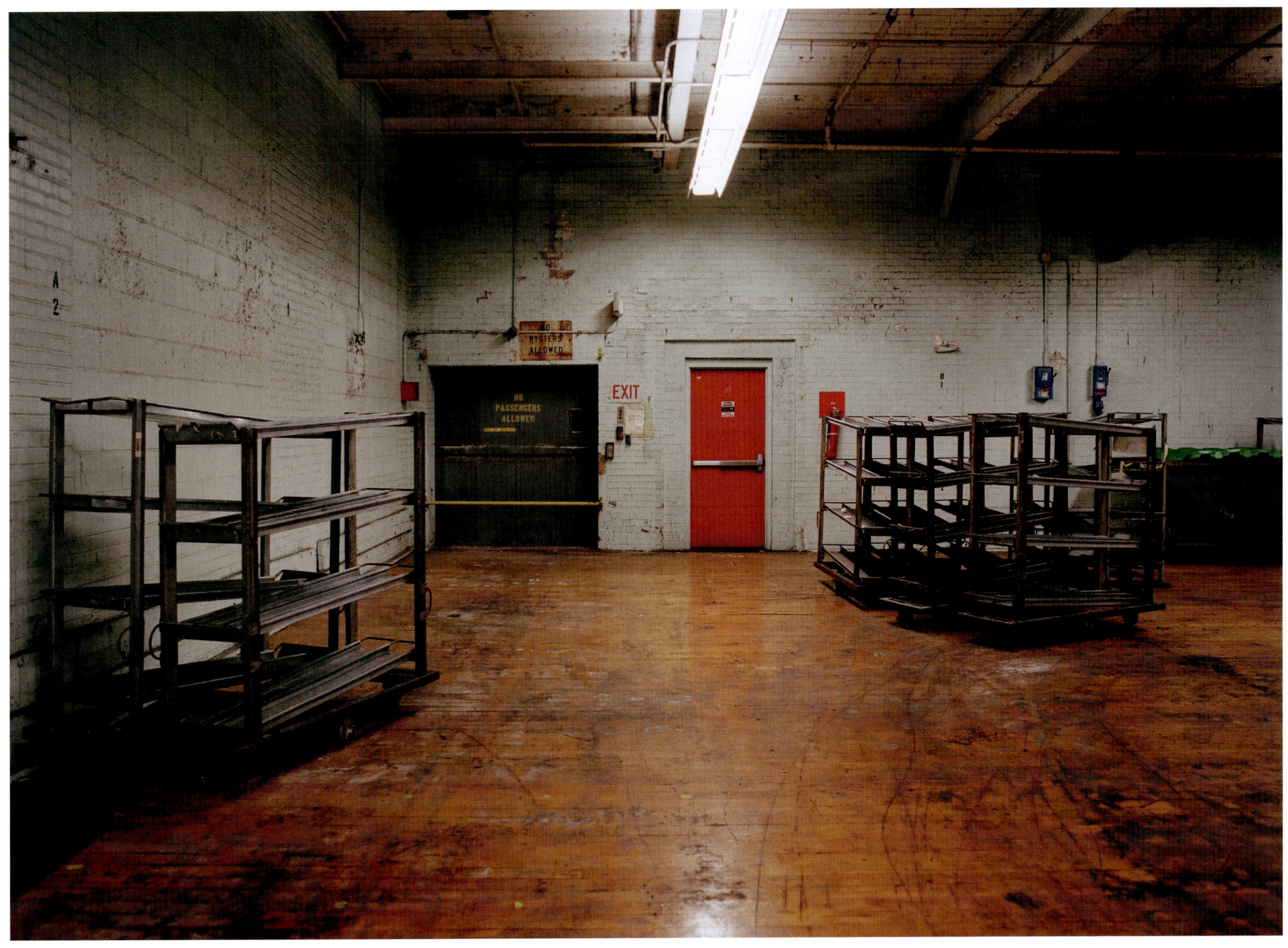

Yarn carts in the 16th Street Annex Building

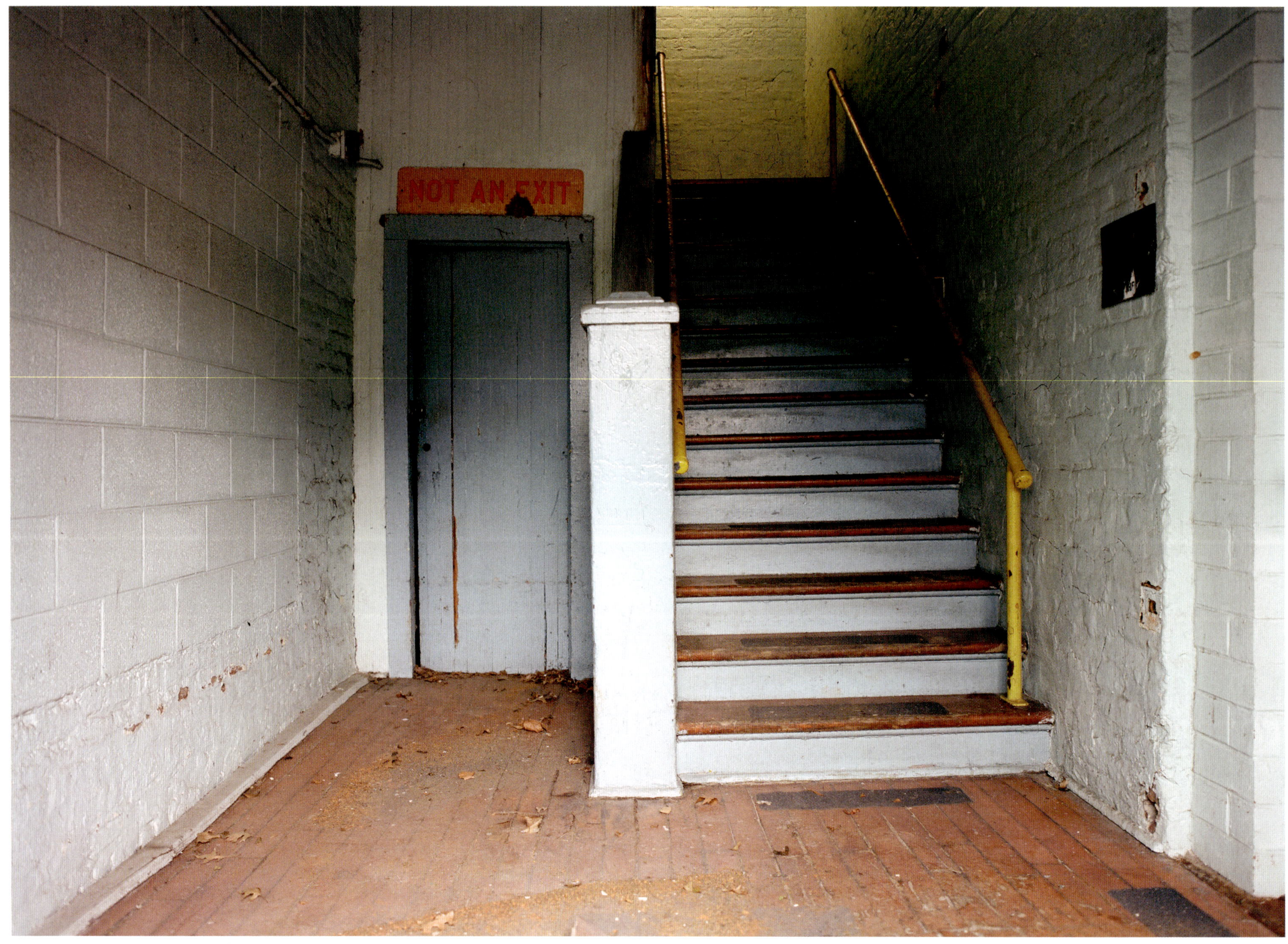

"NOT AN EXIT" sign at the weave room entrance of the 16th Street Annex Building

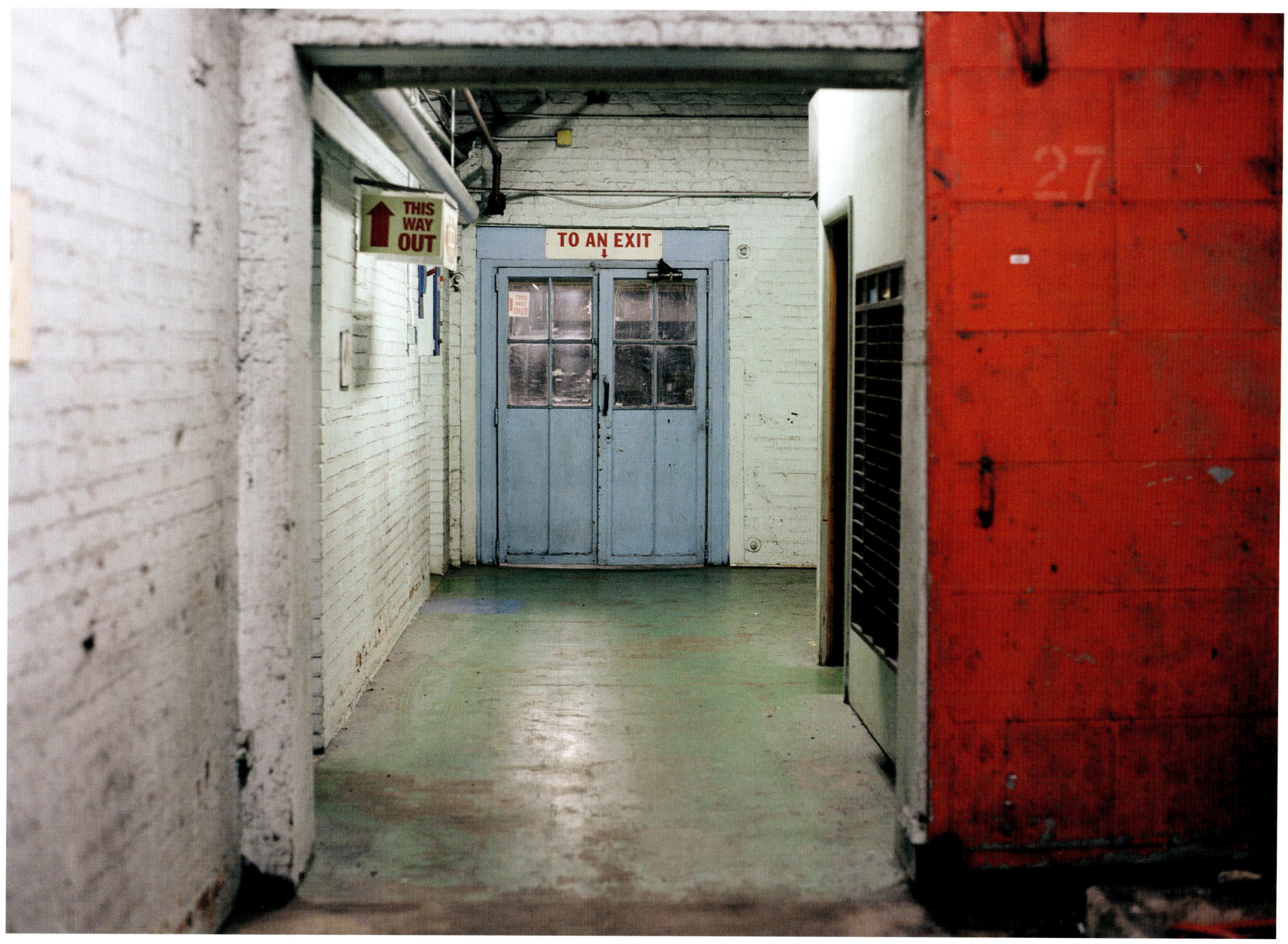

Hallway and doors in the lab area of the Finishing Department

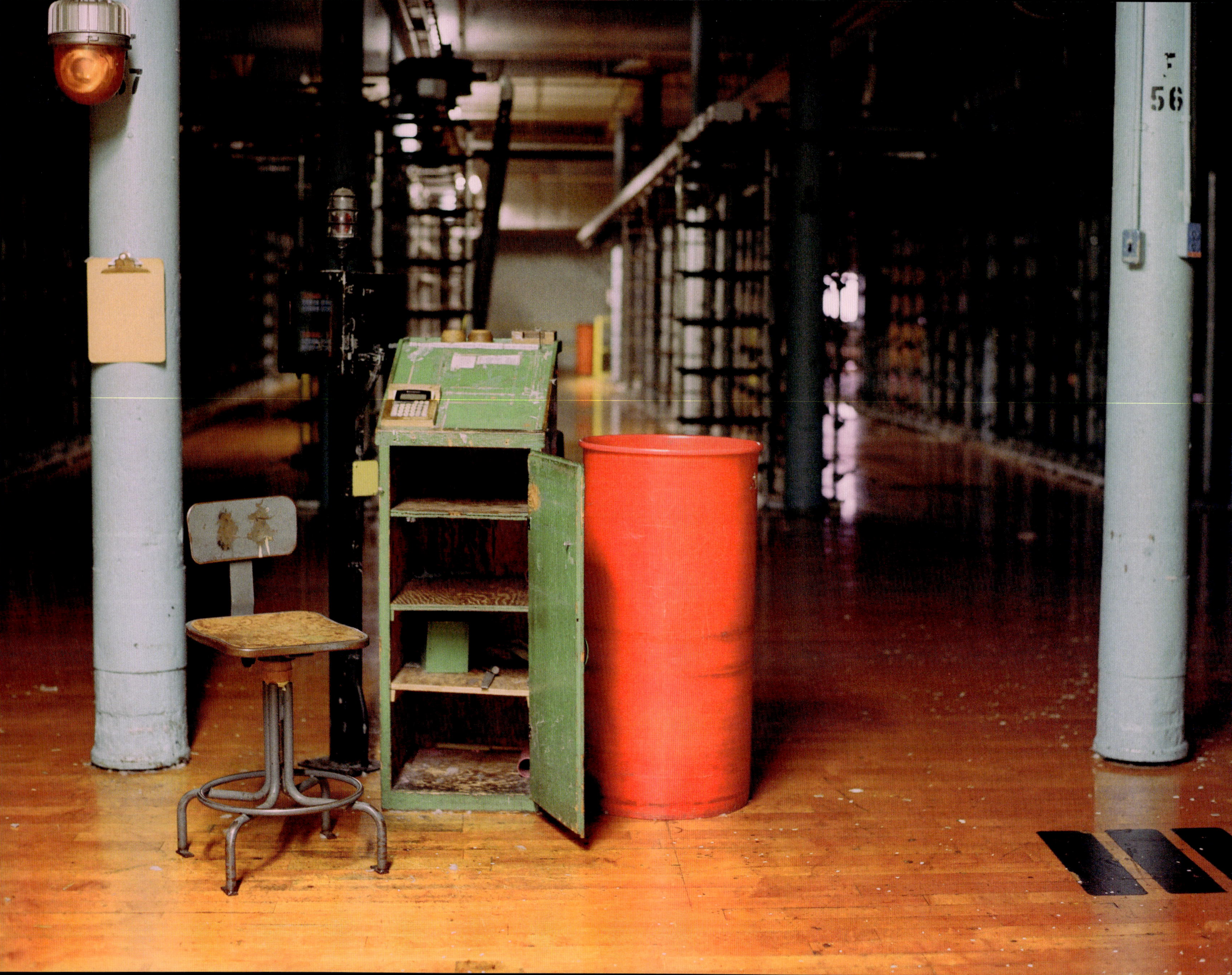

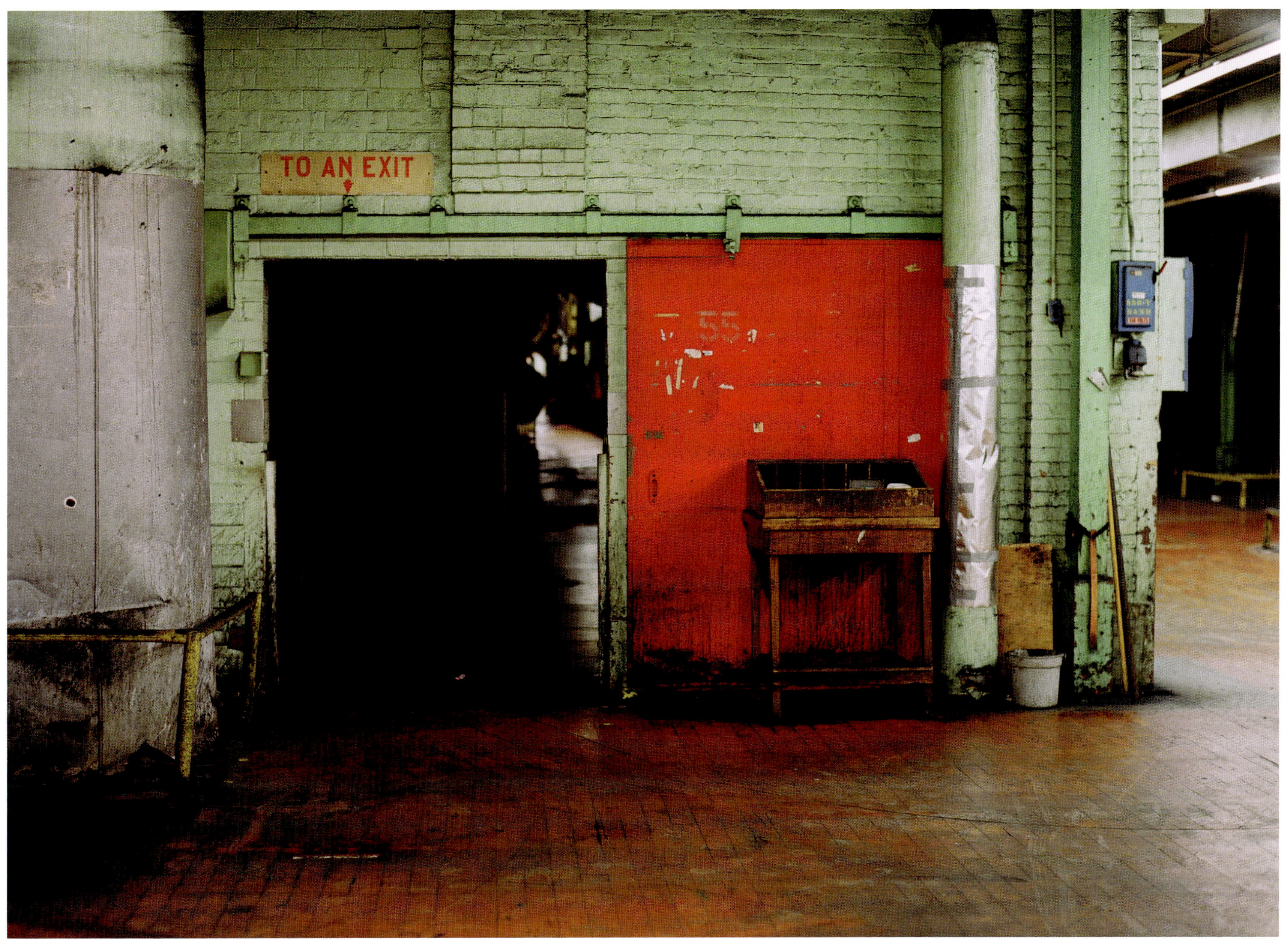

Operator stand in the Warping Department, with empty creels in the background   //   Entryway between rooms in the Finishing Department

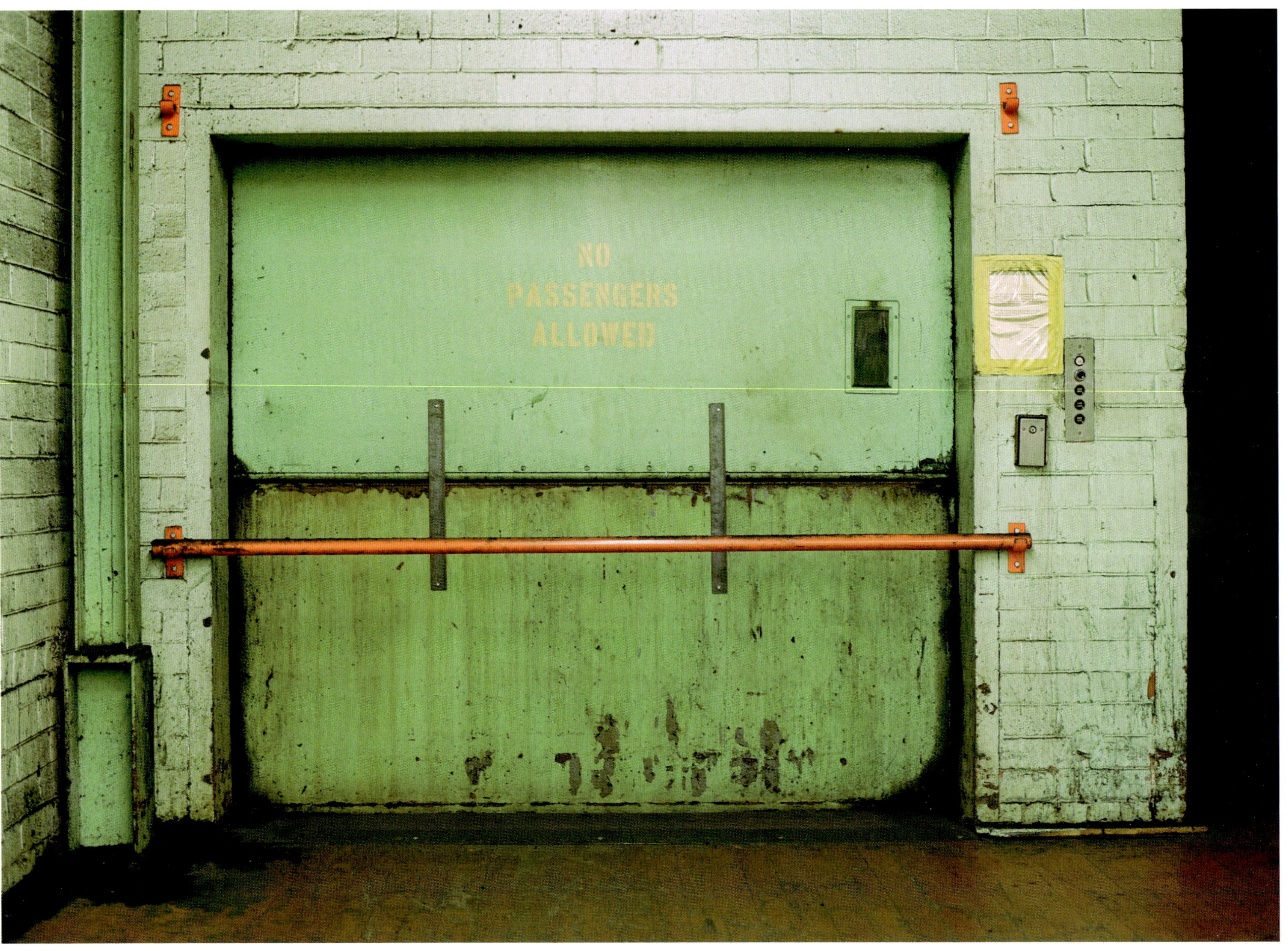

Elevator access in the Finishing Department

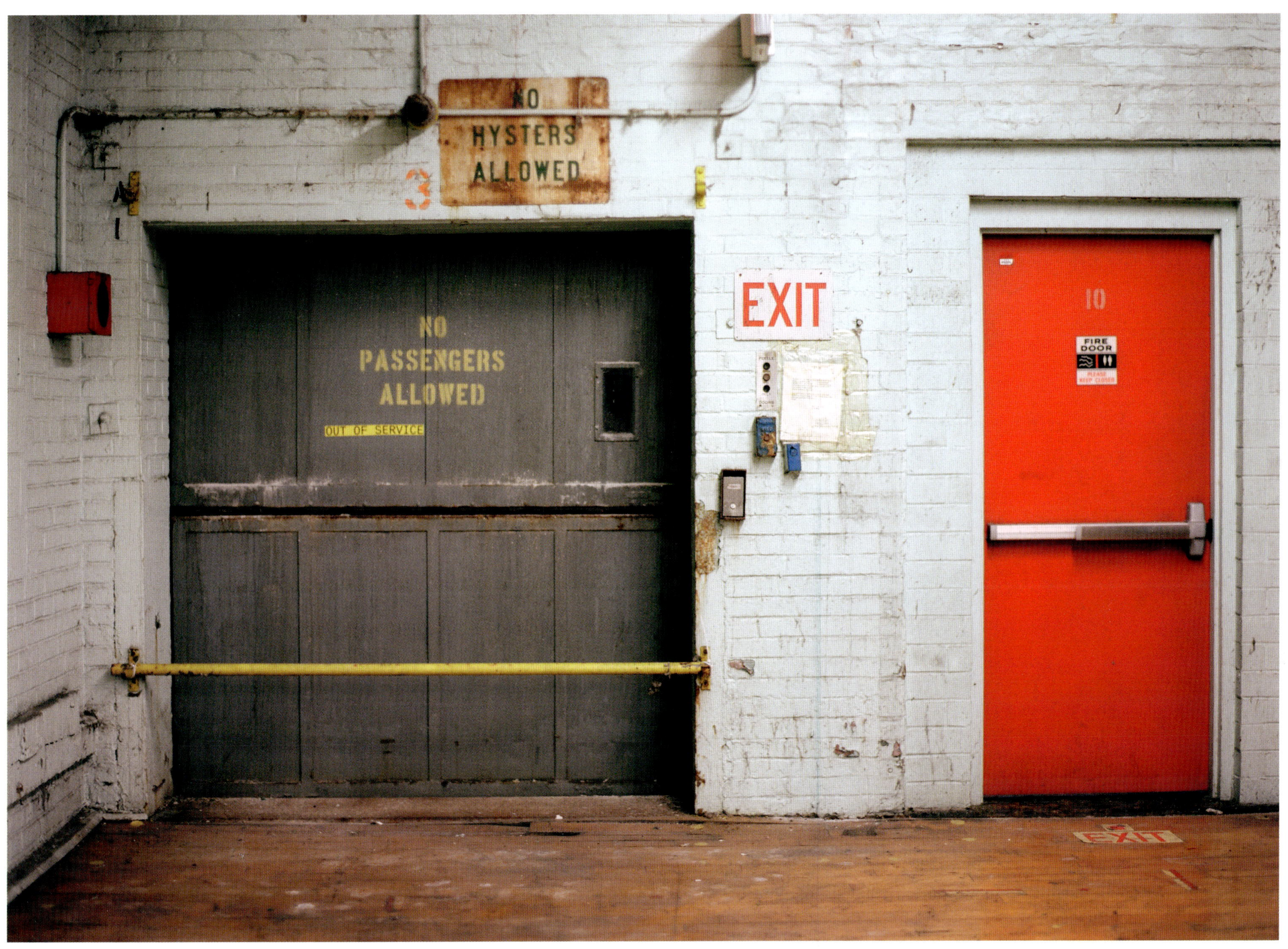

Elevator access, 16th Street Annex Building

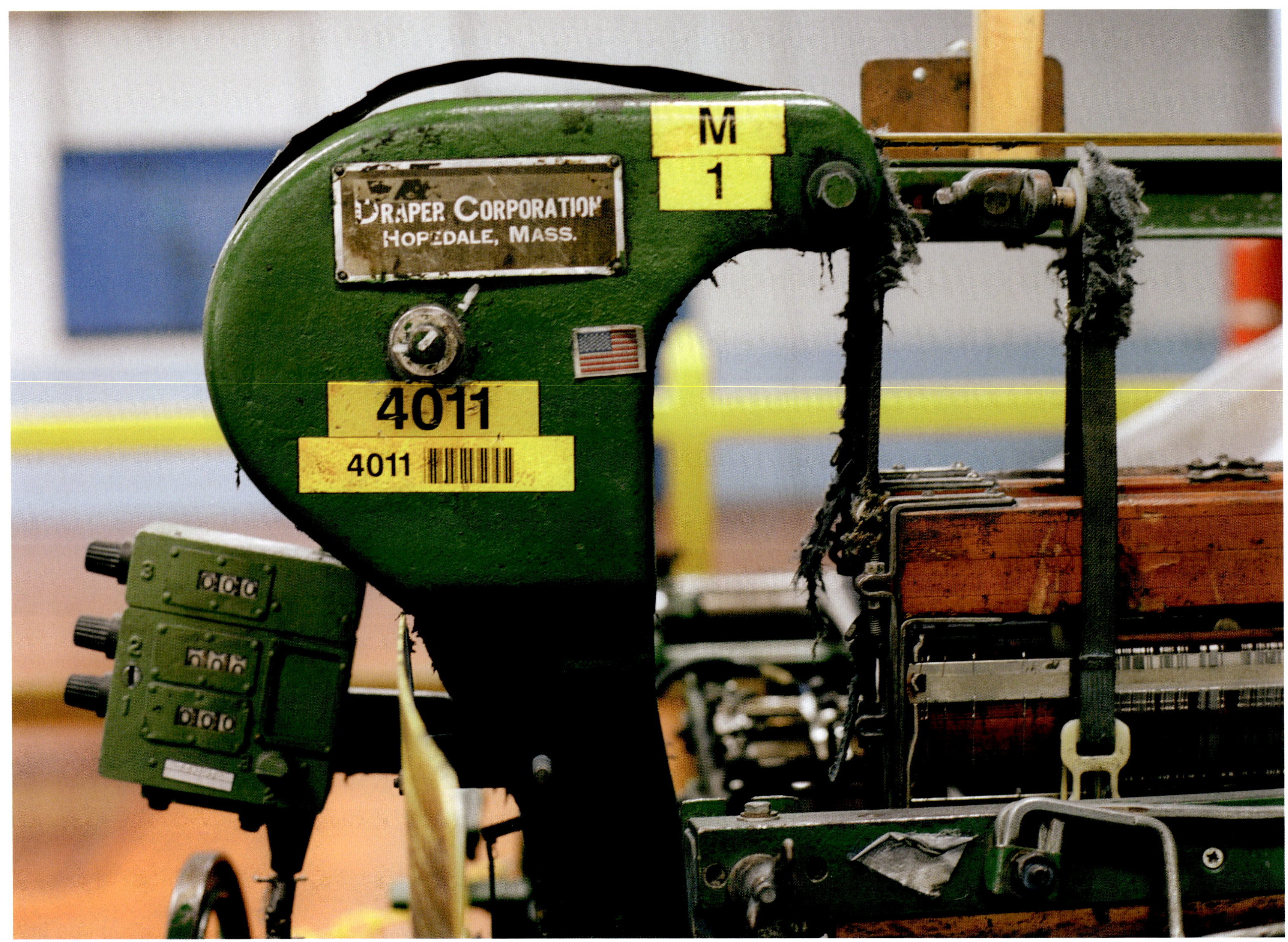

Draper X-3 shuttle loom in the Weaving Department

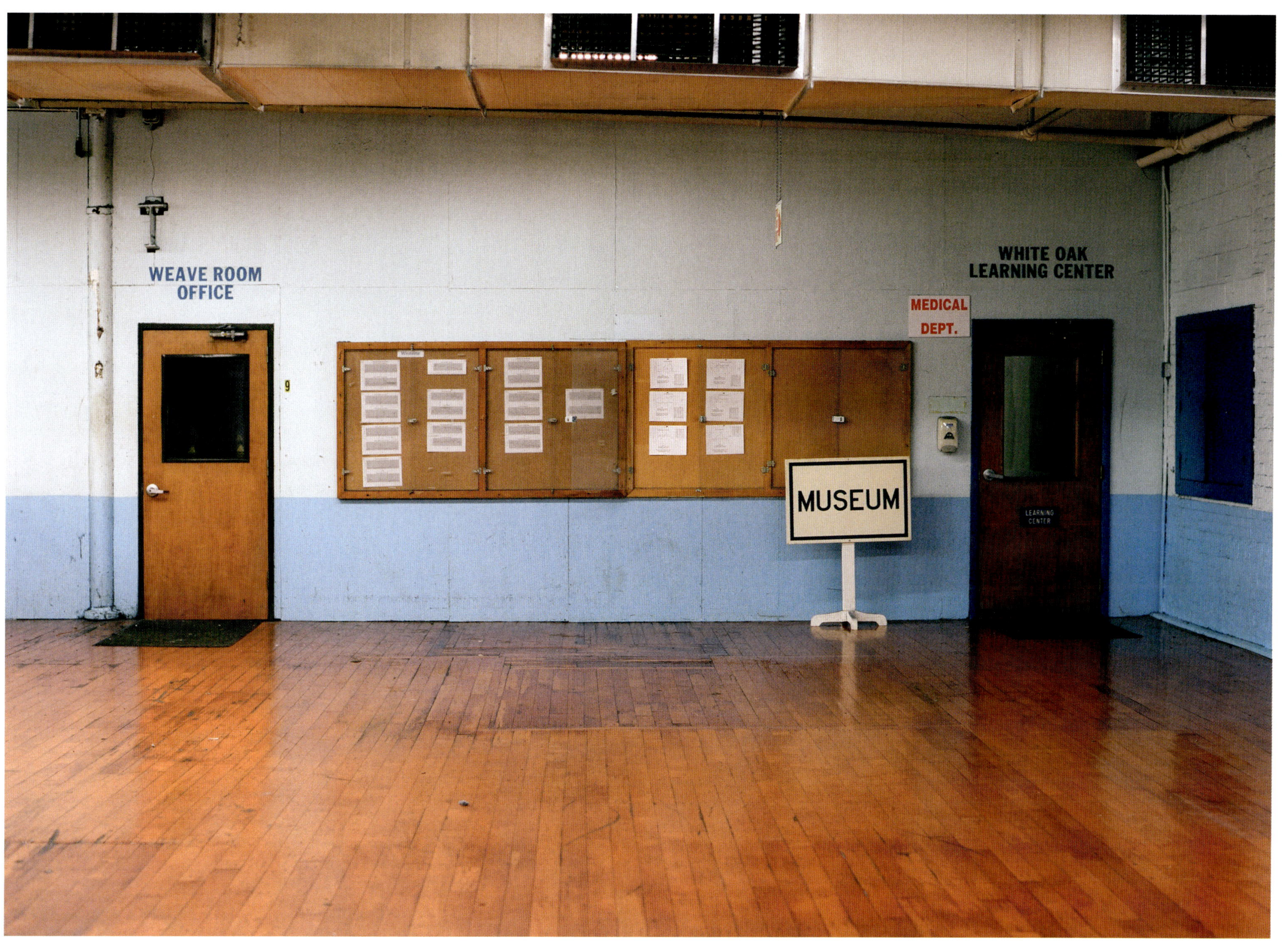

Corner of the Weave Room near the company offices and museum.
Productivity charts on the bulletin board.

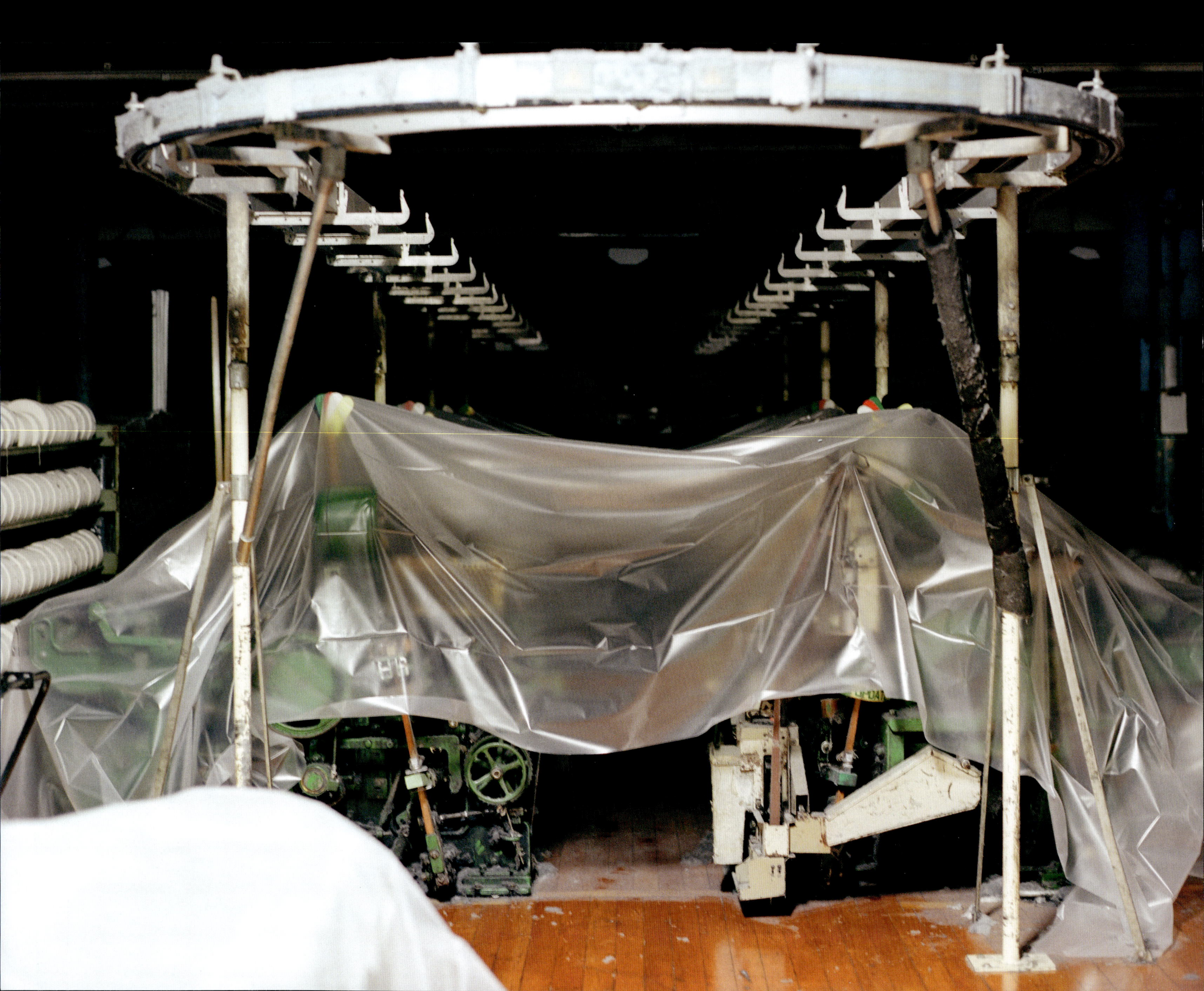

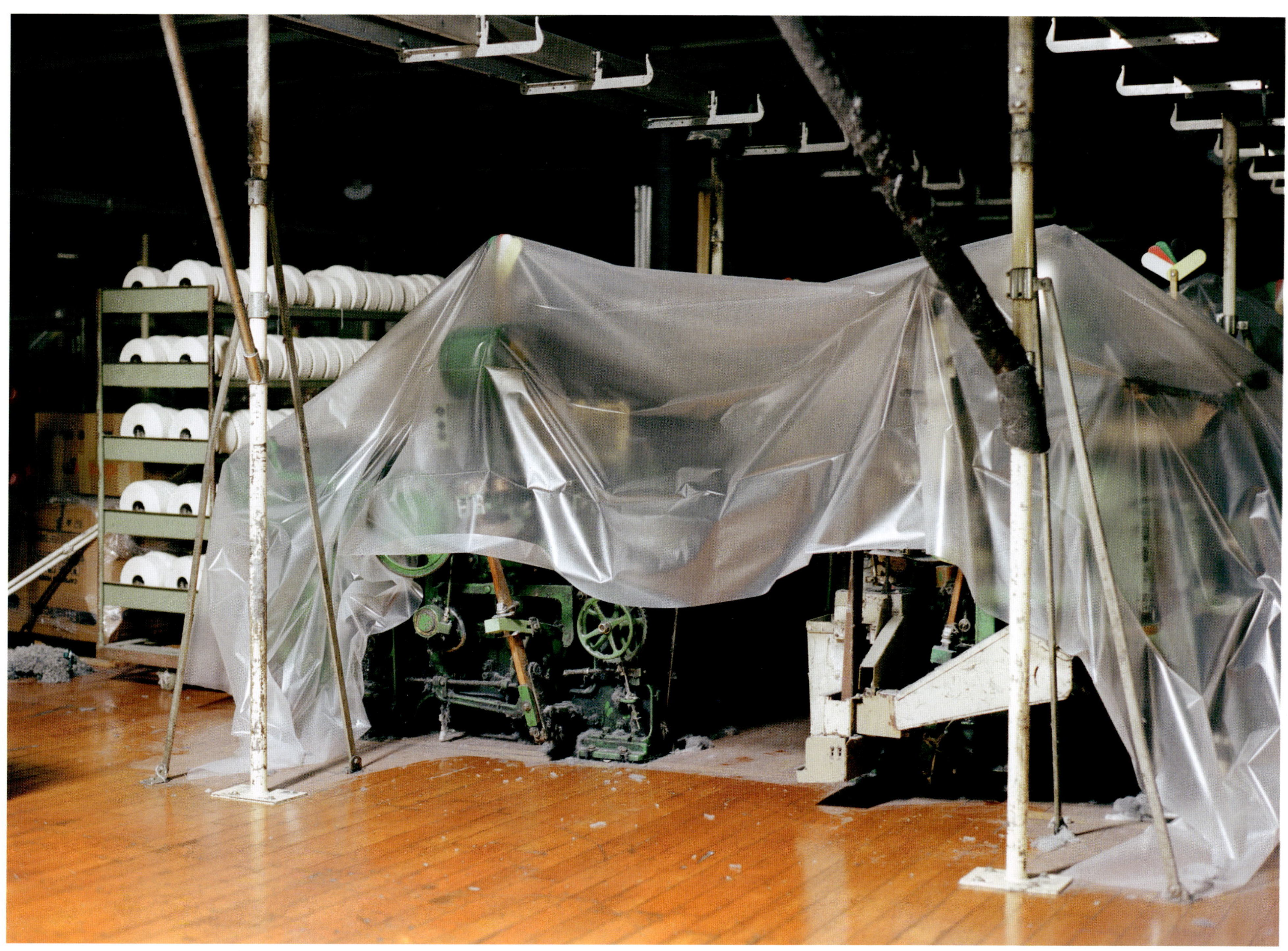

X-3 shuttle looms covered with plastic sheets in the Weave Room

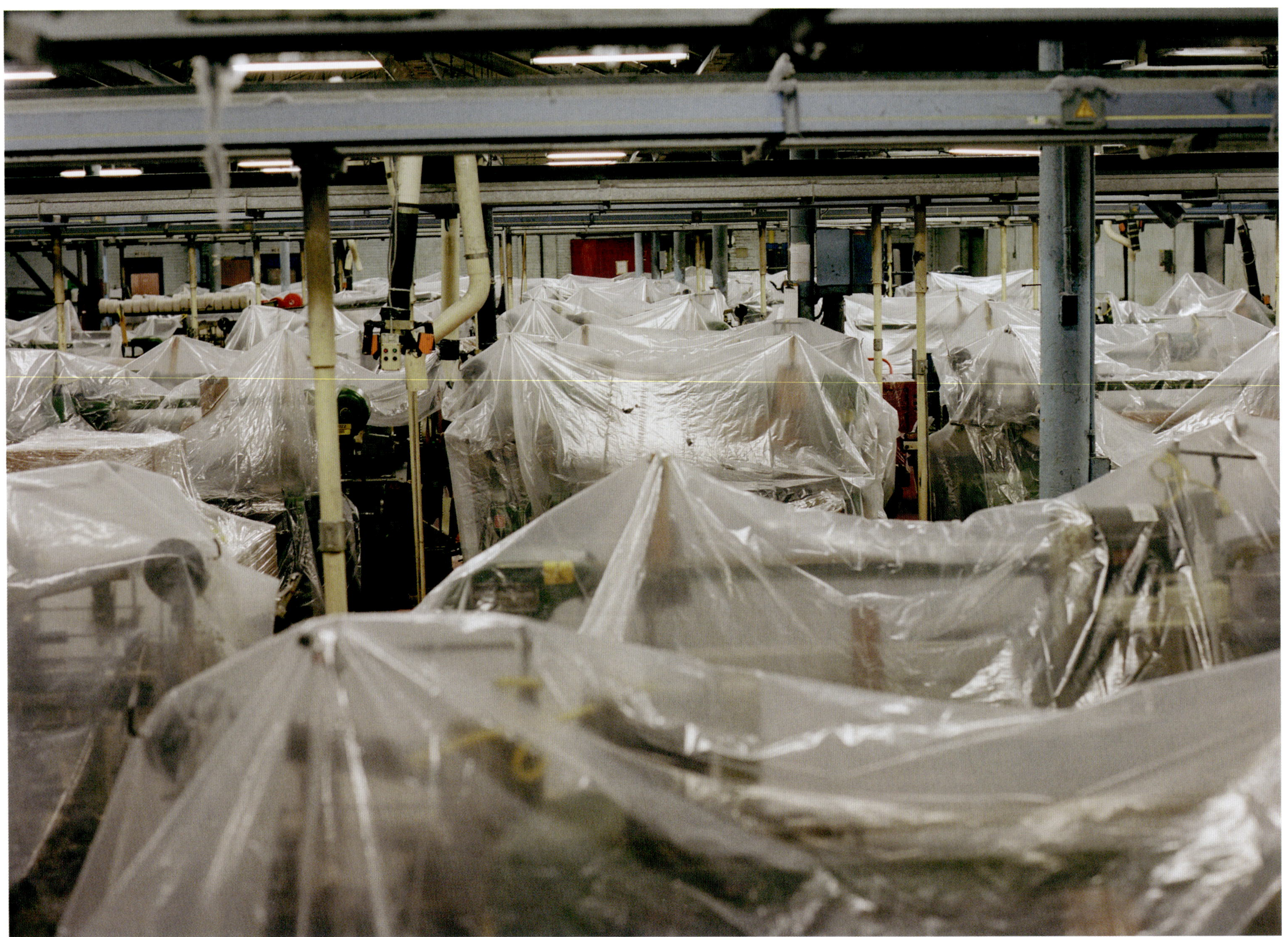

Plastic sheets covering Draper X-3 shuttle looms in the Weaving Department

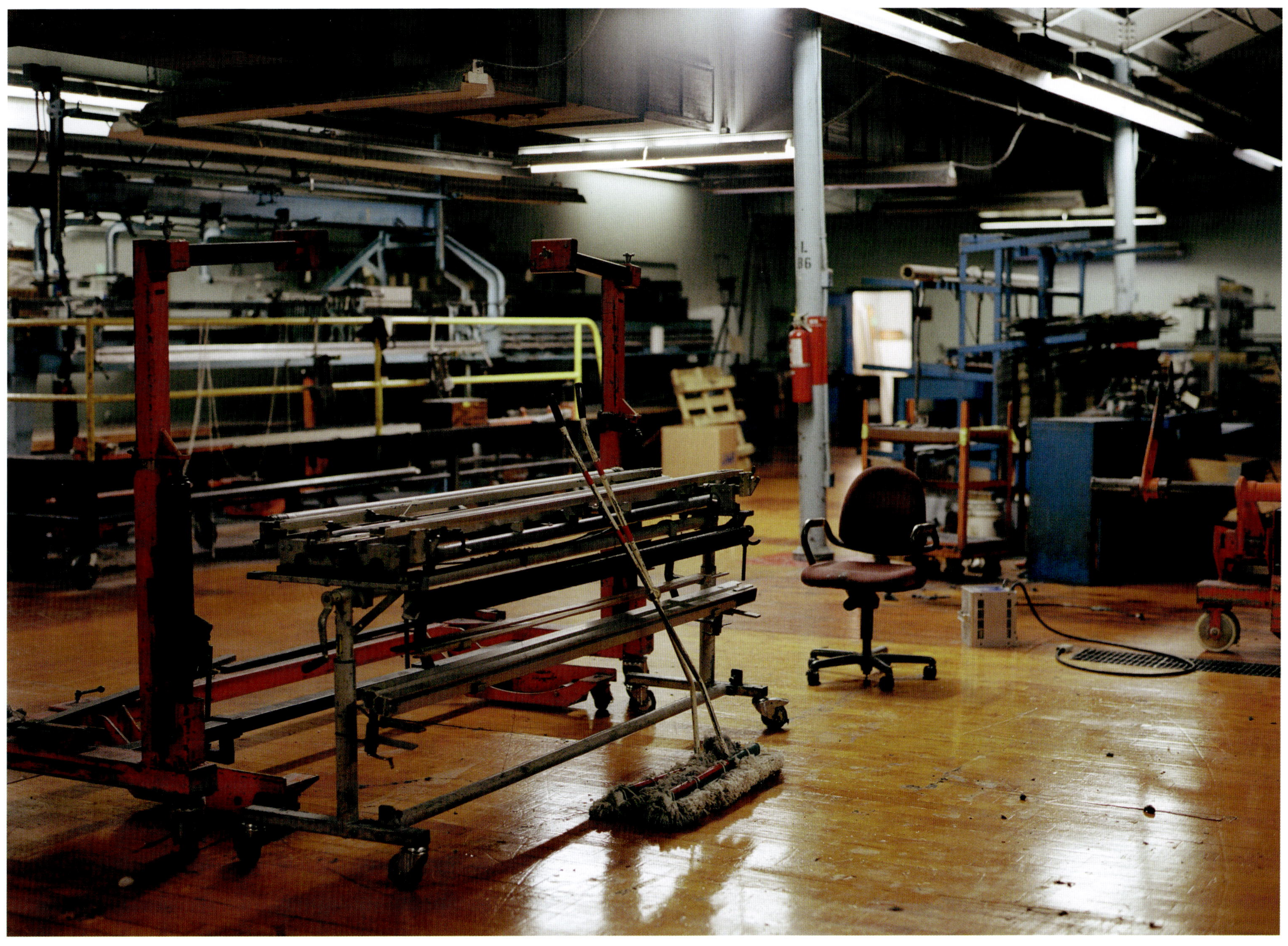

Empty tying-in frame, with broom where it was left

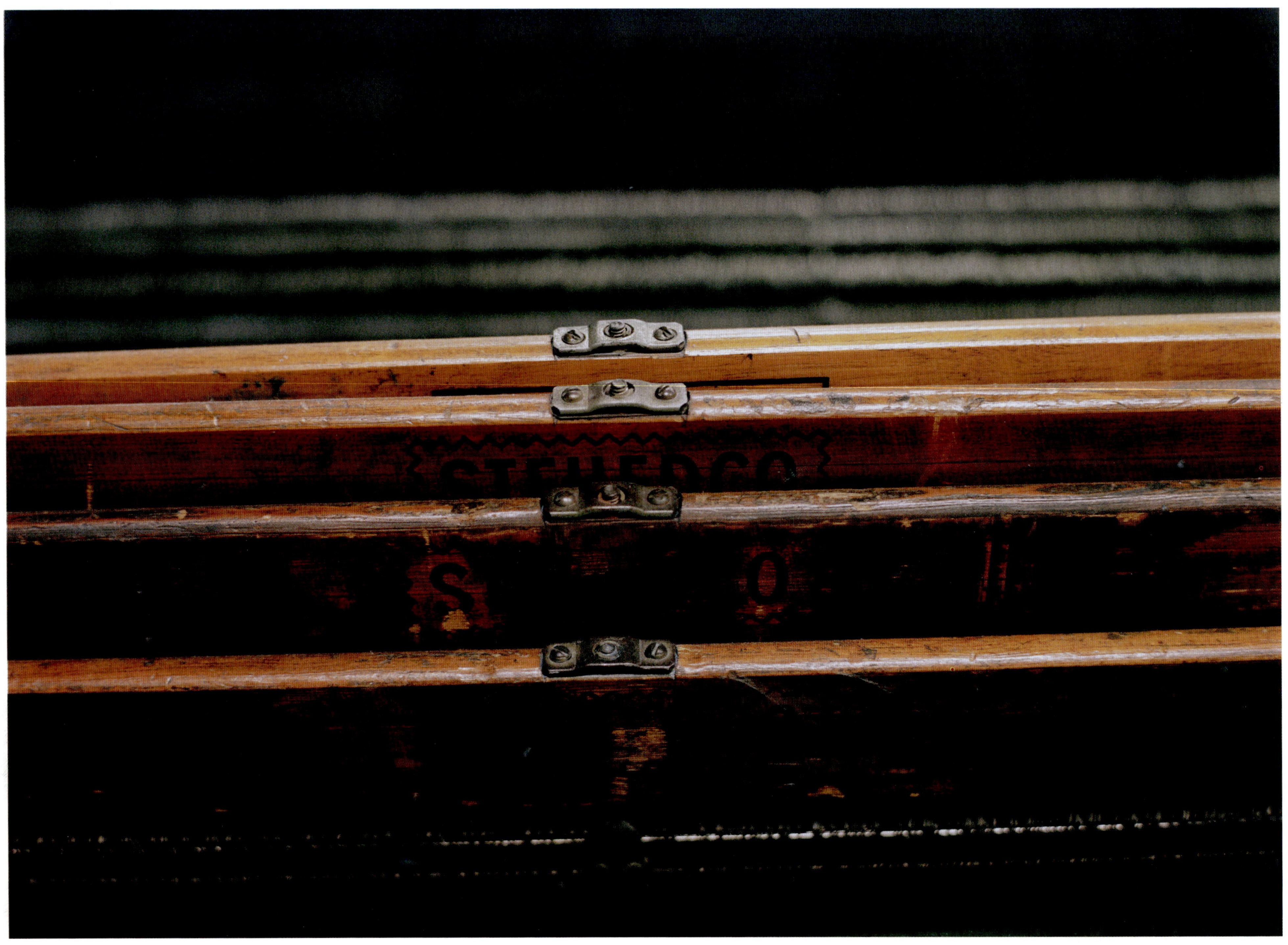

Draper X-3 harnesses

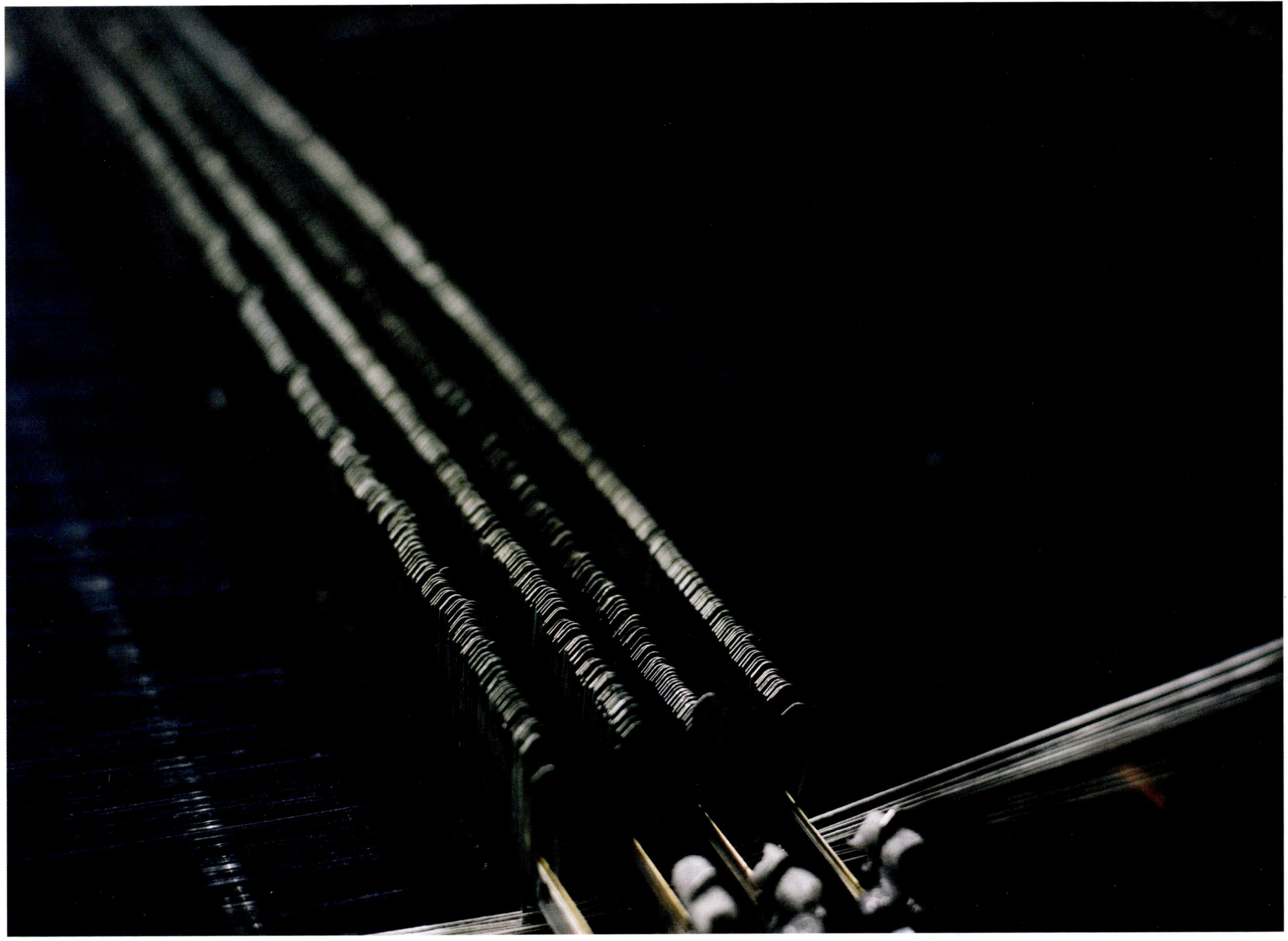

Drop wires on shuttle loom in the Weaving Department

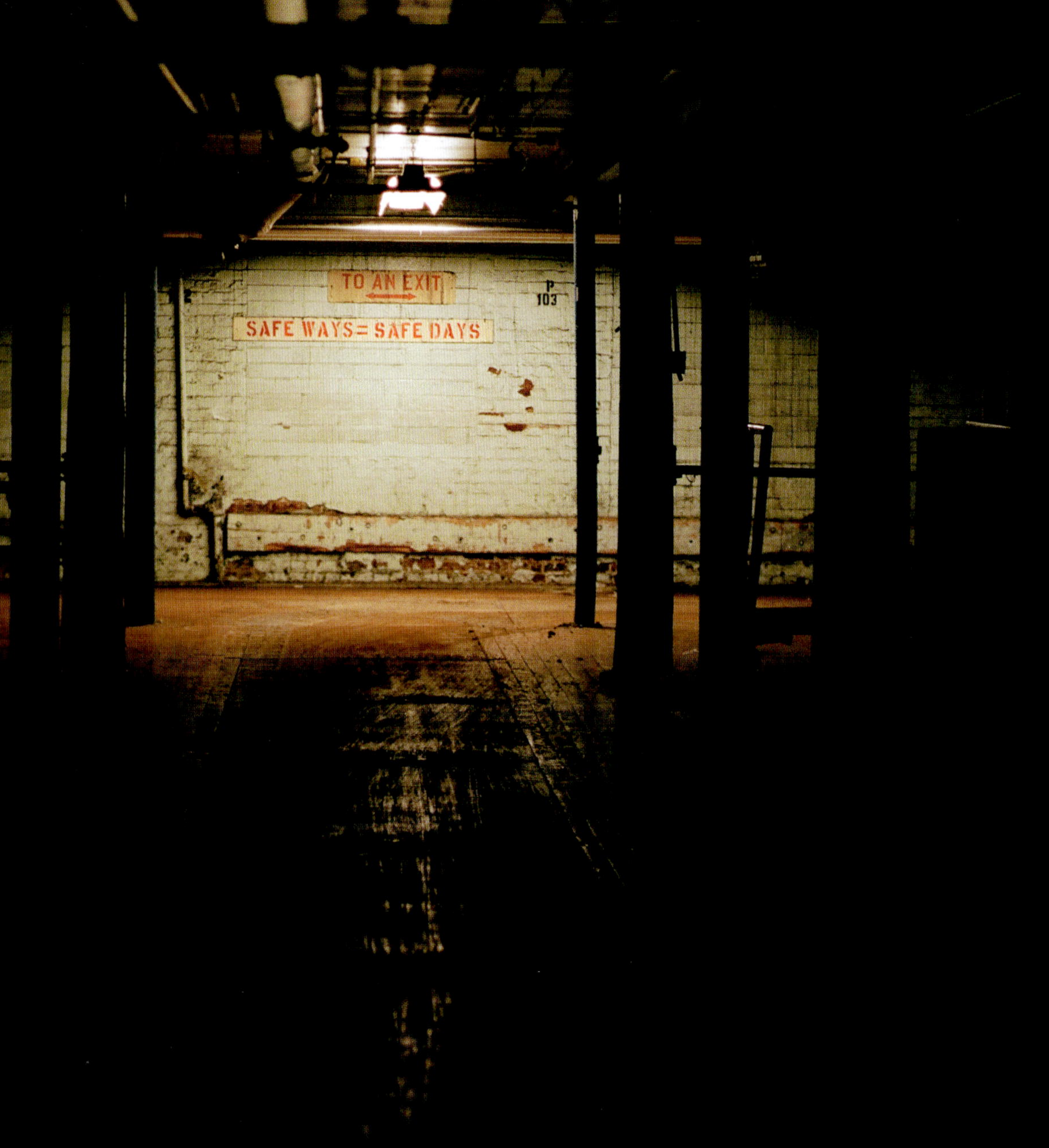

TO AN EXIT
SAFE WAYS = SAFE DAYS
P
103

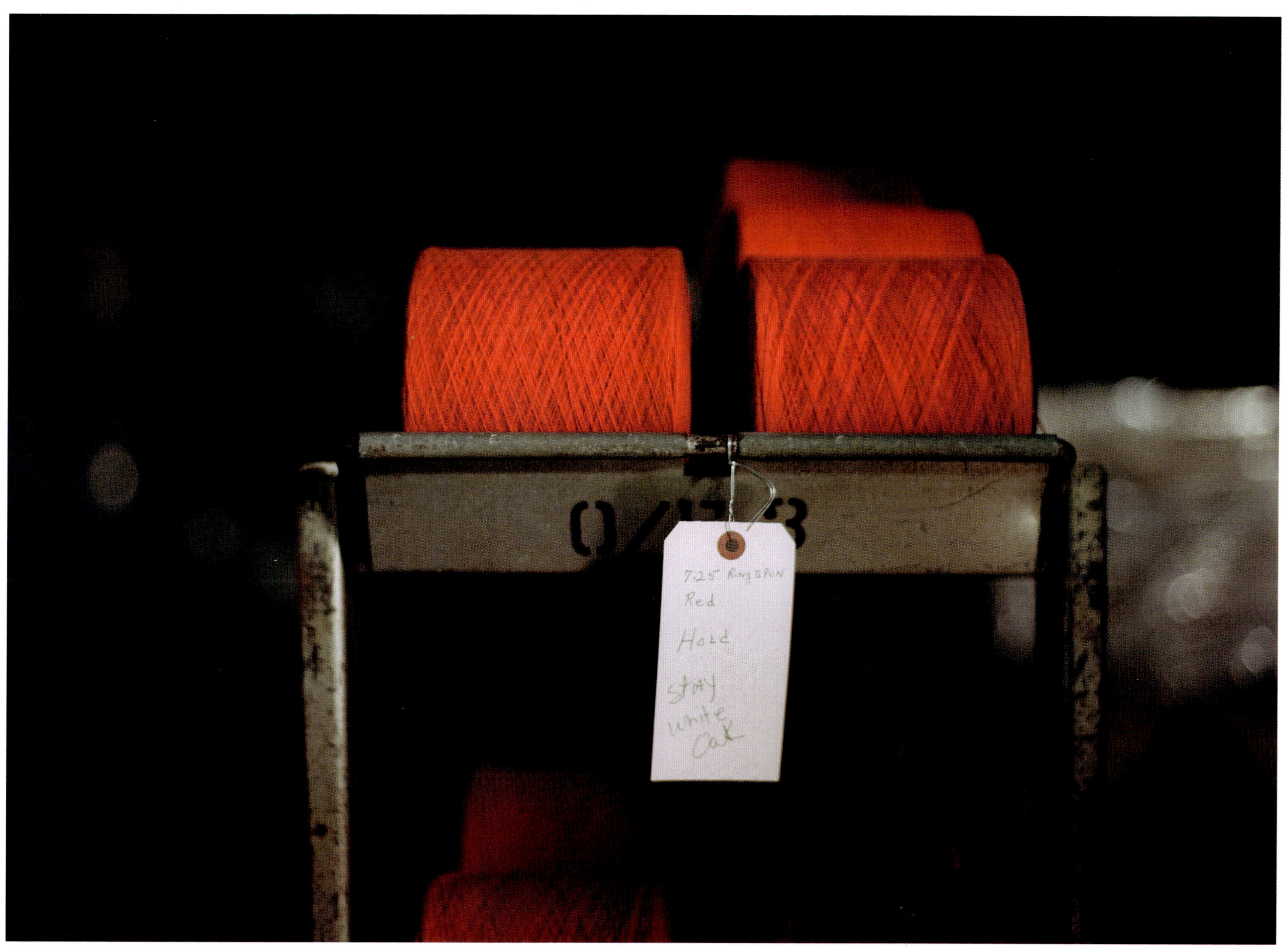

A view down a long corridor on the mezzanine floor of the Weave Building   //   XX Red Selvage ID yarn for Draper X-3 shuttle looms on yarn cart in the Weaving Department

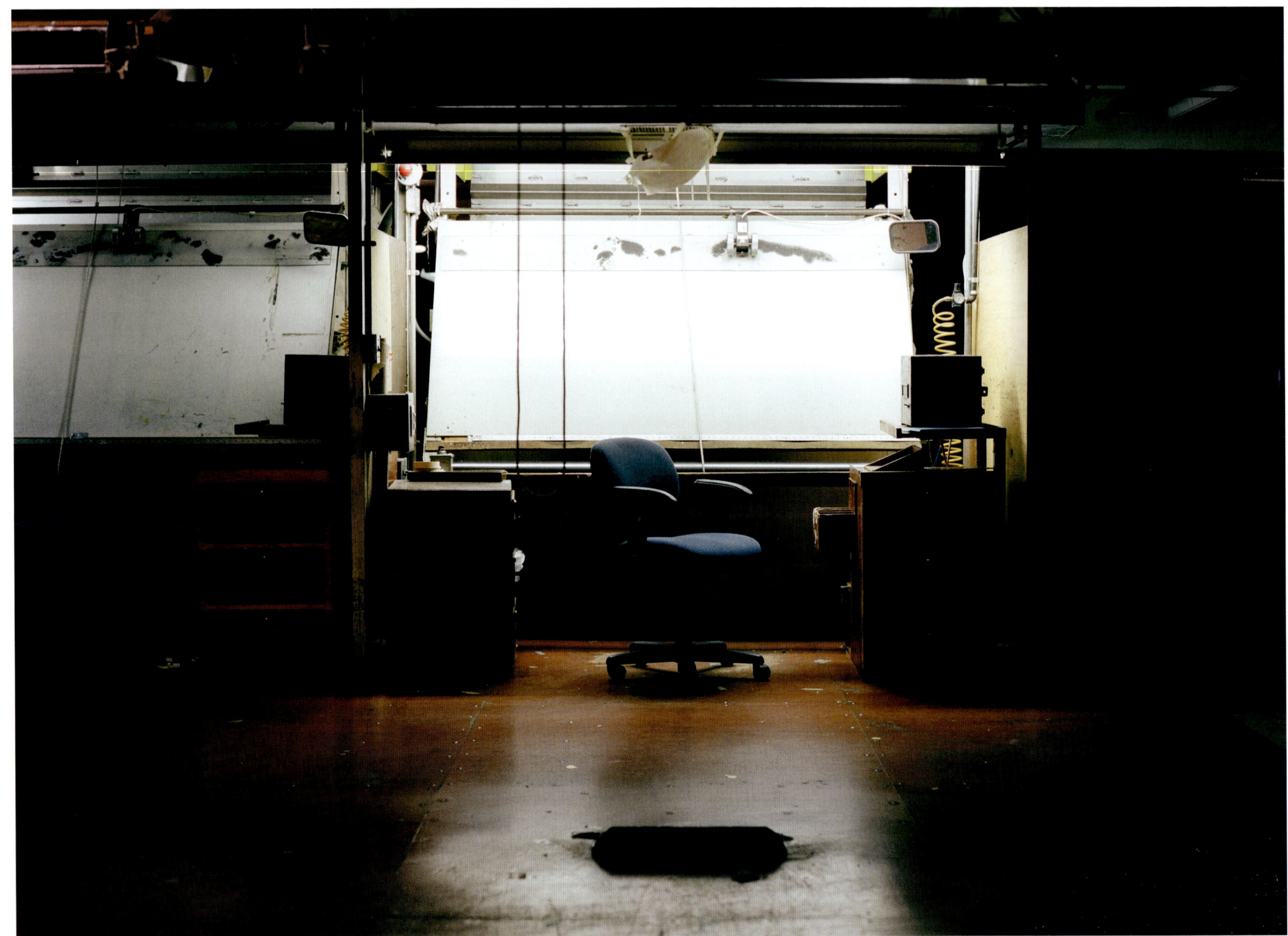

Empty inspection station in the Quality Control Department

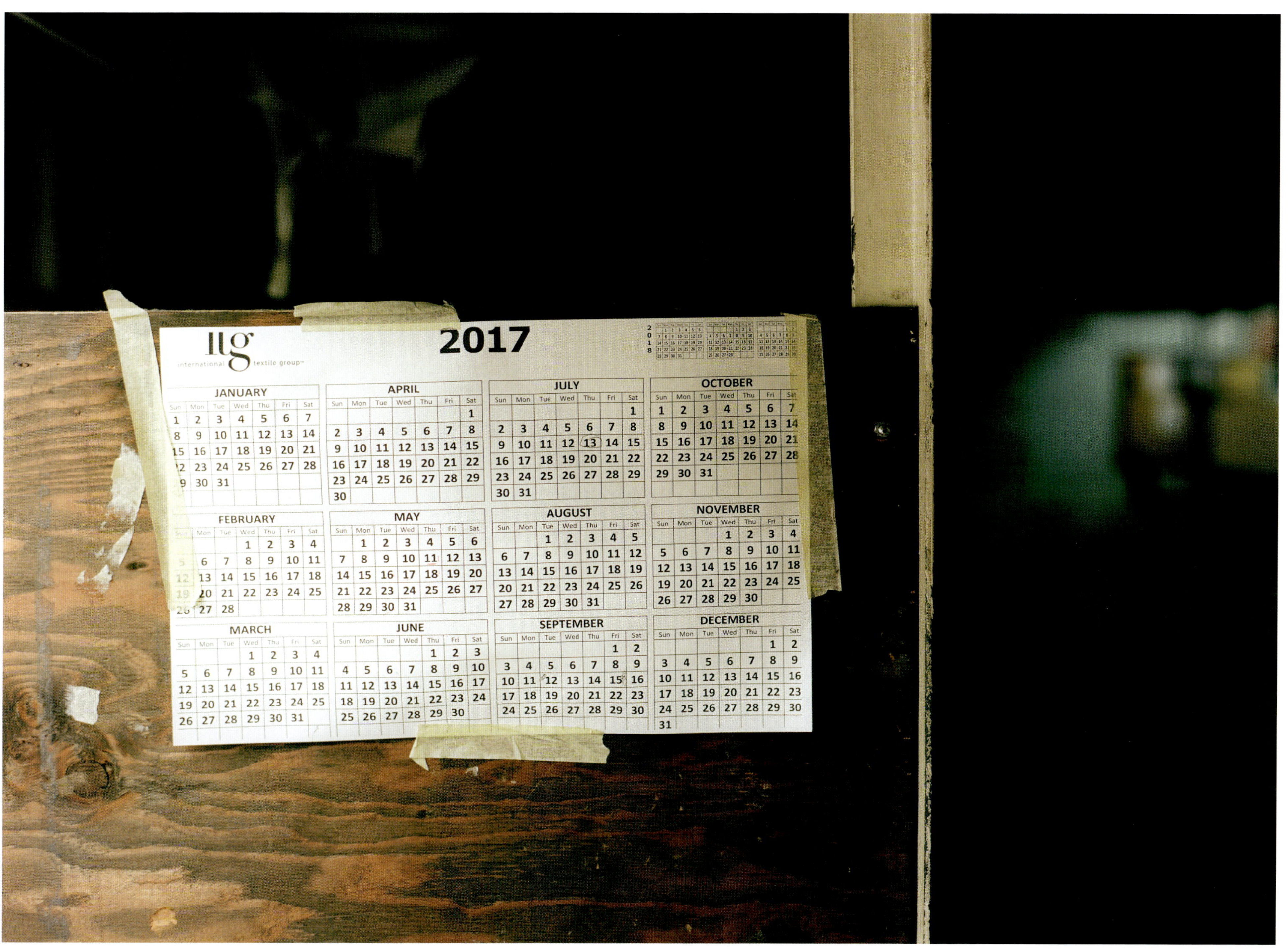

Company calendar at an inspection station in the Quality Control Department,
showing the final year and month that Cone denim was made at White Oak

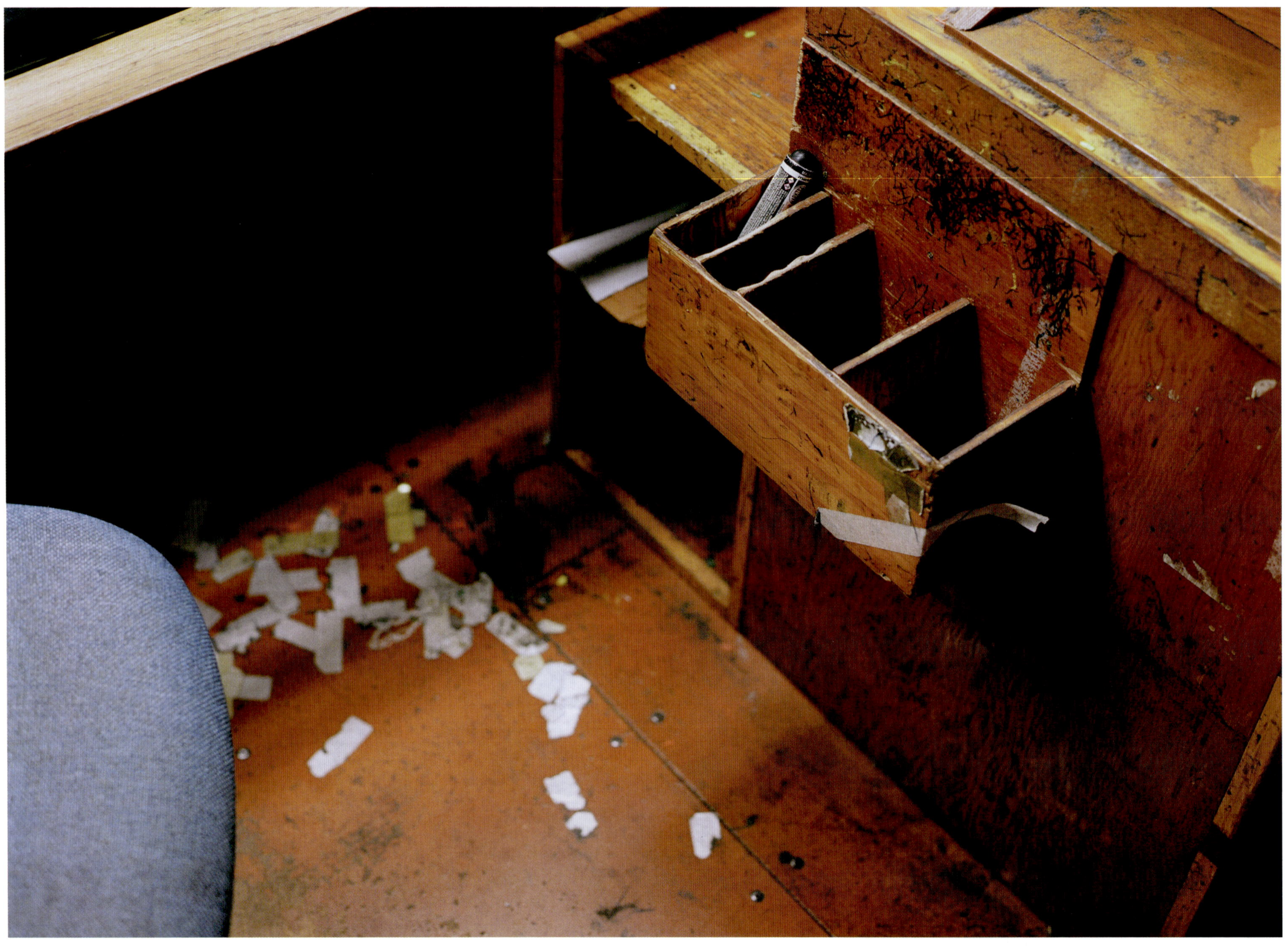

Toolbox at inspection booth in the Quality Control Department // Cloth inspection stations in the Quality Control Department

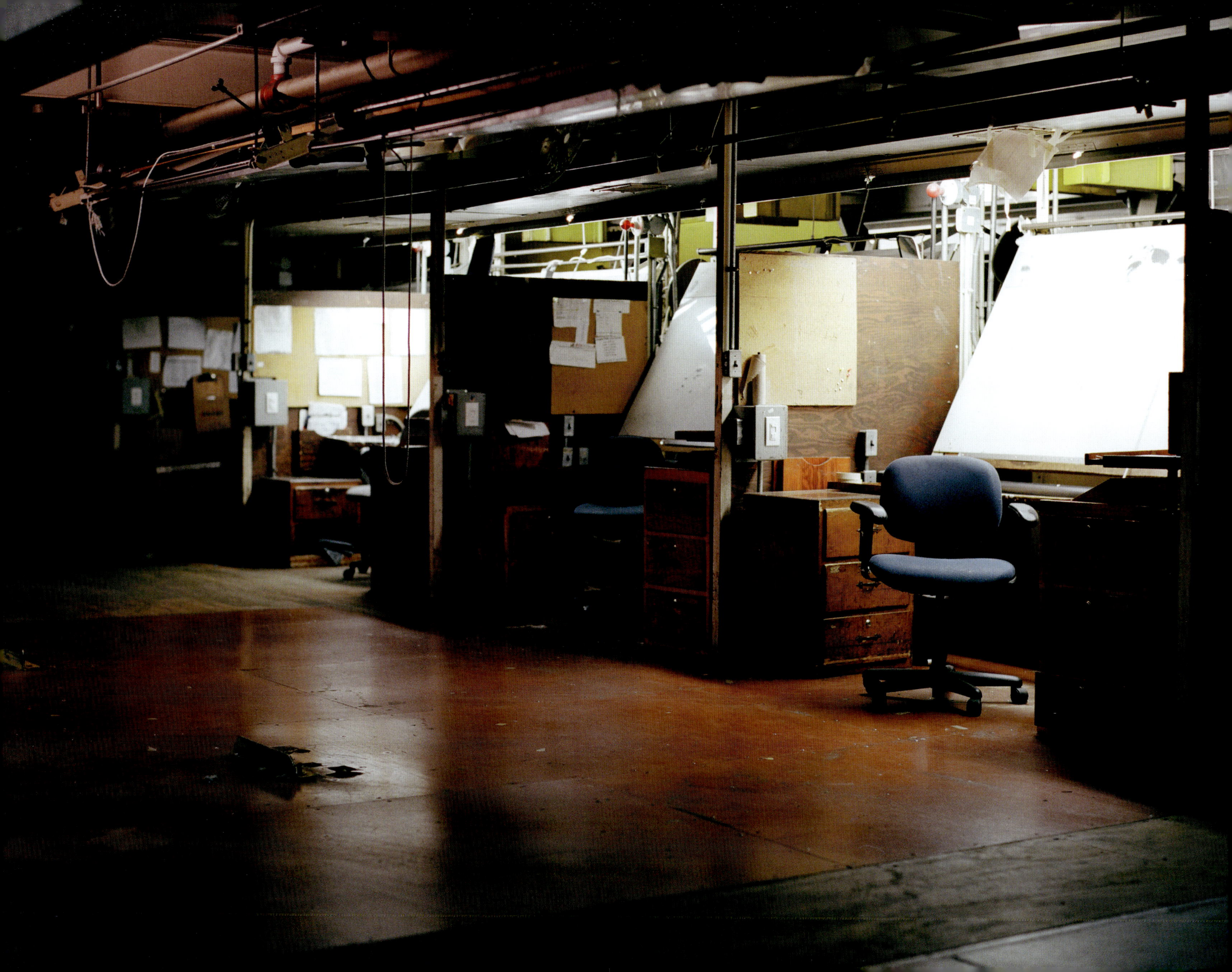

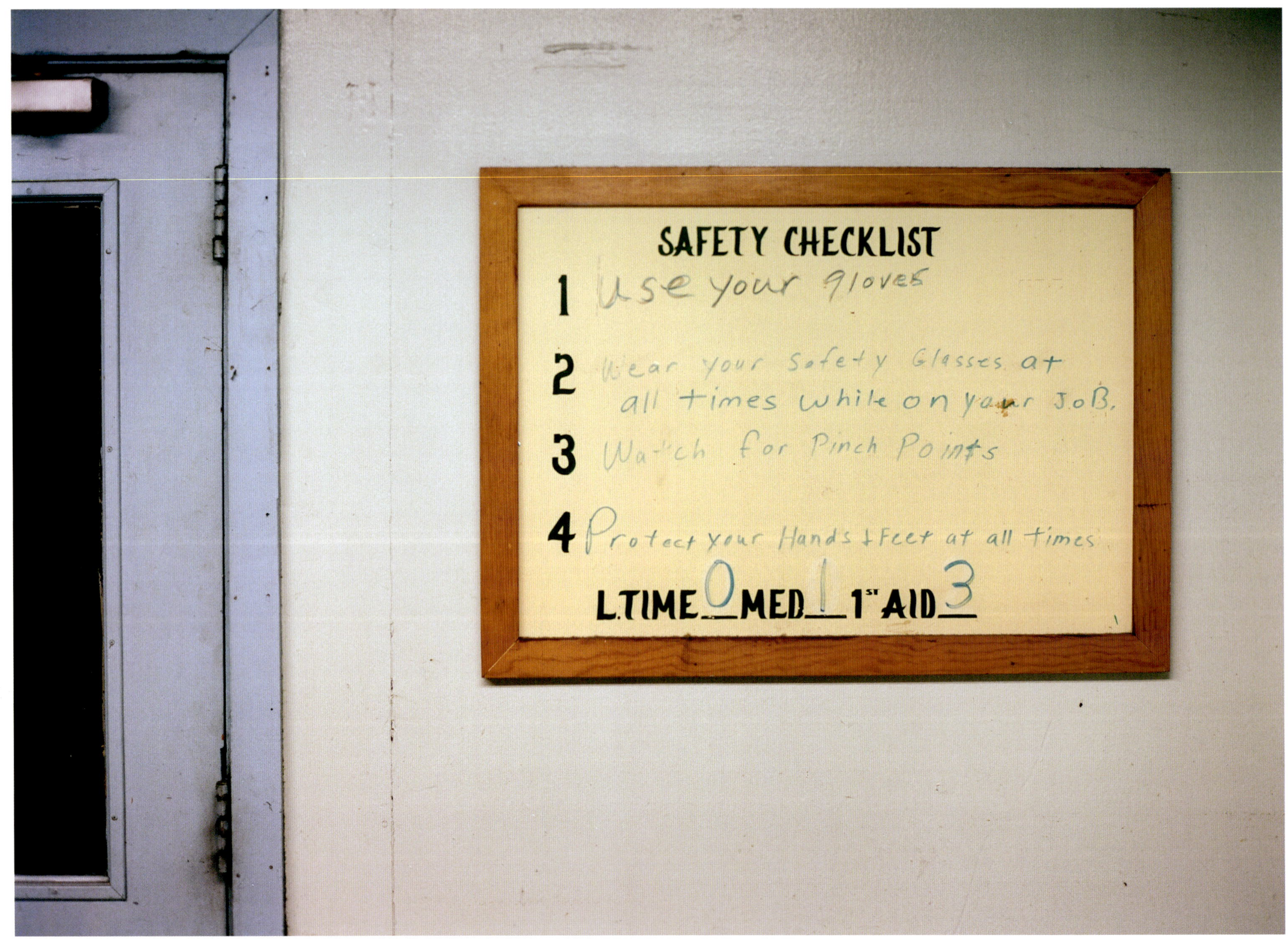

Safety checklist at the Testing Laboratory entrance

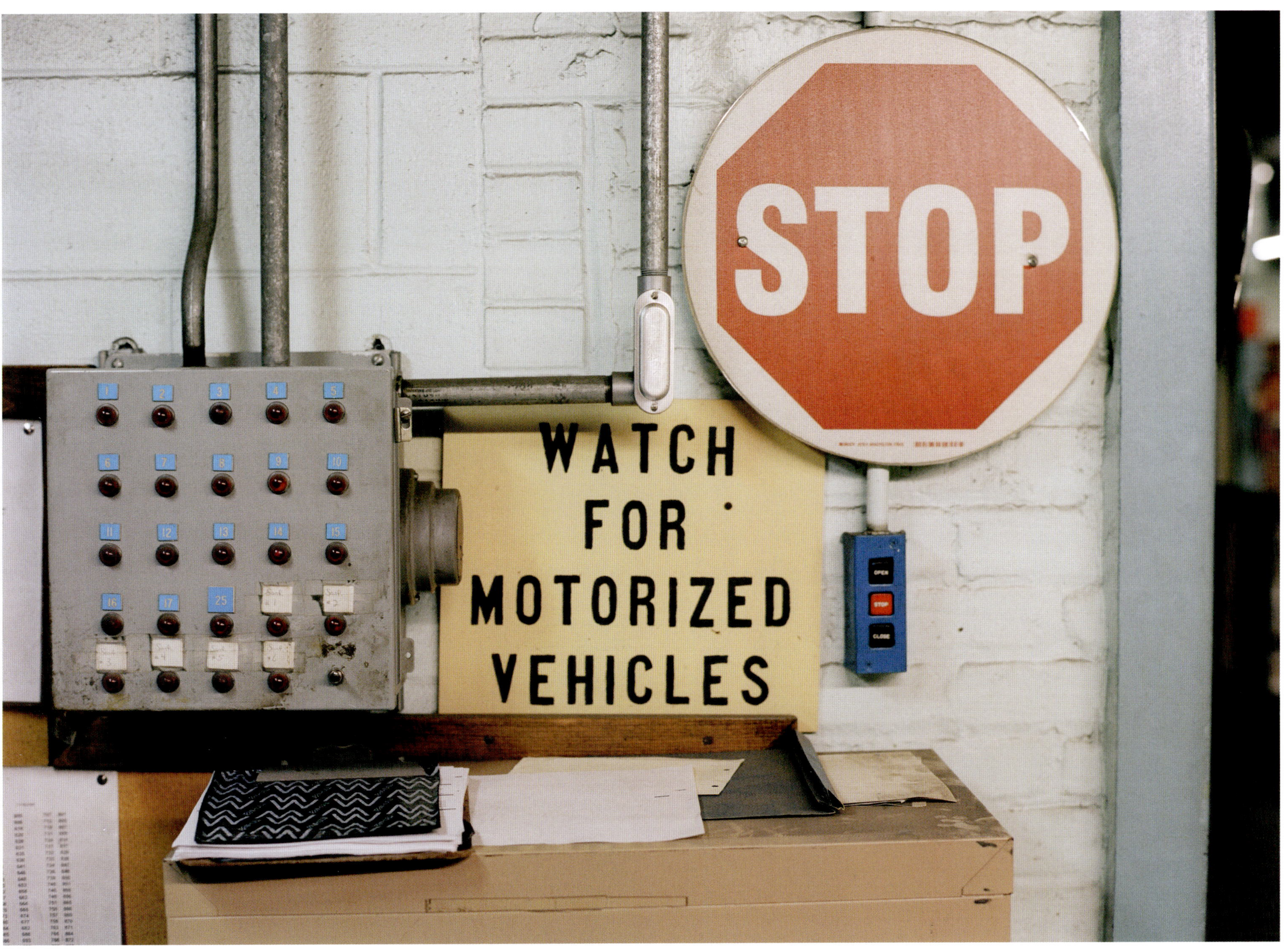

Control panel in passway between the Weave Room and Slashing Department

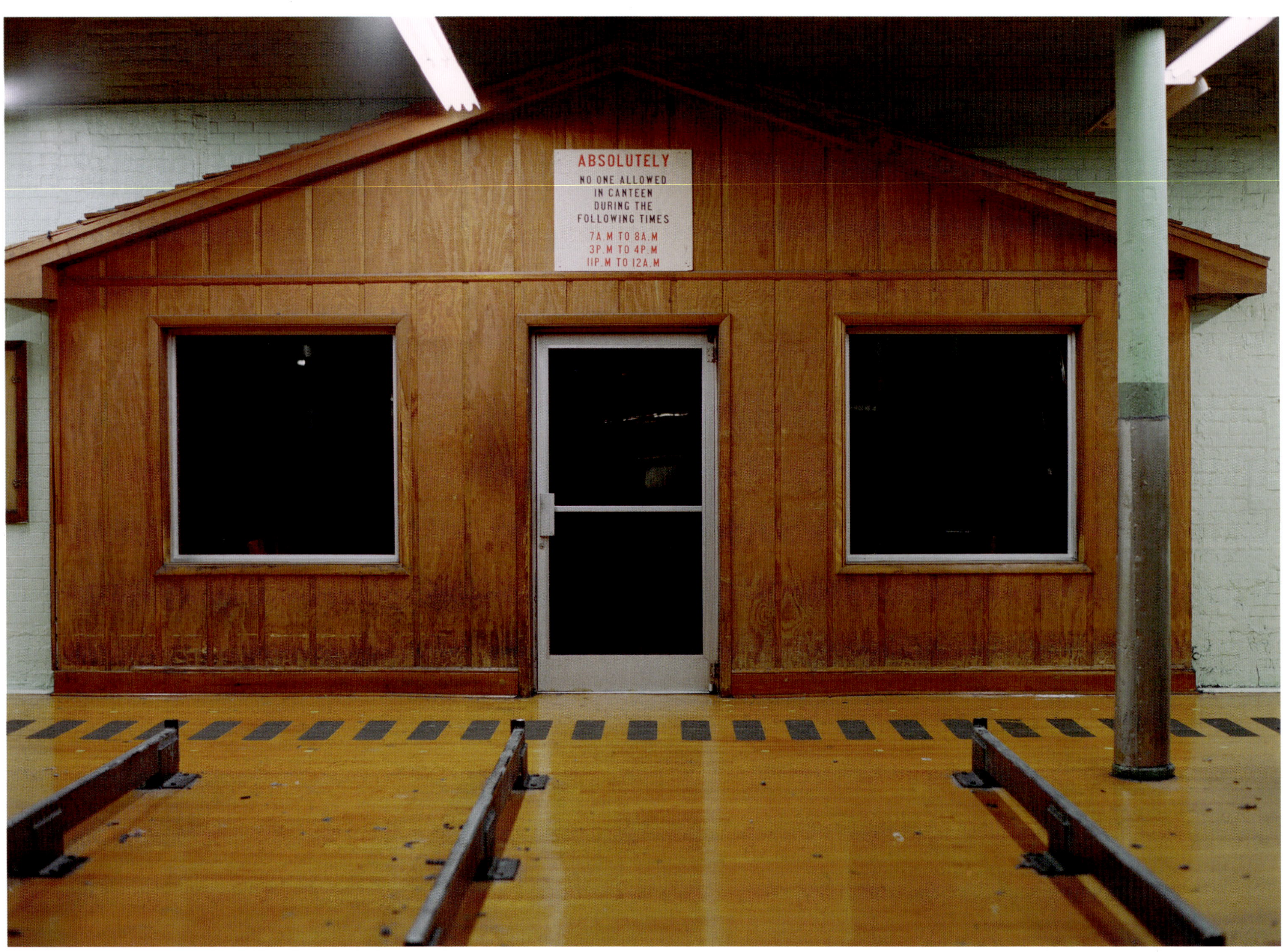

Outside the Company Canteen in the Slashing Department // Inside the Canteen

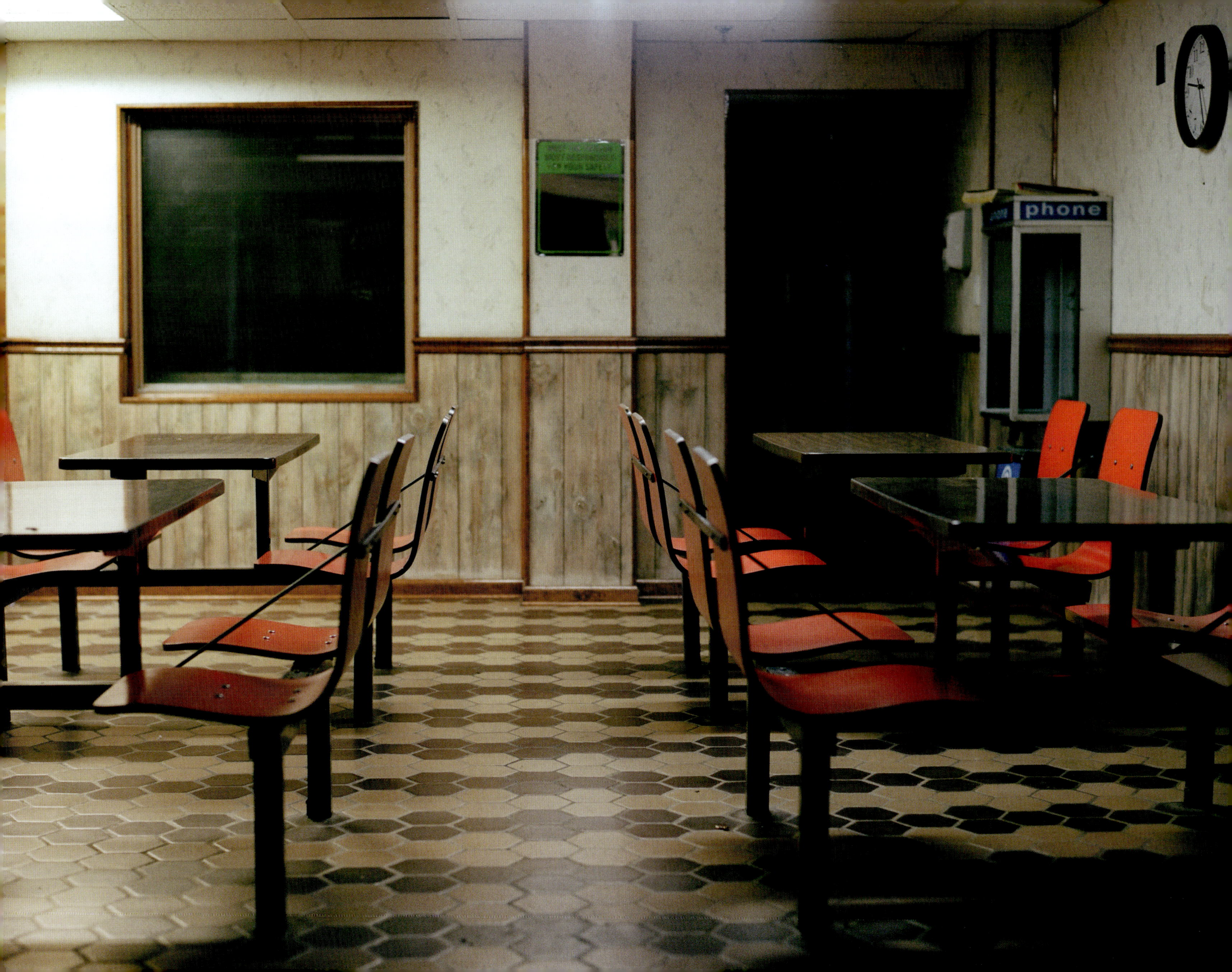

phone

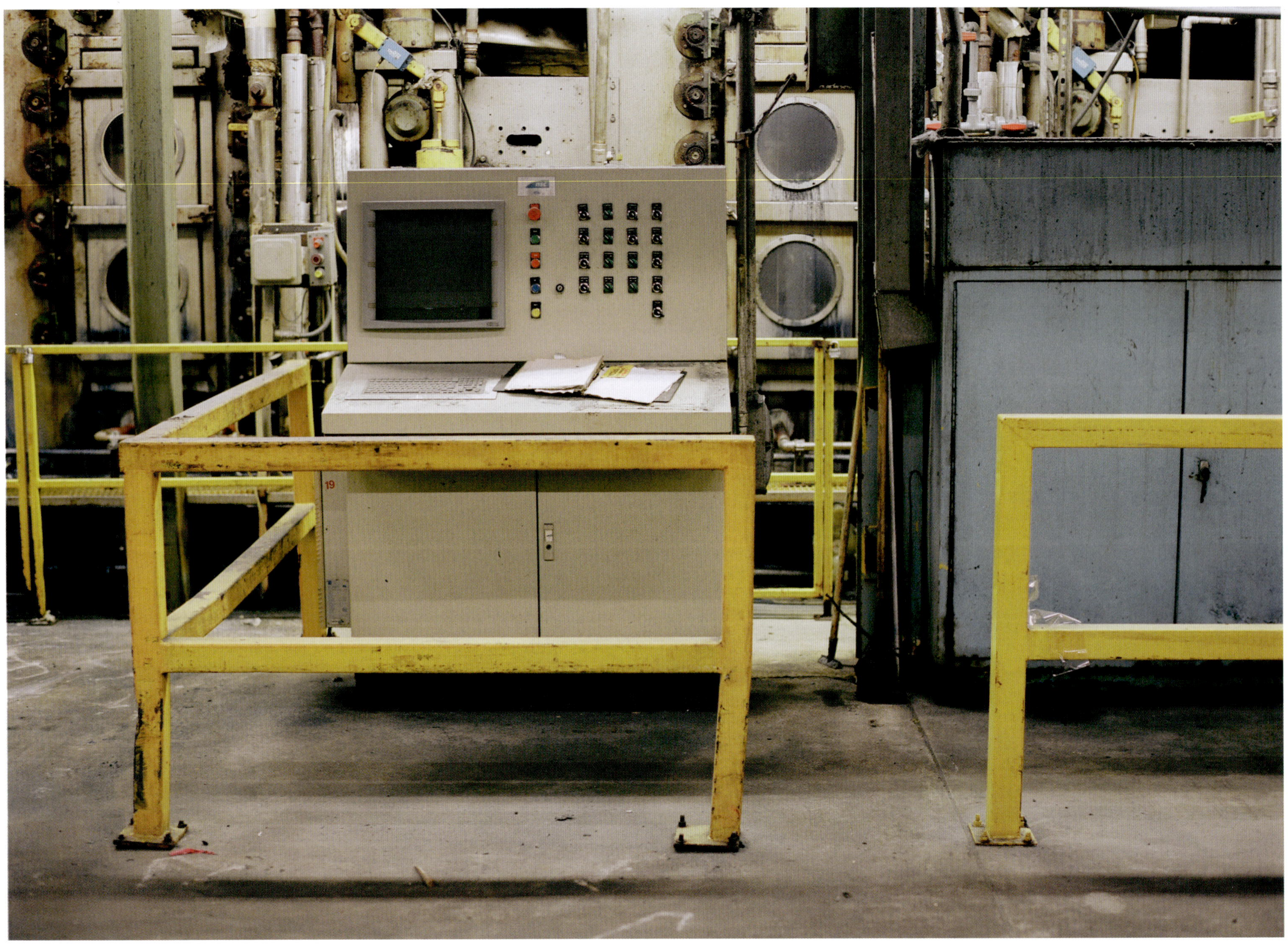

Control panel in the Finishing Department

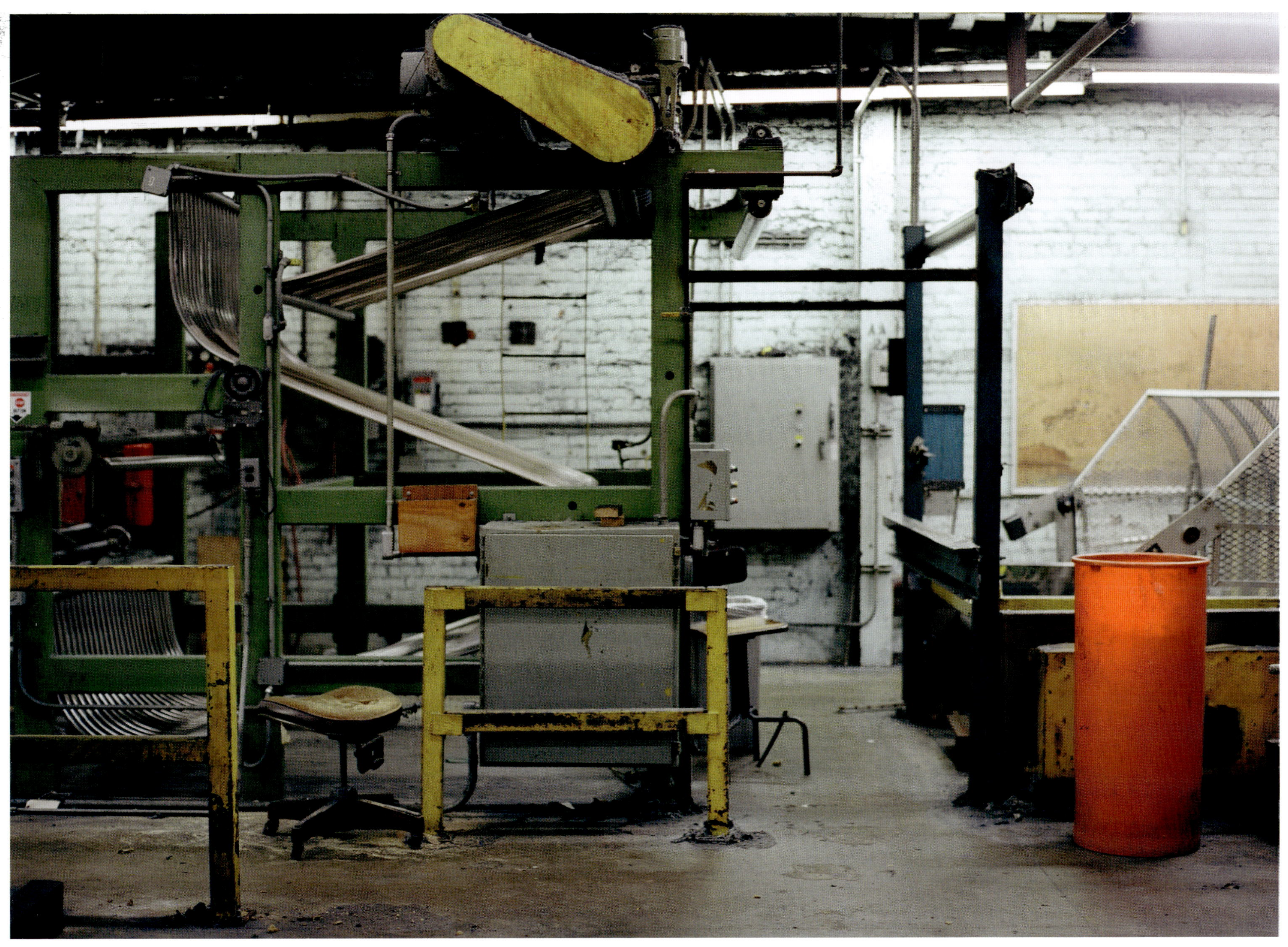

Workstation in the Finishing Department

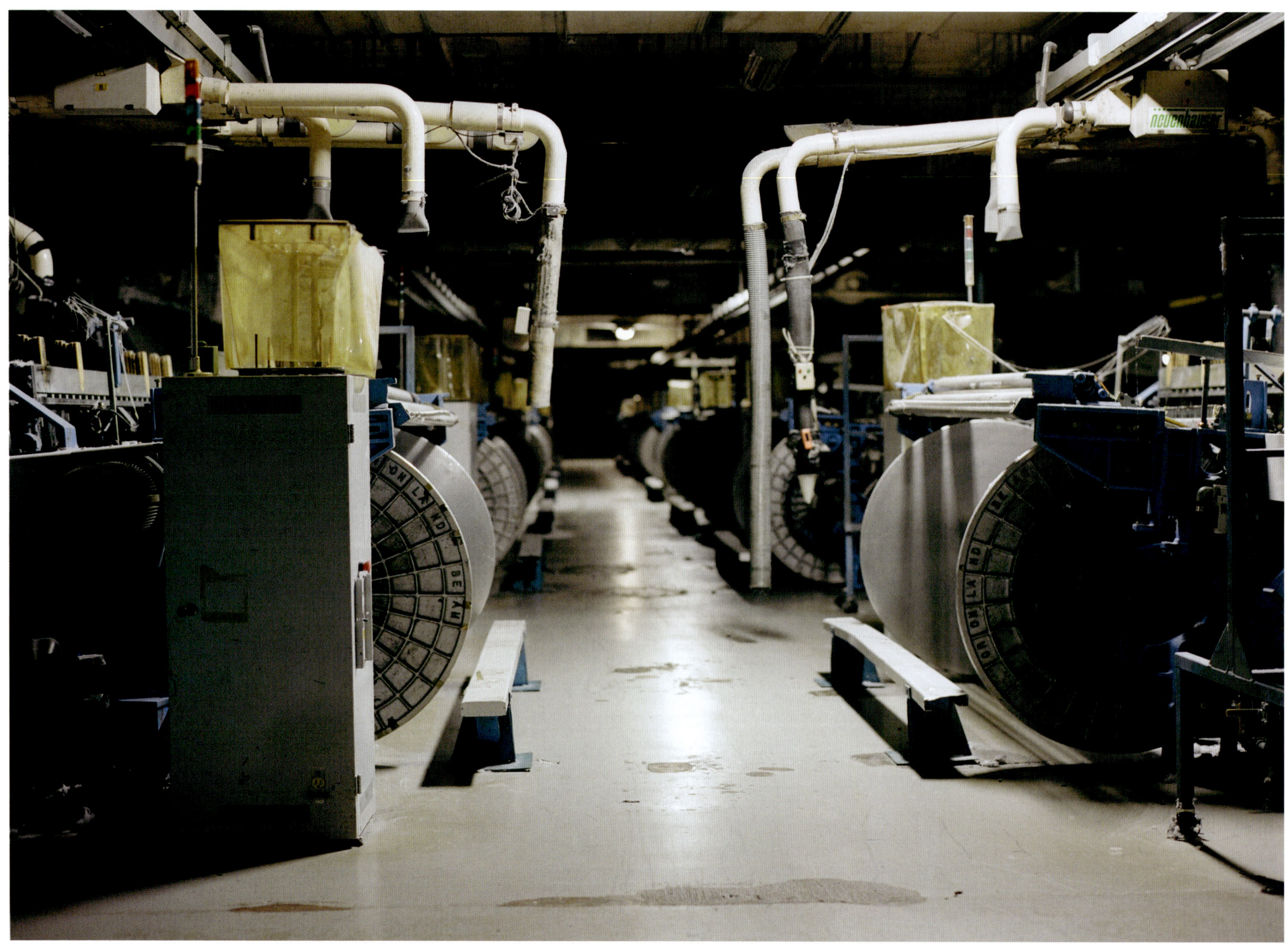

Weaver's alley in an air jet weave room in the Weaving Department

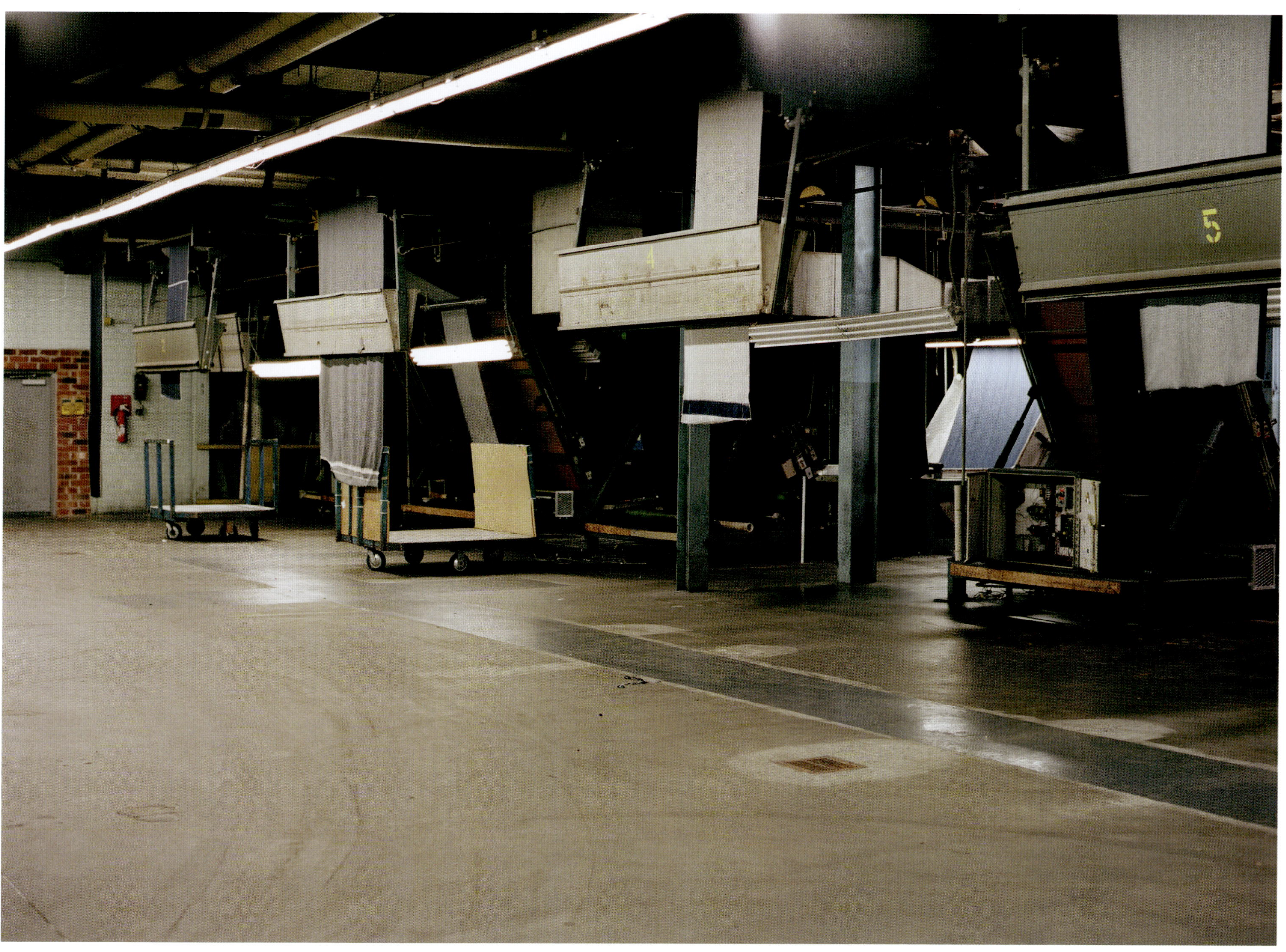

Empty carts at sanforizer stations in the Finishing Department. Cloth still in place.

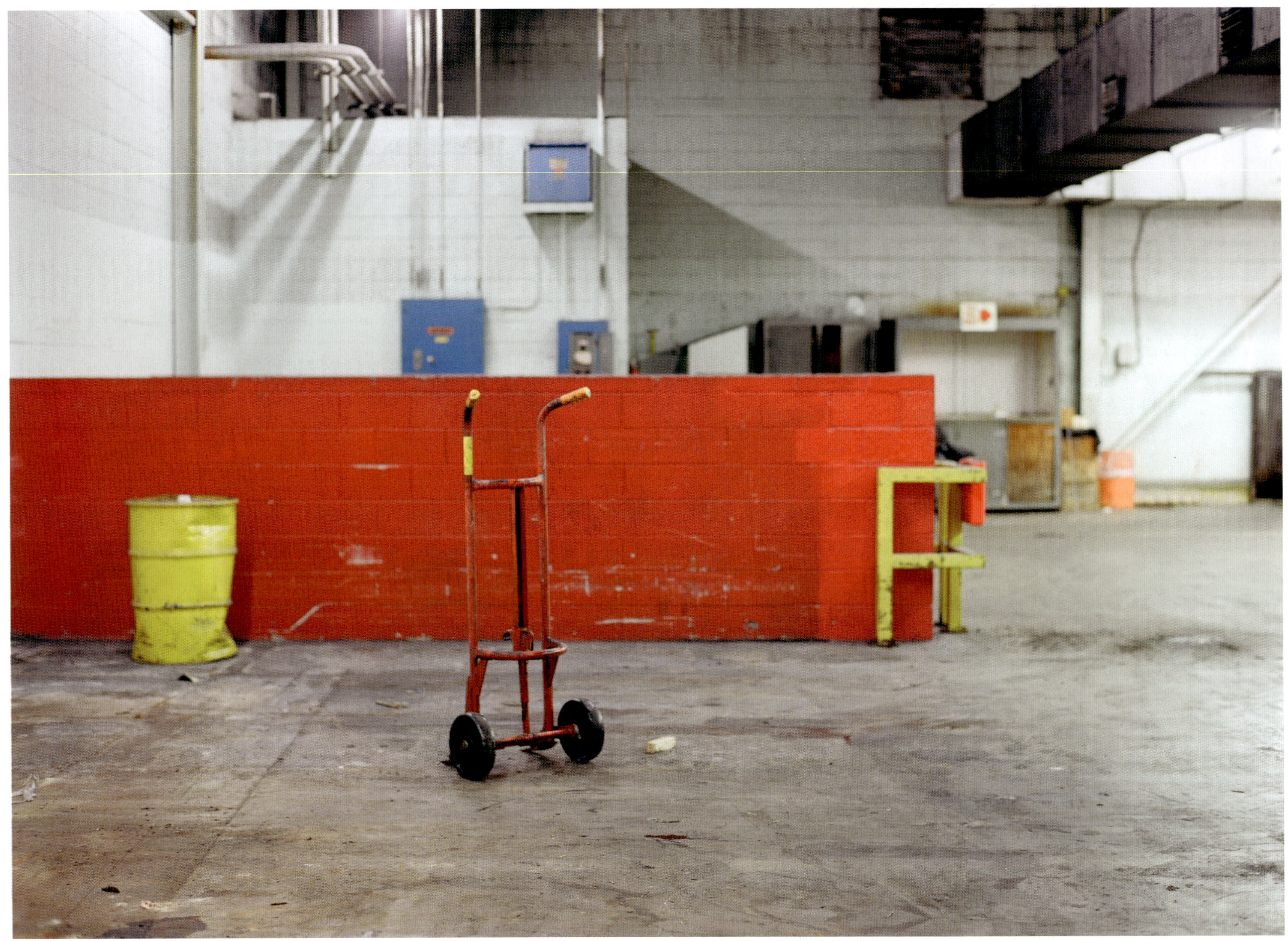

160

Hand truck in the Finishing Department

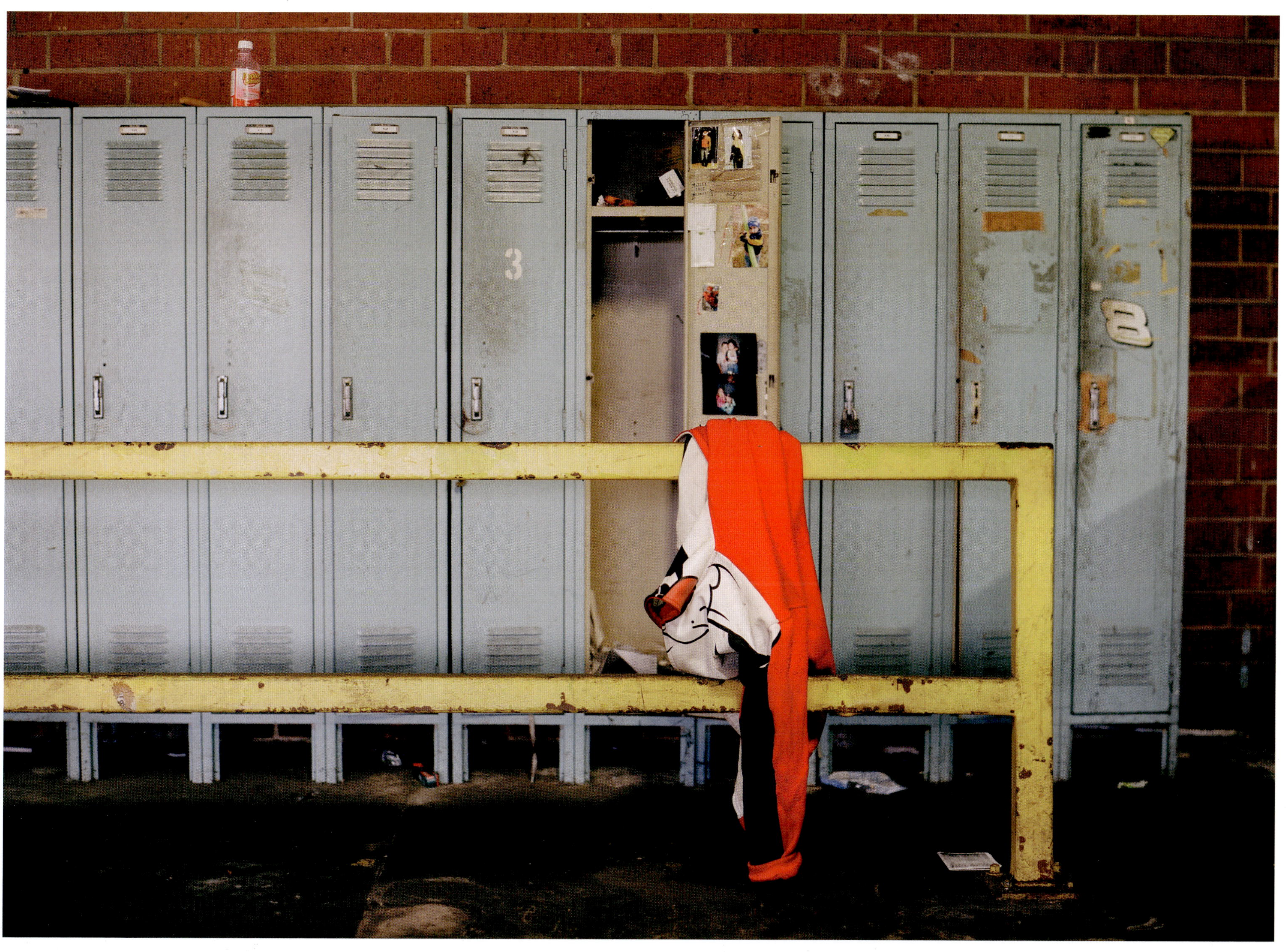

Vacated employee lockers in the Slashing Department

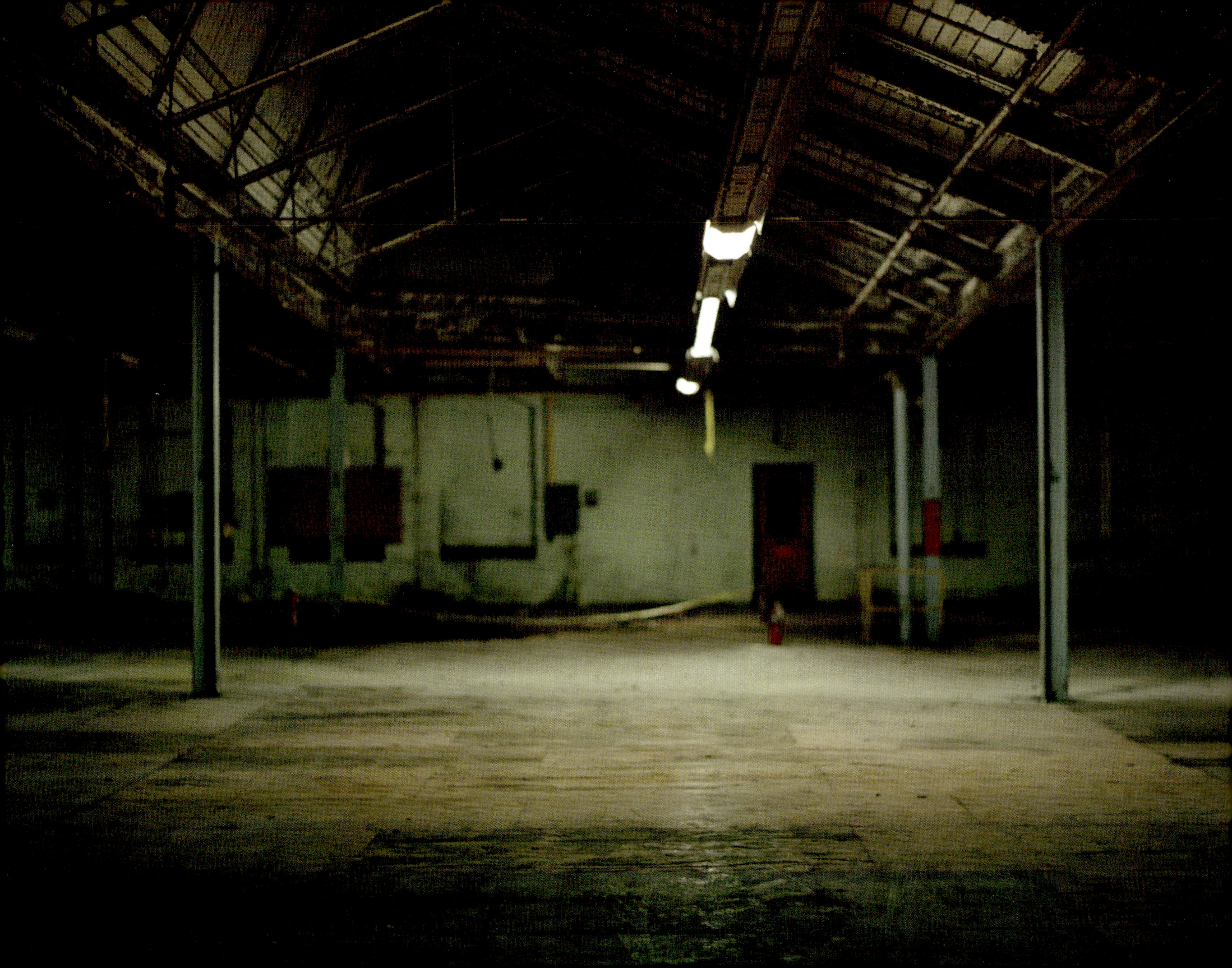

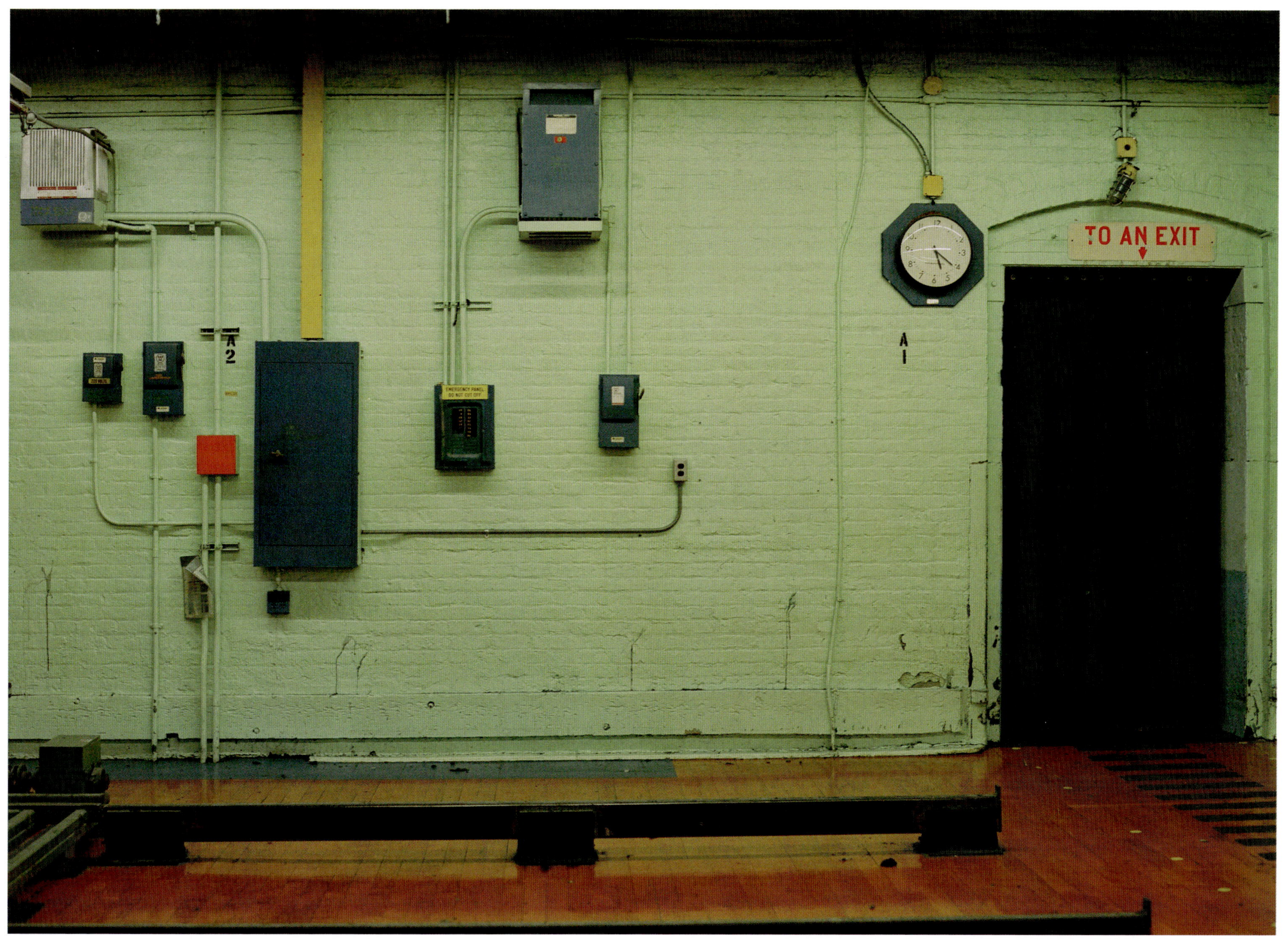

South end of the original Weave Room　//　Passway between the Weave Floor and Slashing Room in the Slashing Department

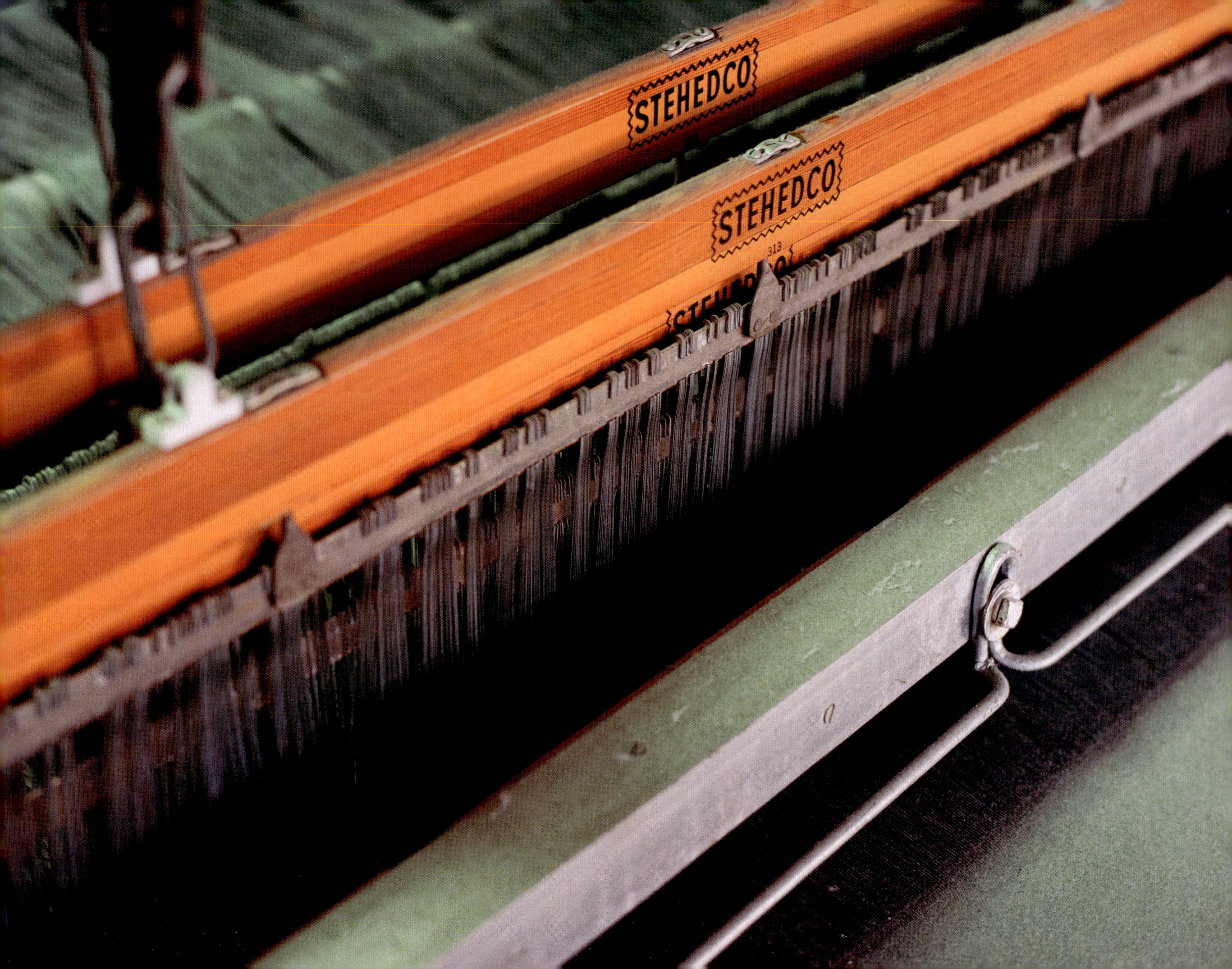
STEHEDCO
STEHEDCO
313

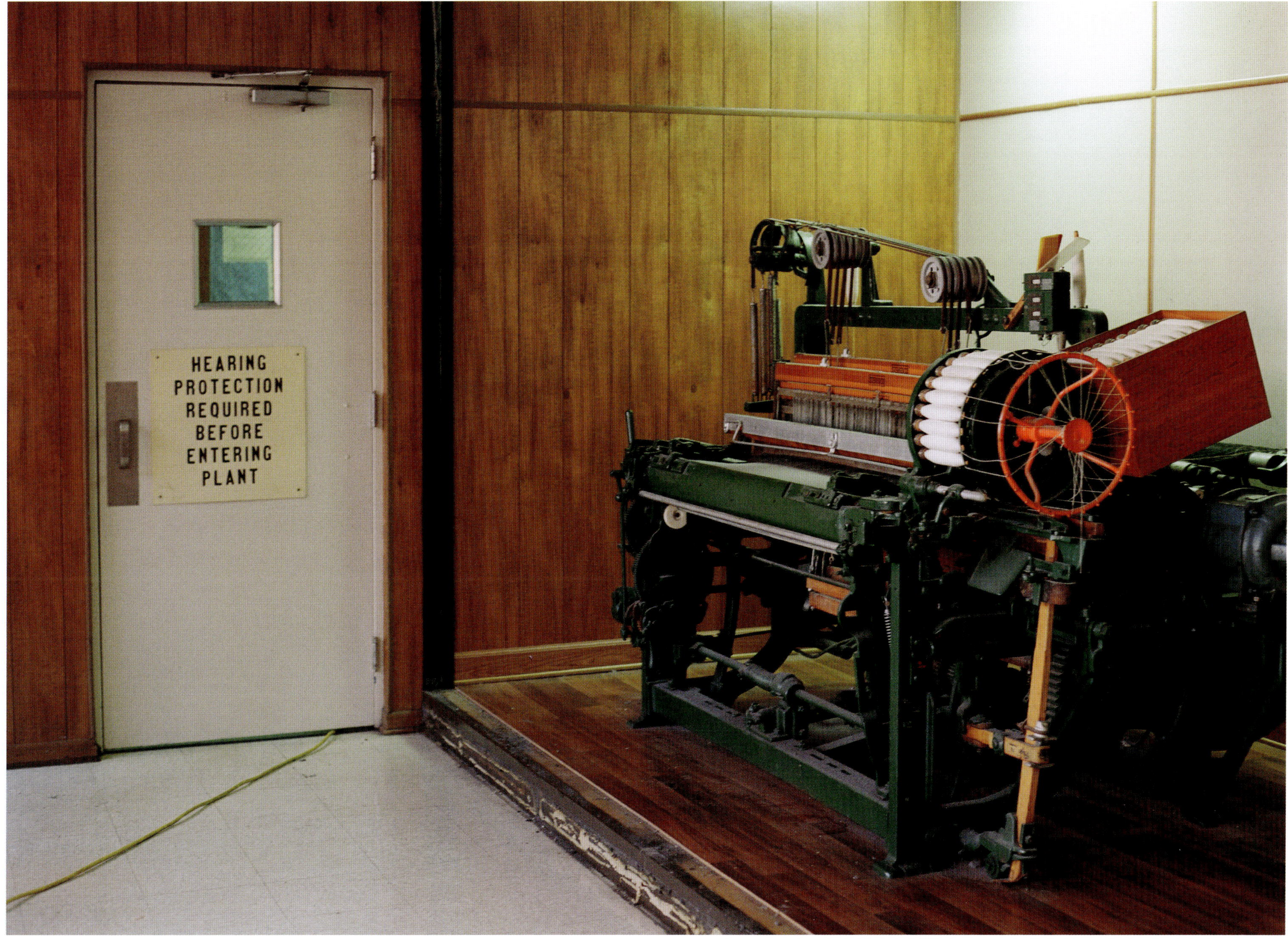

165

Draper X-2 shuttle loom in the front lobby of White Oak Mill entrance to the Weave Room

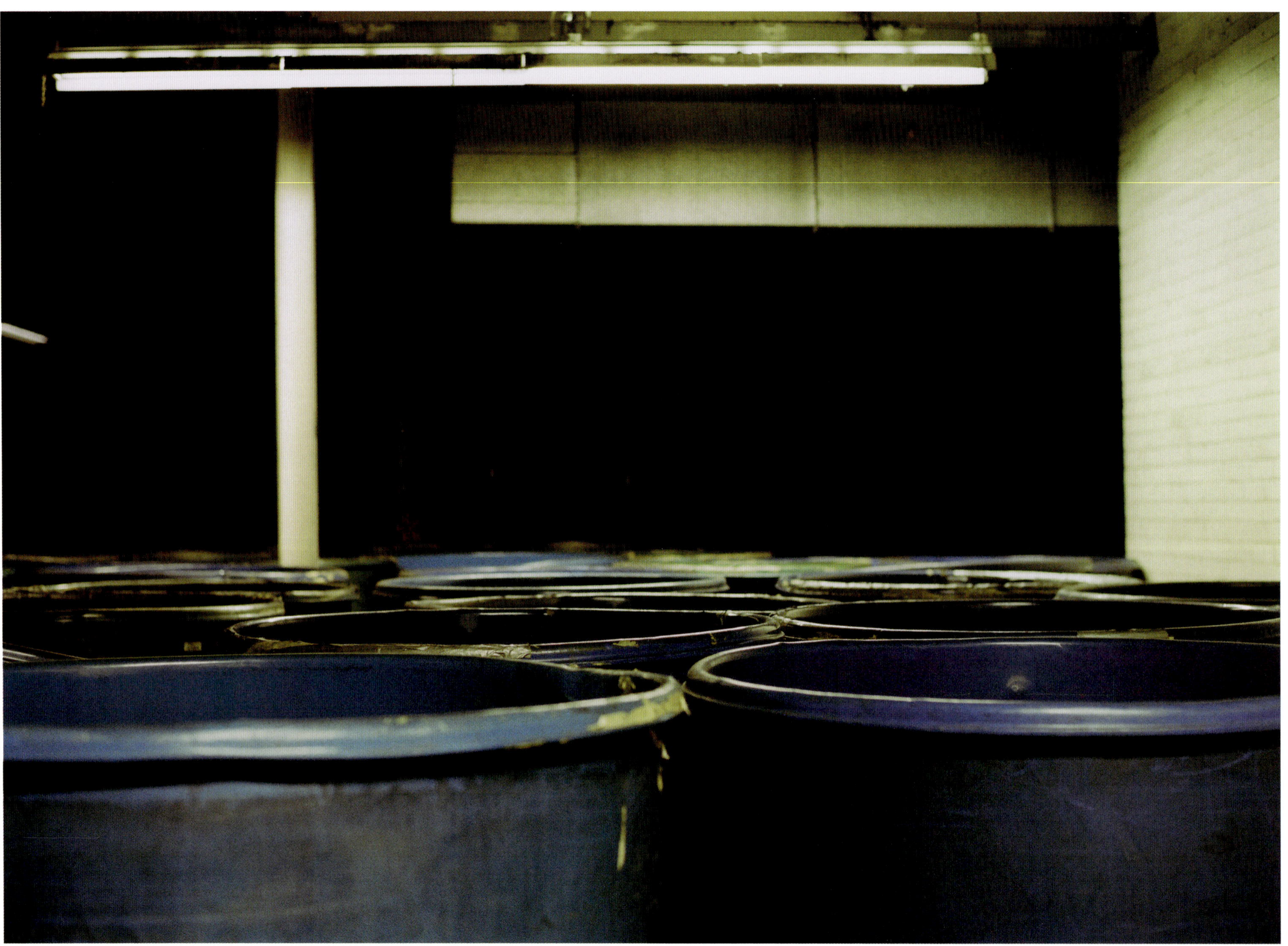

Empty tubs for dyed warps in the Beaming Department

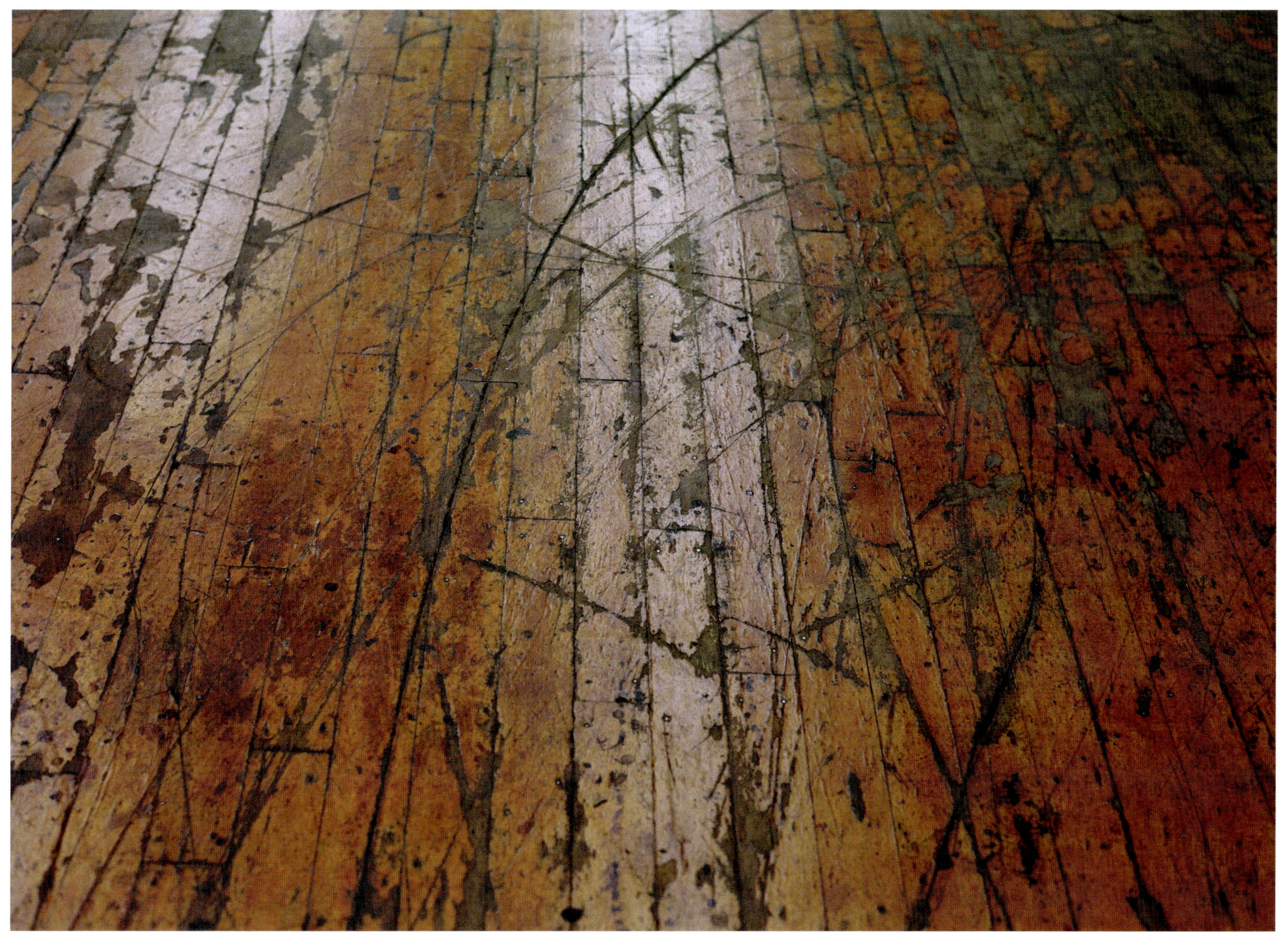

Material bin wheel marks on maple flooring in the Spinning Department

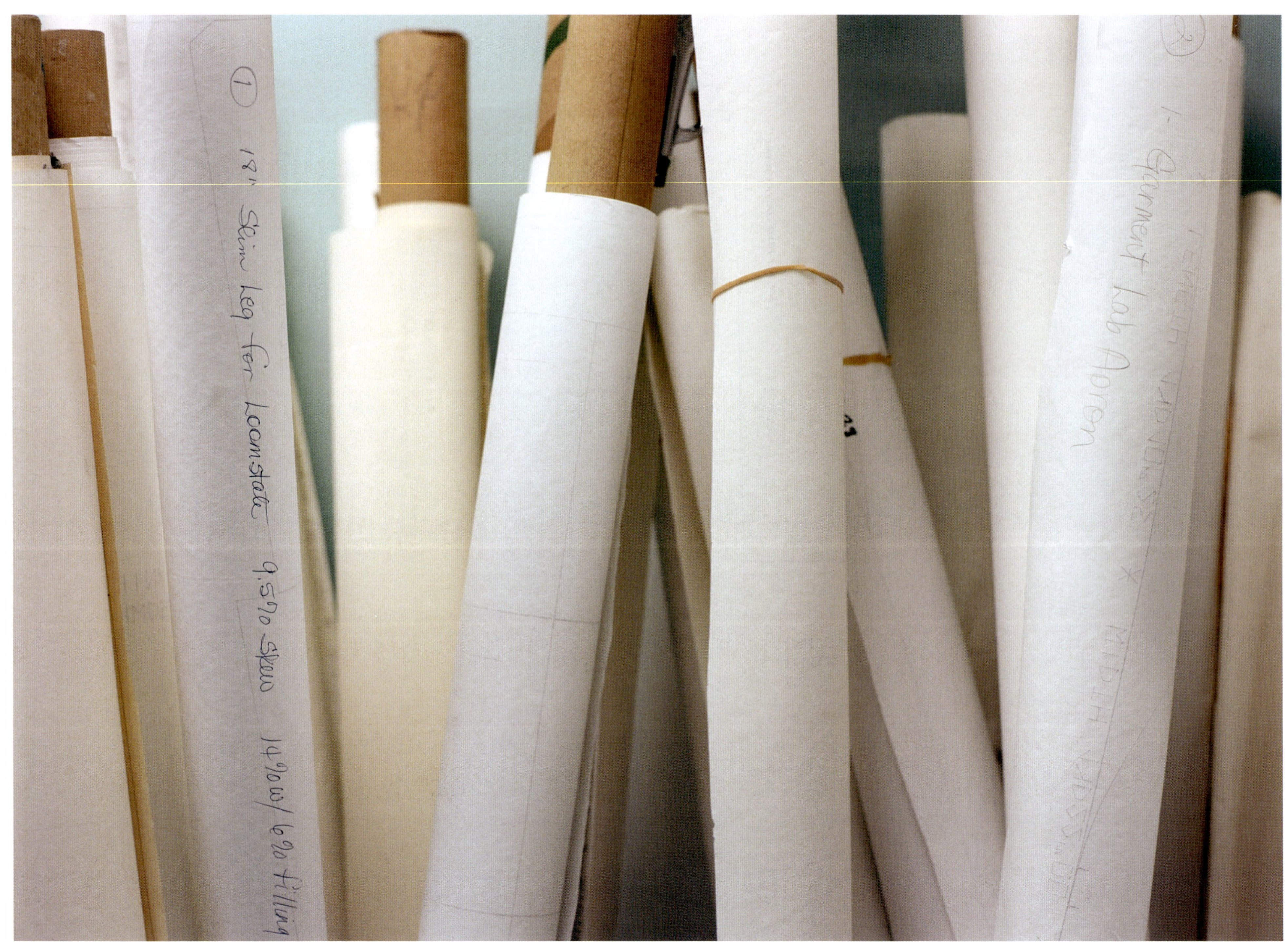

Rolled-up sample garment patterns in the Garment Lab. Handwriting of Lynda Layton, former Head of the Garment Lab, is on the scroll on the left.

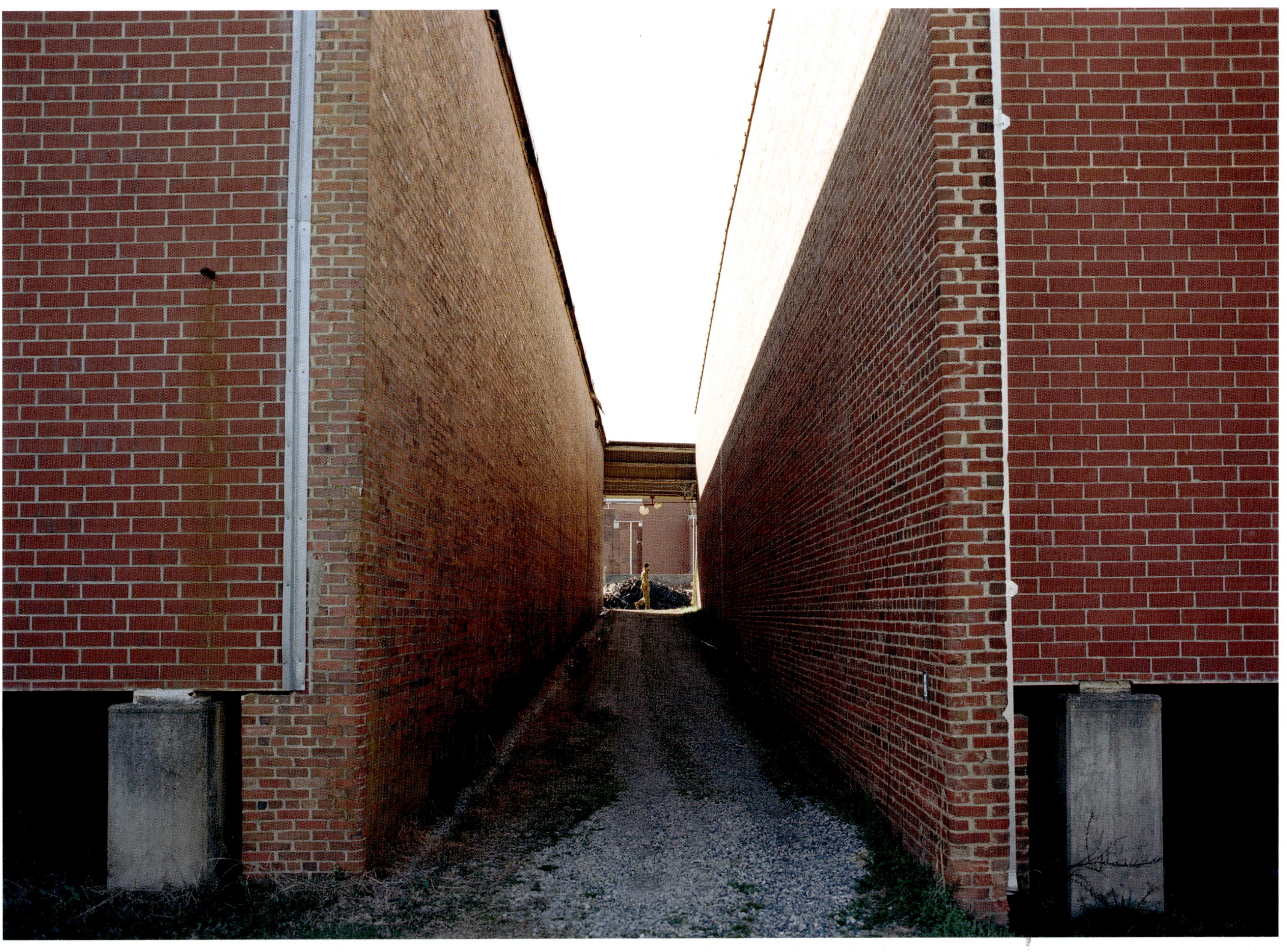

Evan Morrison walking past stripped copper remnants from cotton warehouses

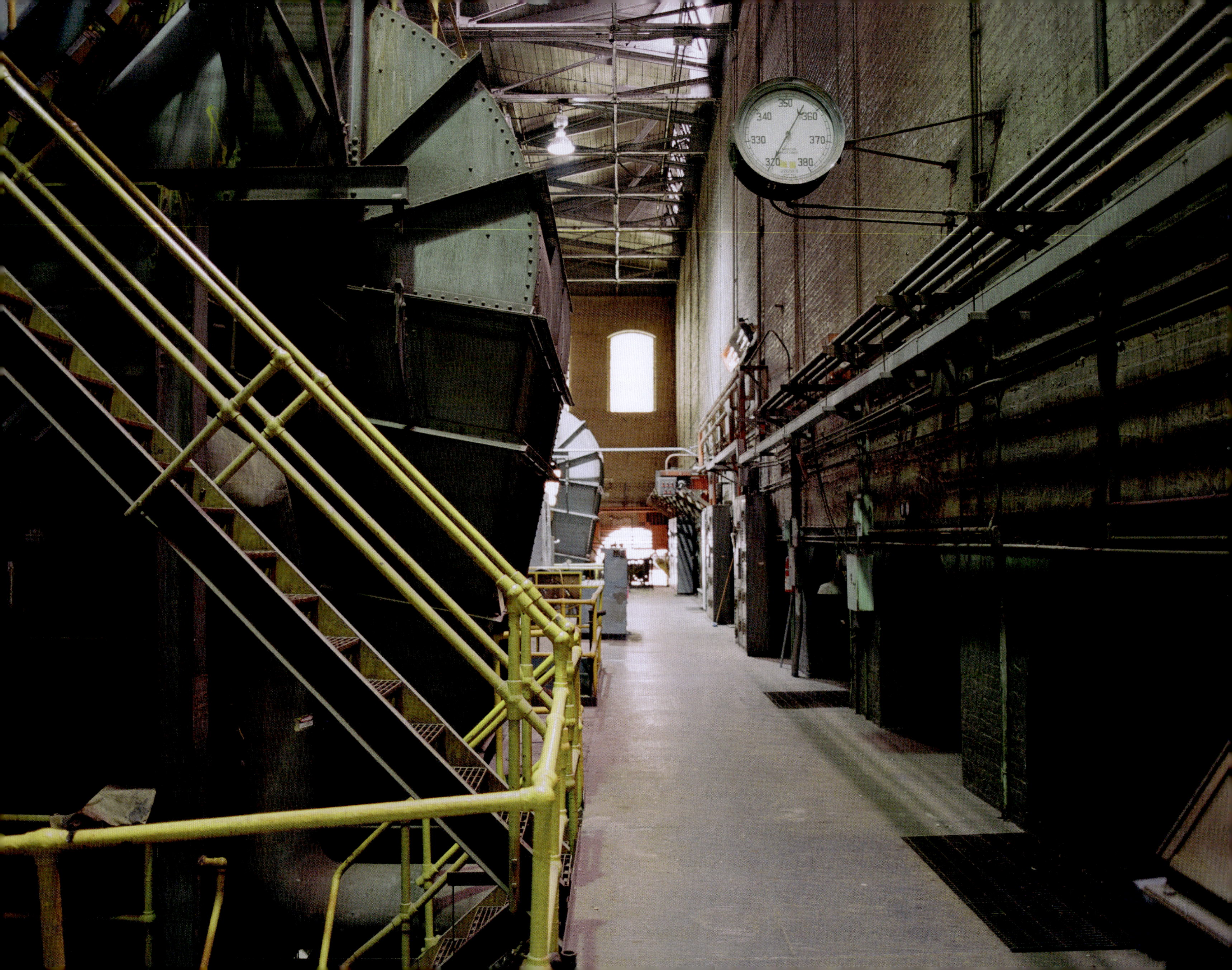

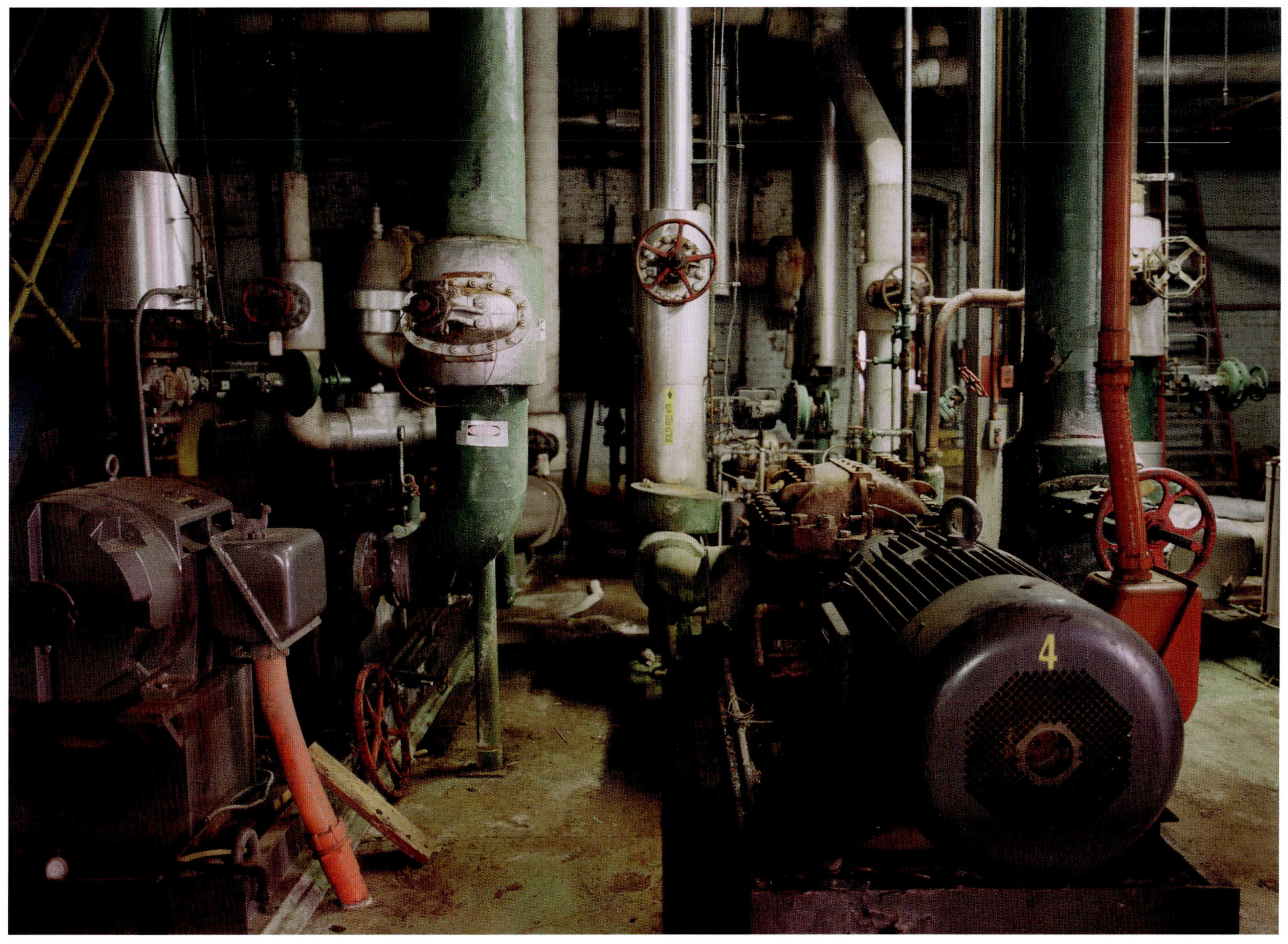

Control Room platform in the Powerhouse   //   Generator room full of steam pipes in the Powerhouse

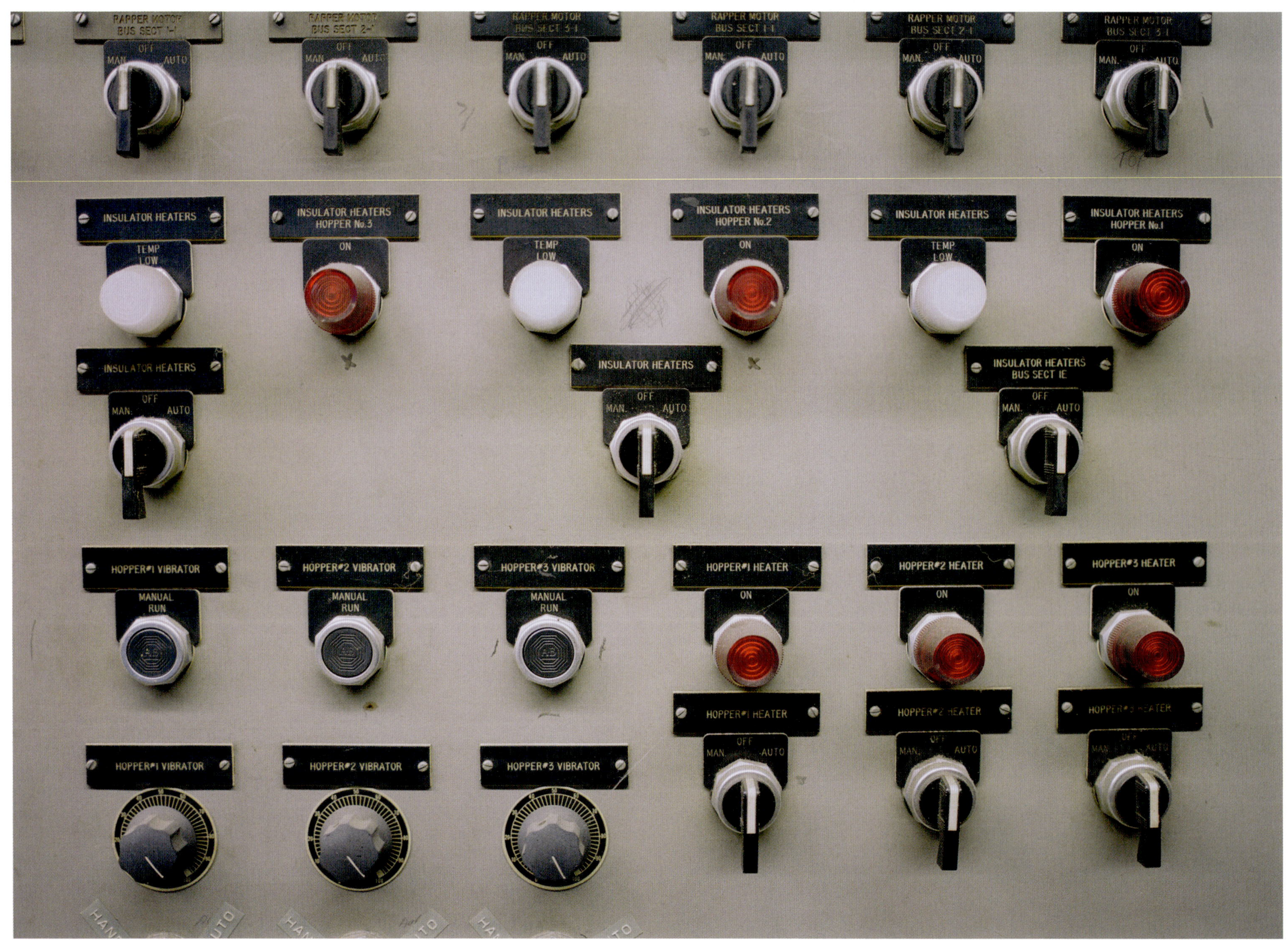

172

Hi/Low steam pressure control panel in the Powerhouse

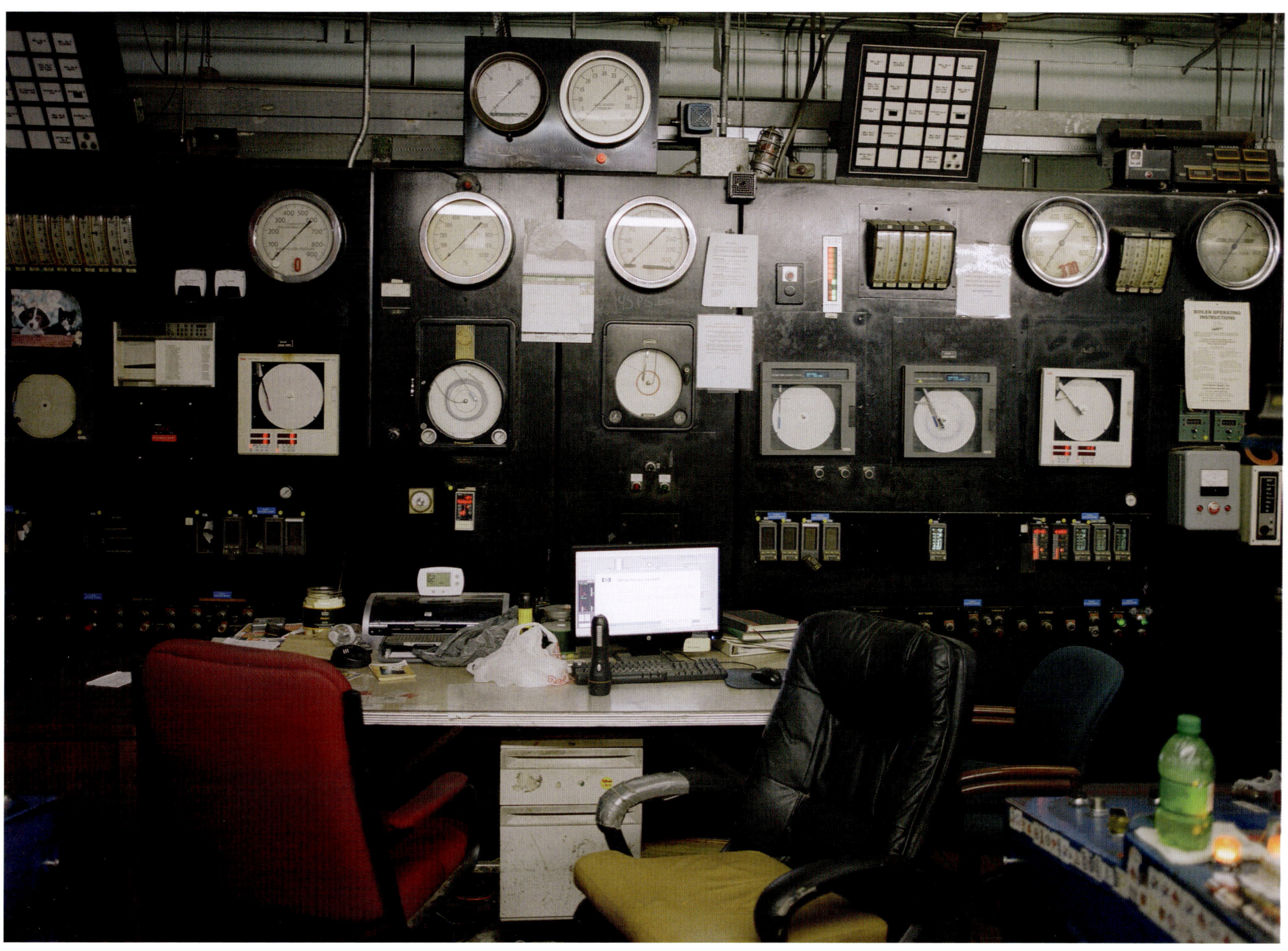

Vacated main office of the Control Room in the Powerhouse

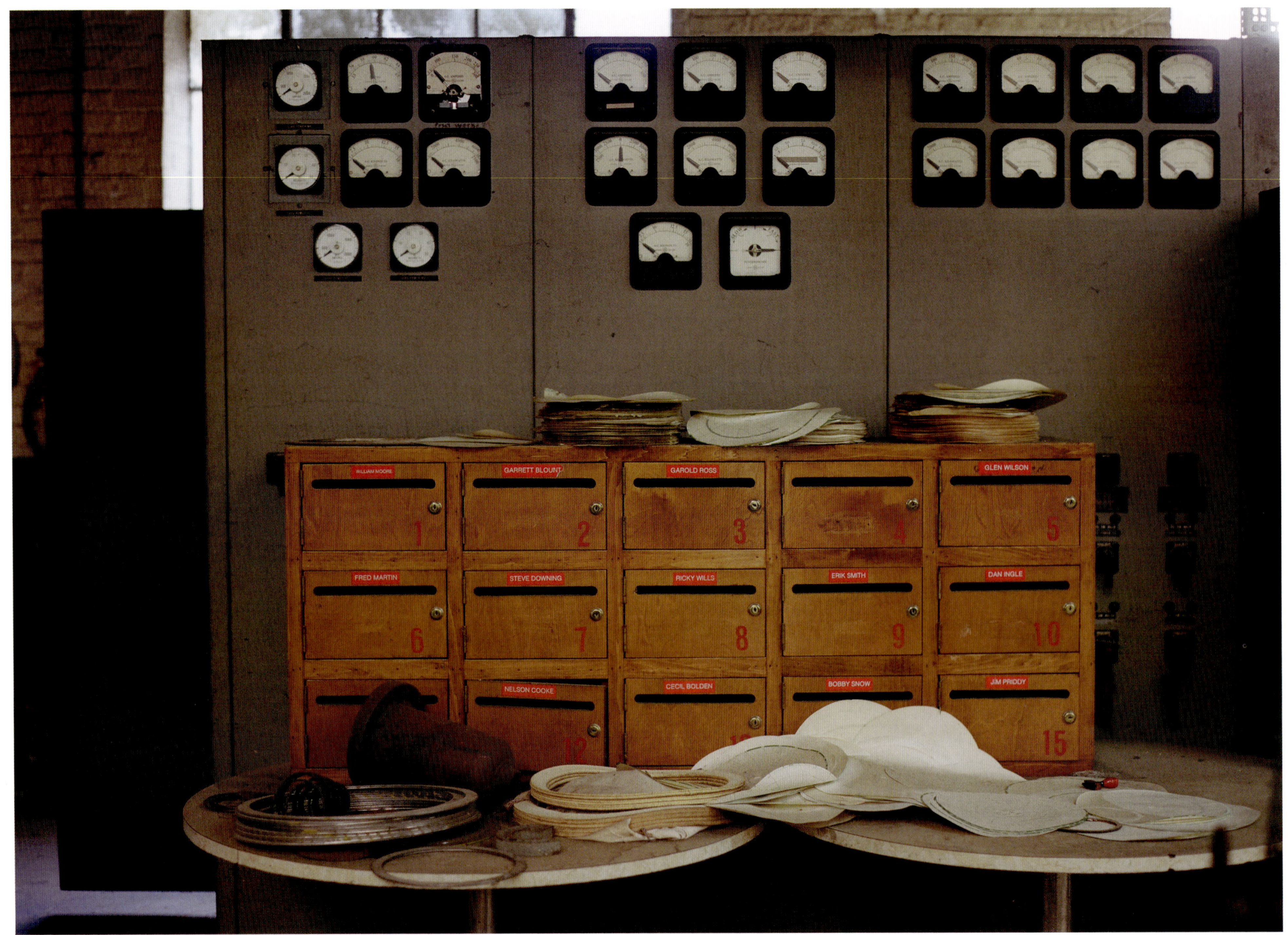

Control panel and employee drop boxes in the Generator Room in the Powerhouse

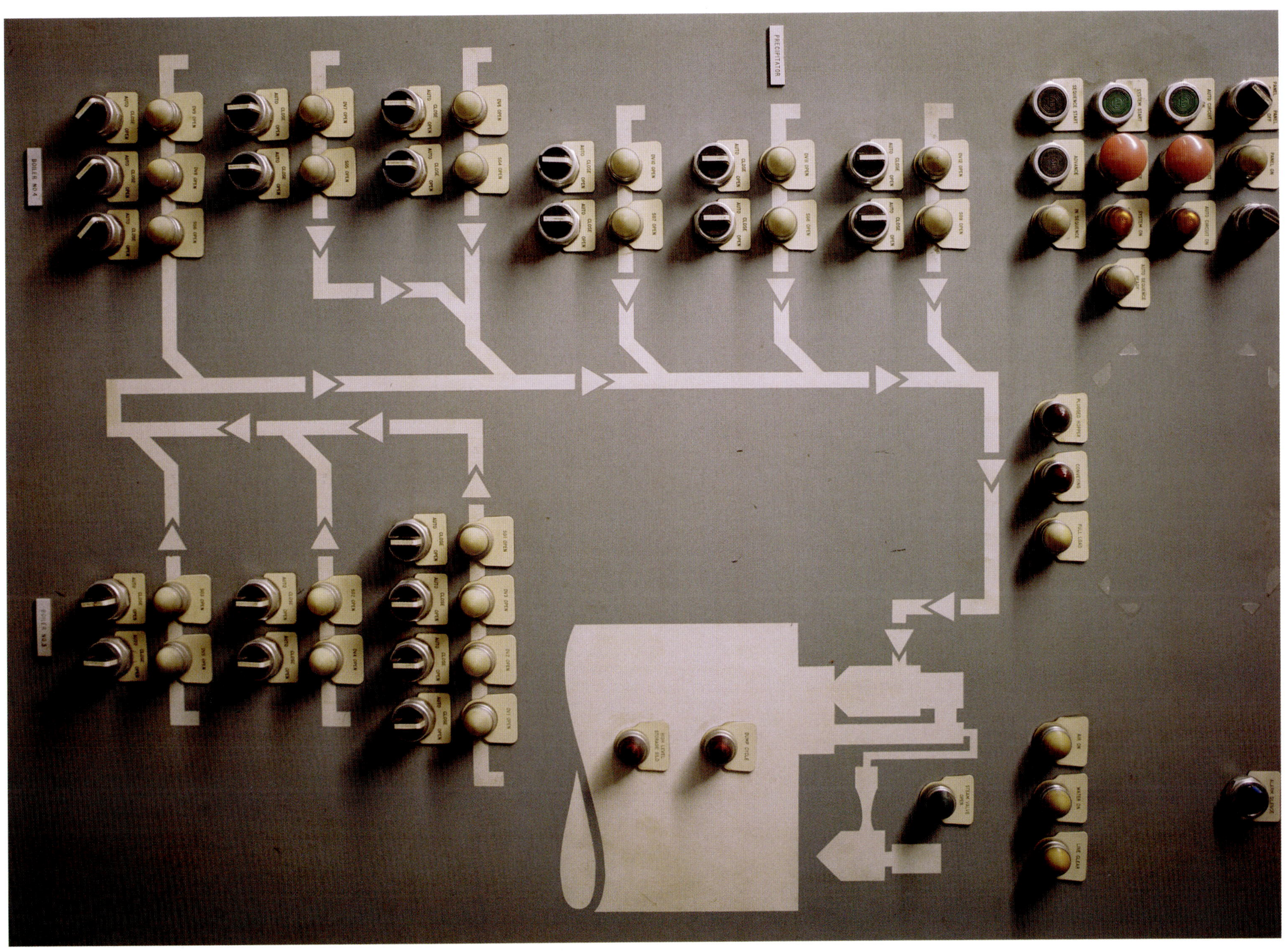

Hi/Low-pressure steam flow chart on the control panel in the Powerhouse

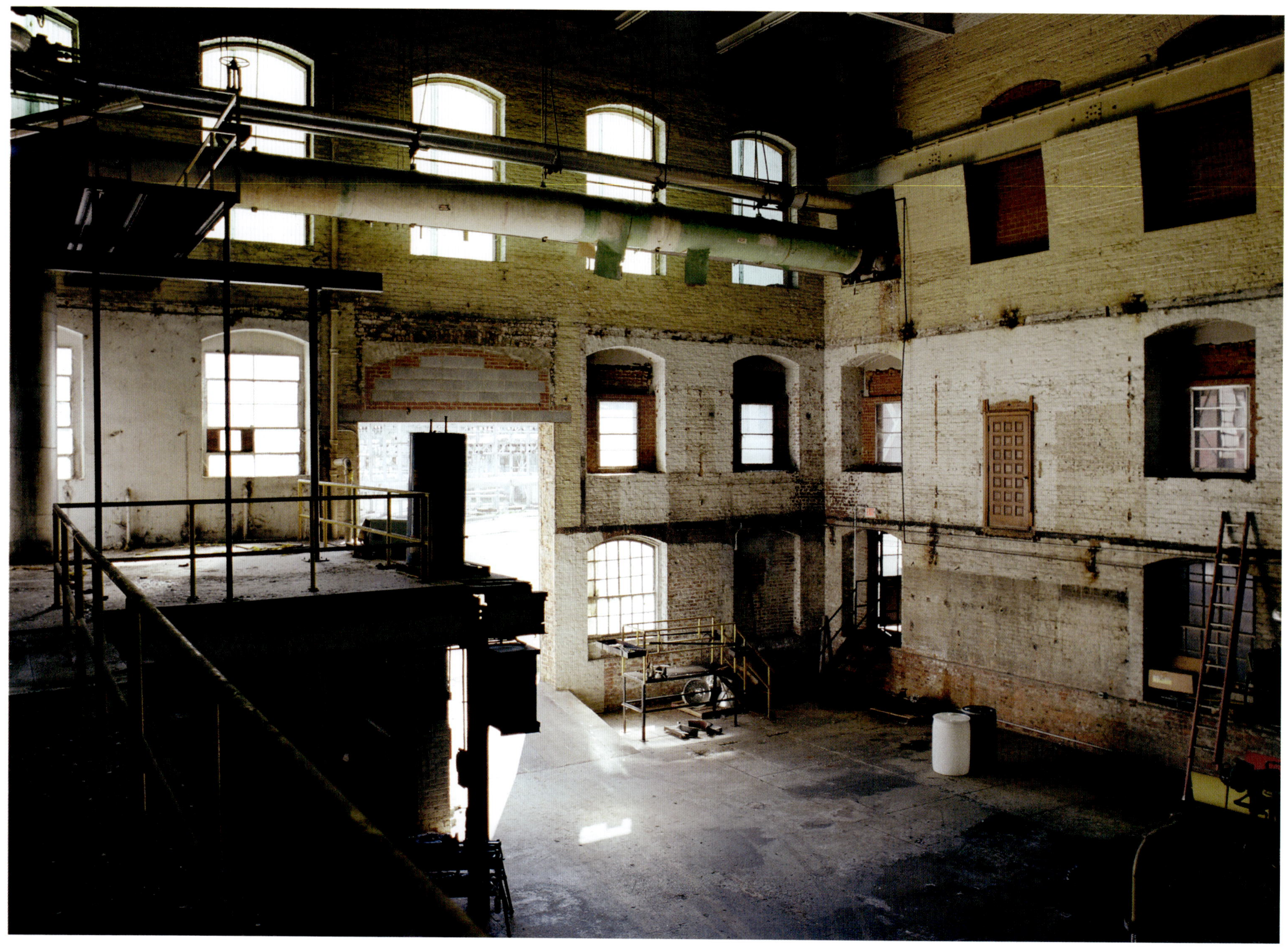

Doorway without a floor on the second floor of the Powerhouse

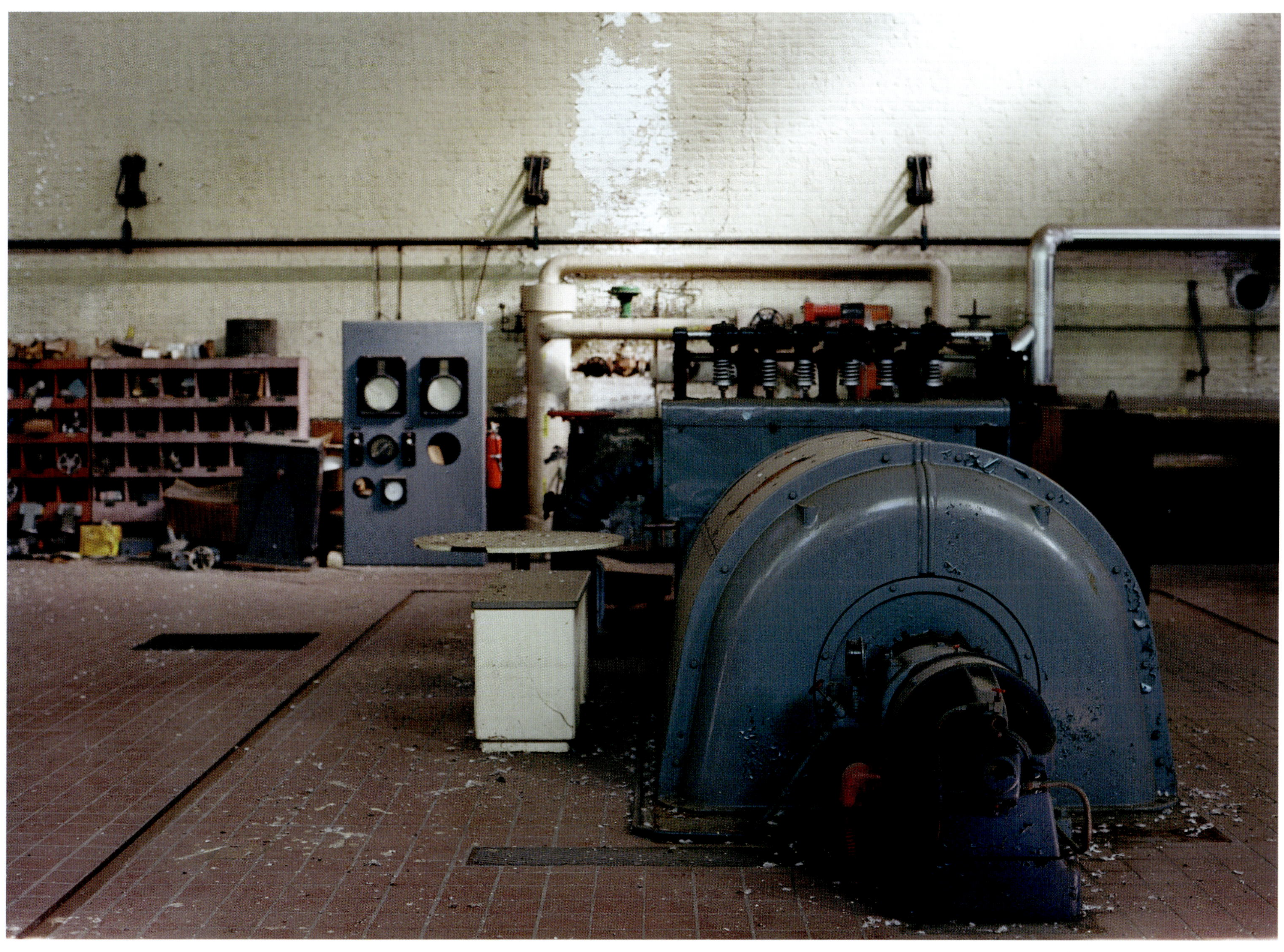

Electrical conversion generator in the Powerhouse

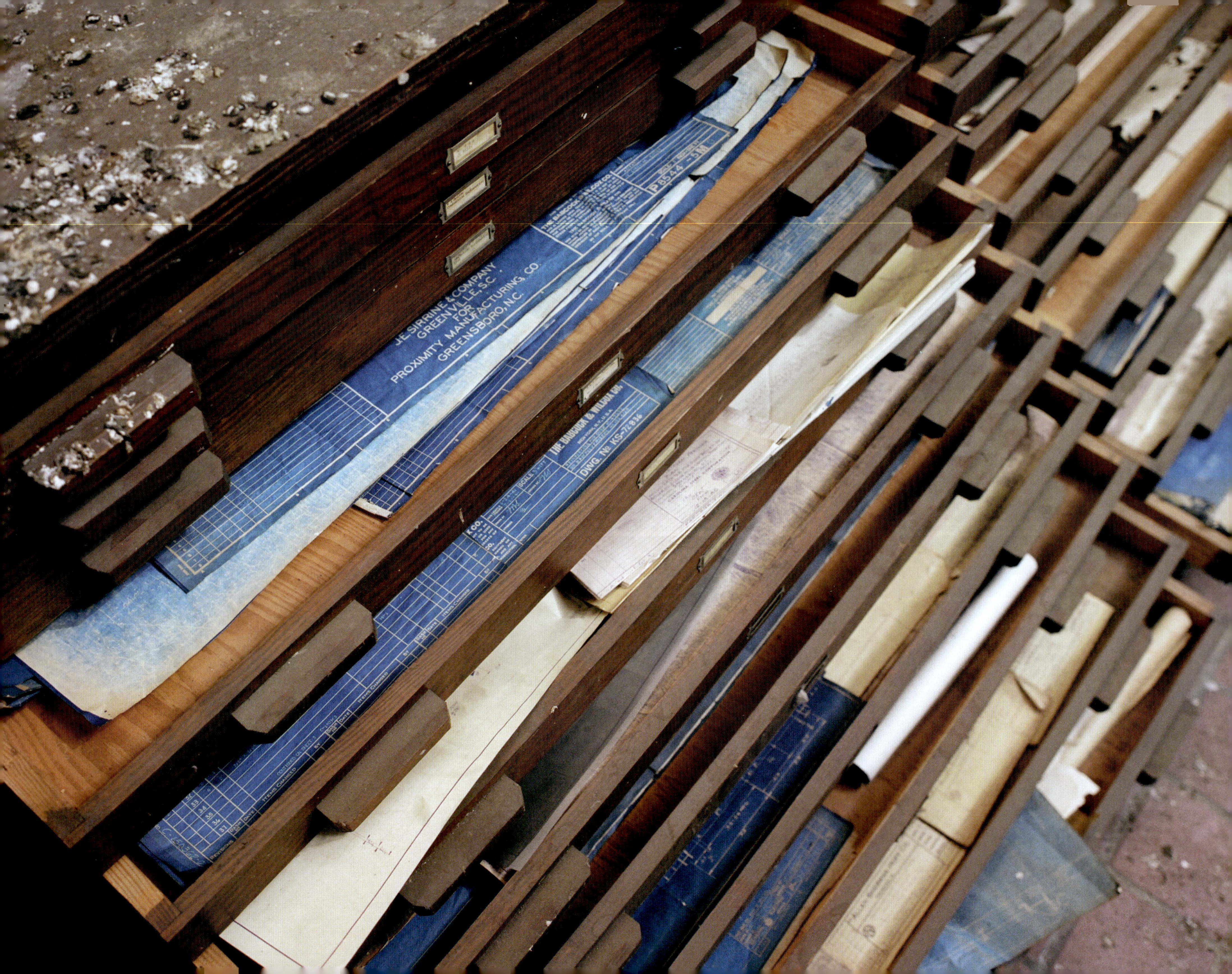

J.E. SIRRINE & COMPANY
GREENVILLE, S.C.
FOR
PROXIMITY MANUFACTURING CO.
GREENSBORO, N.C.
P8544 - 3N
DWG. Nº KG-7193

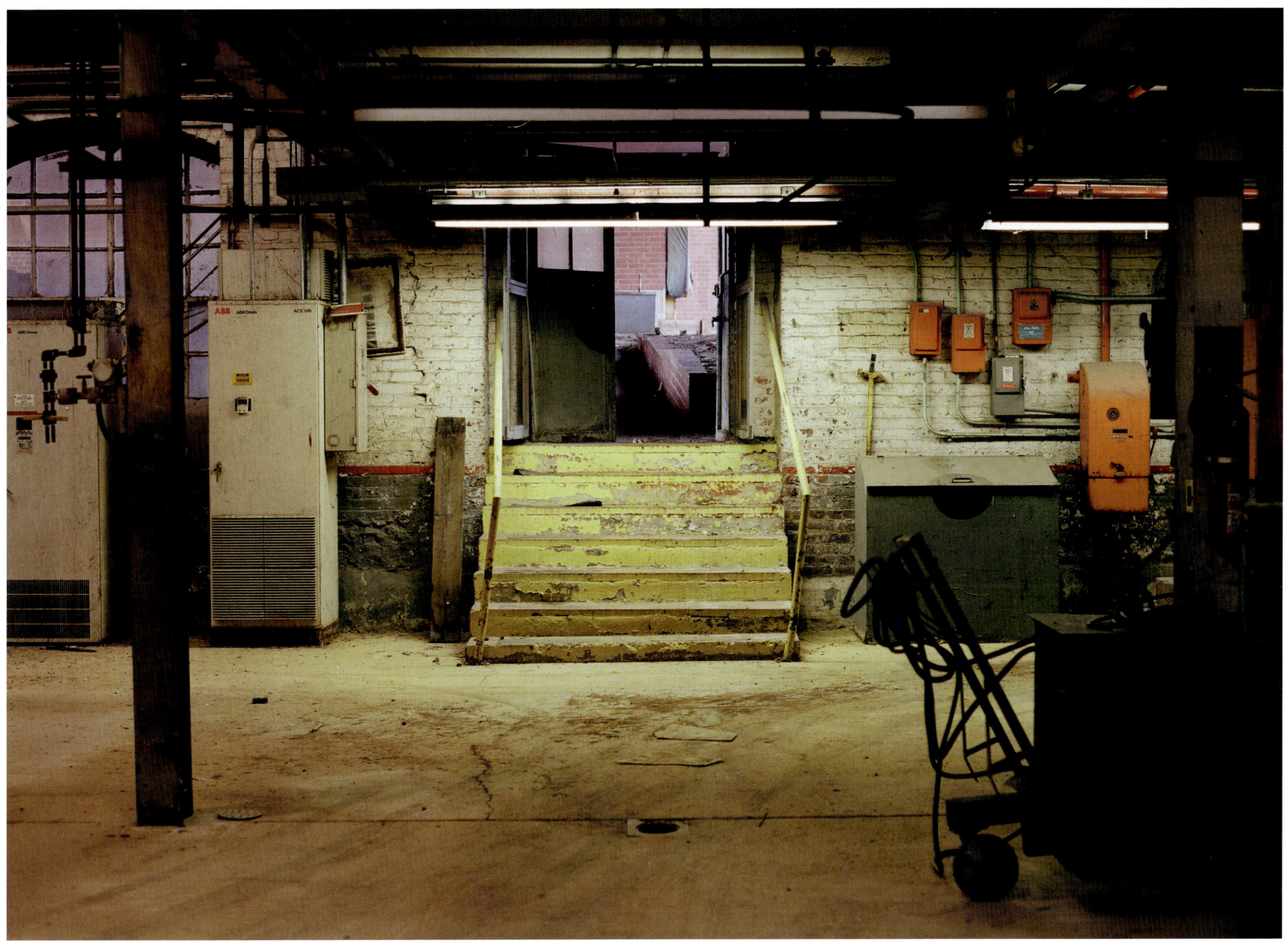

Blueprint cabinet, found by local birds, in the Powerhouse   //   Entryway to the first floor of Generator Room in the Powerhouse

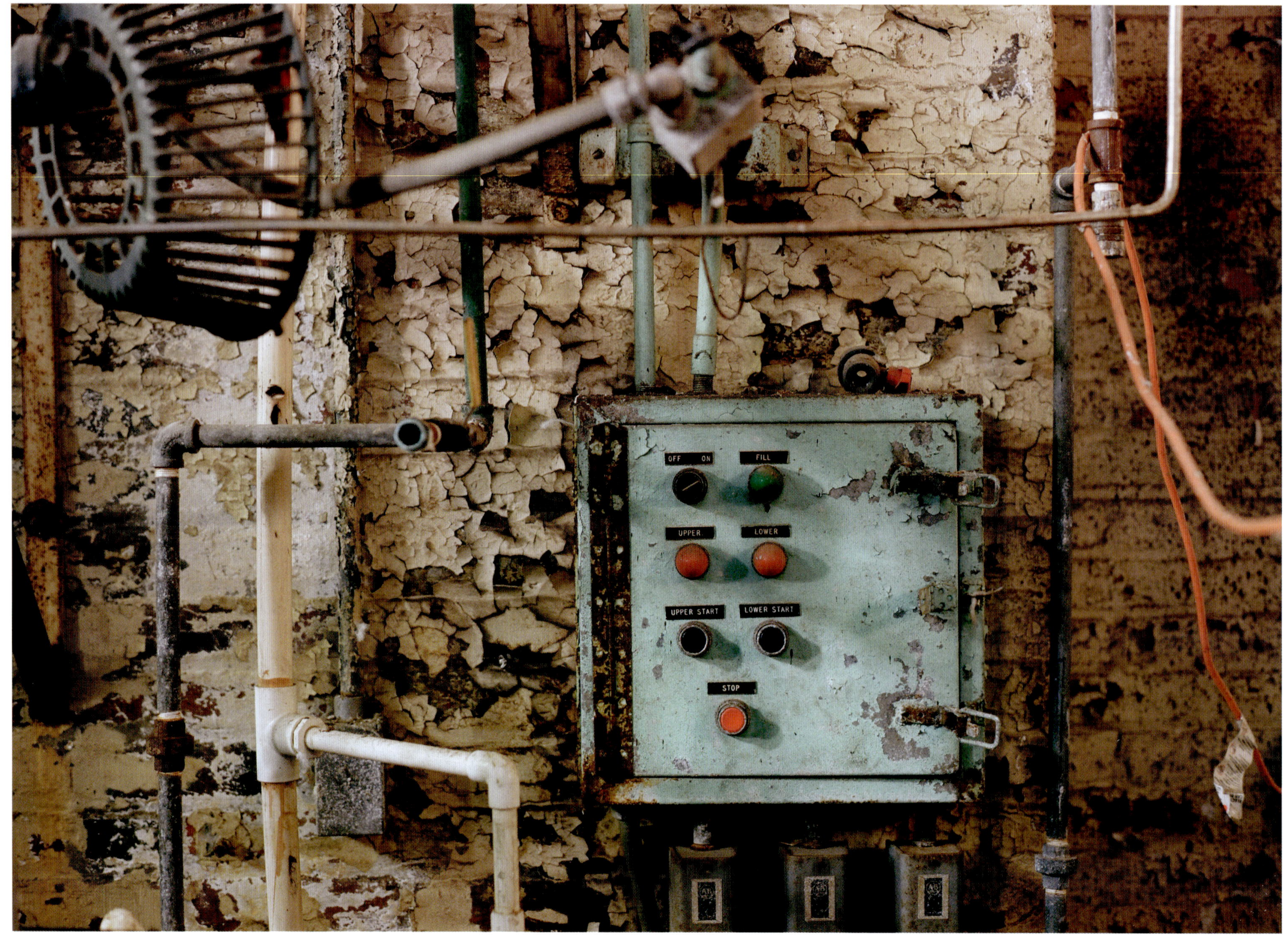

Control panel on the first floor of the Generator Room in the Powerhouse

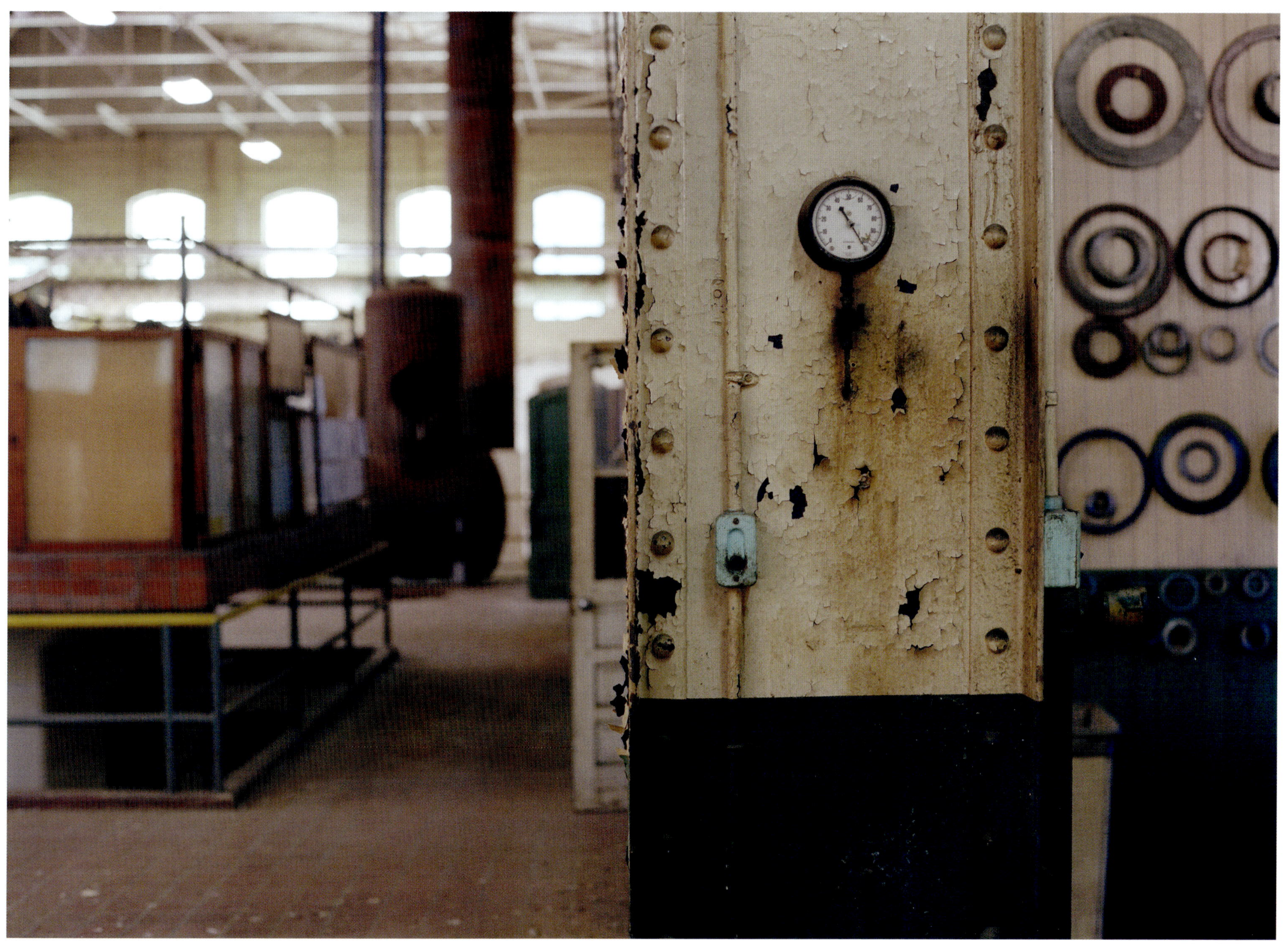

Second floor of the Generator Room in the Powerhouse

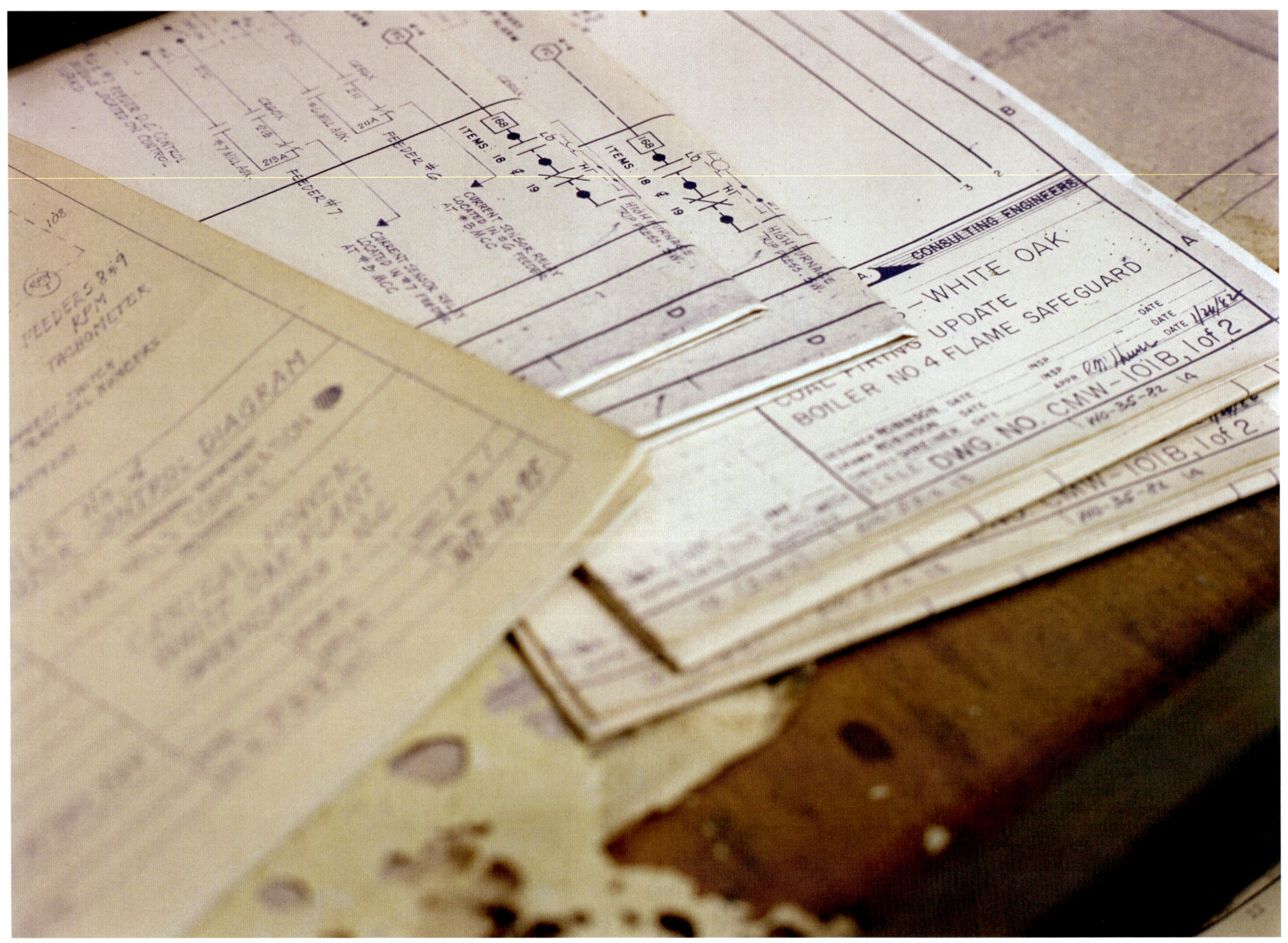

Blueprints in the Powerhouse

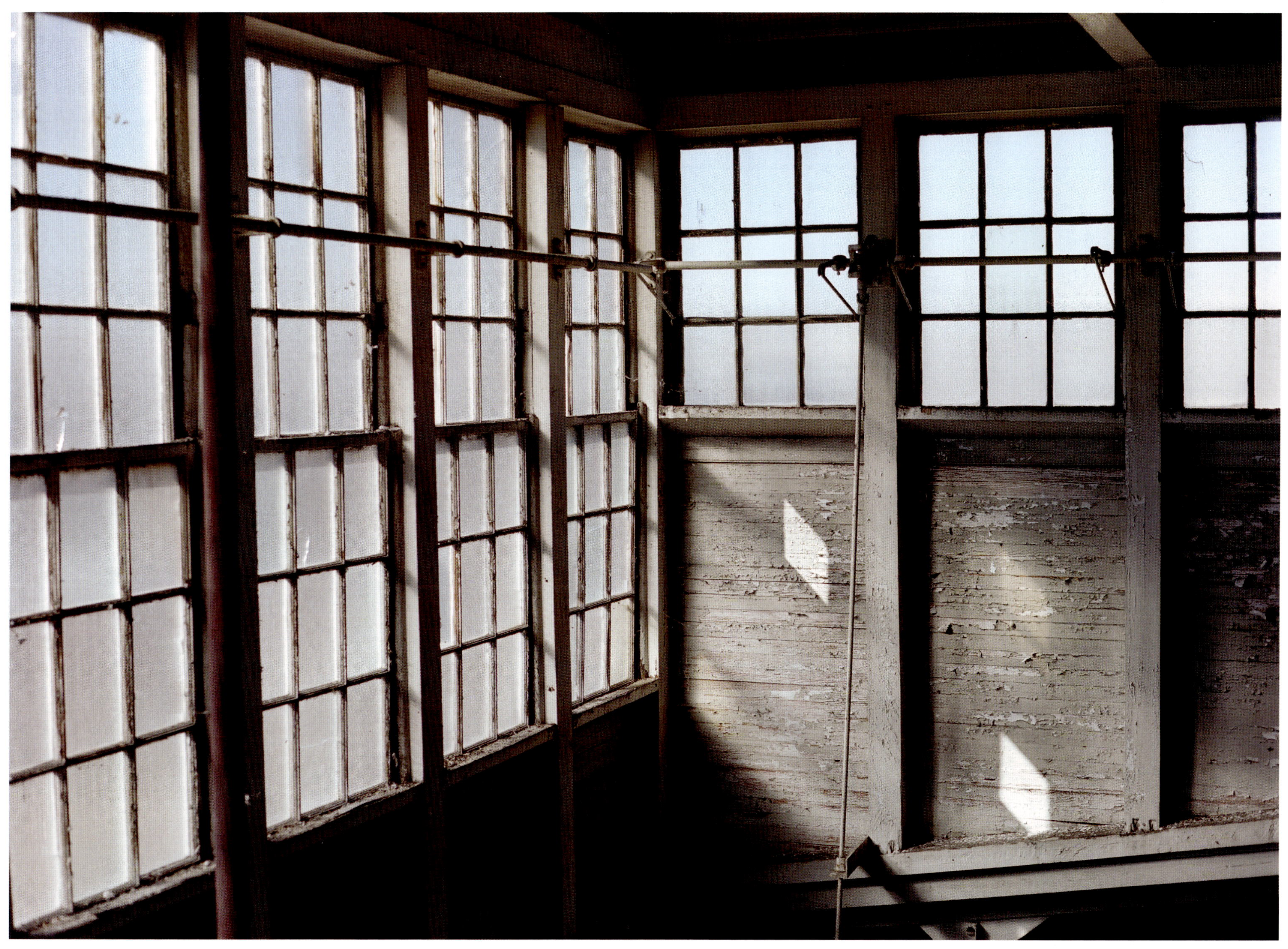

Skylight windows in the Generator Room in the Powerhouse

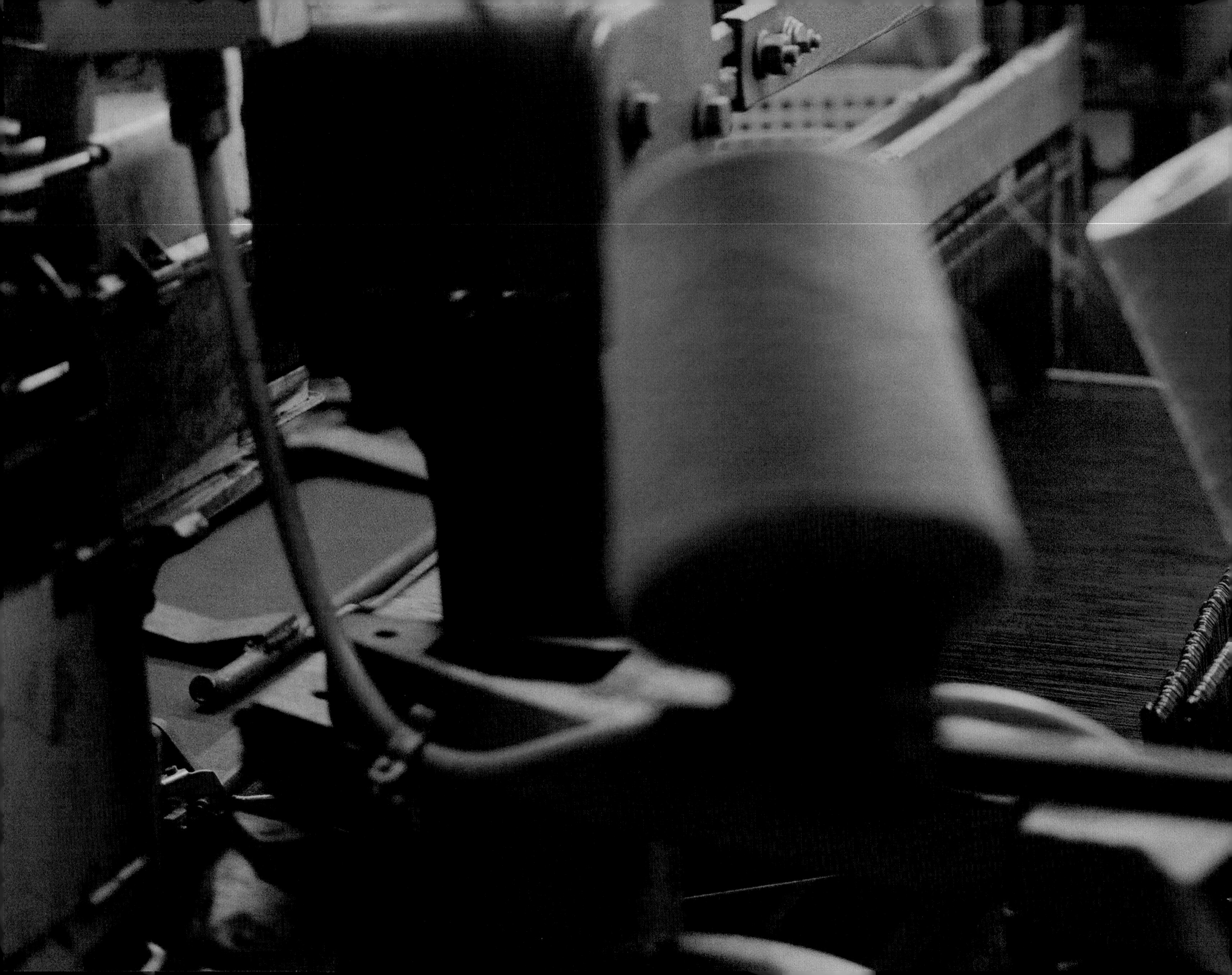

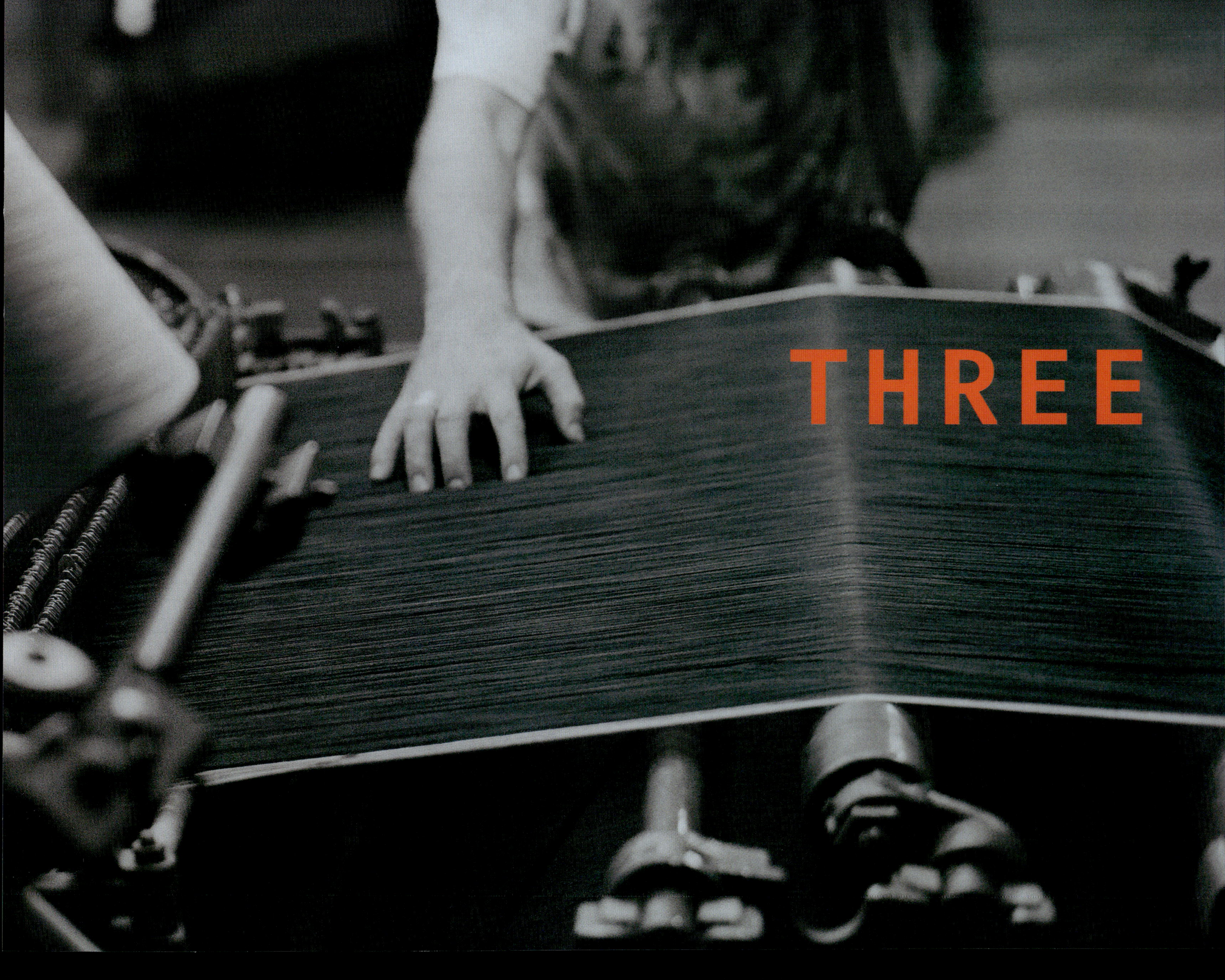

THREE

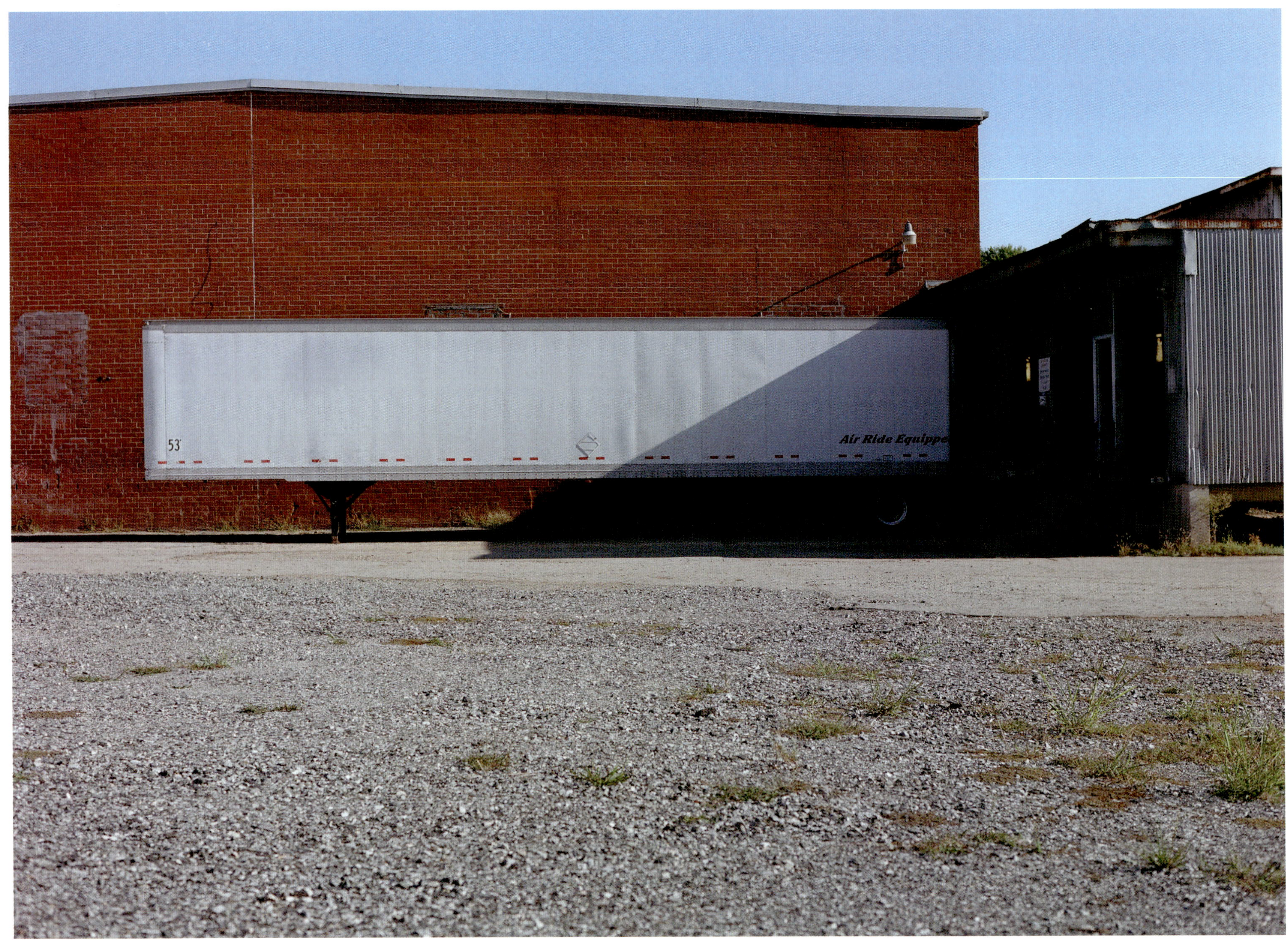

Loading dock at the Cotton Warehouses

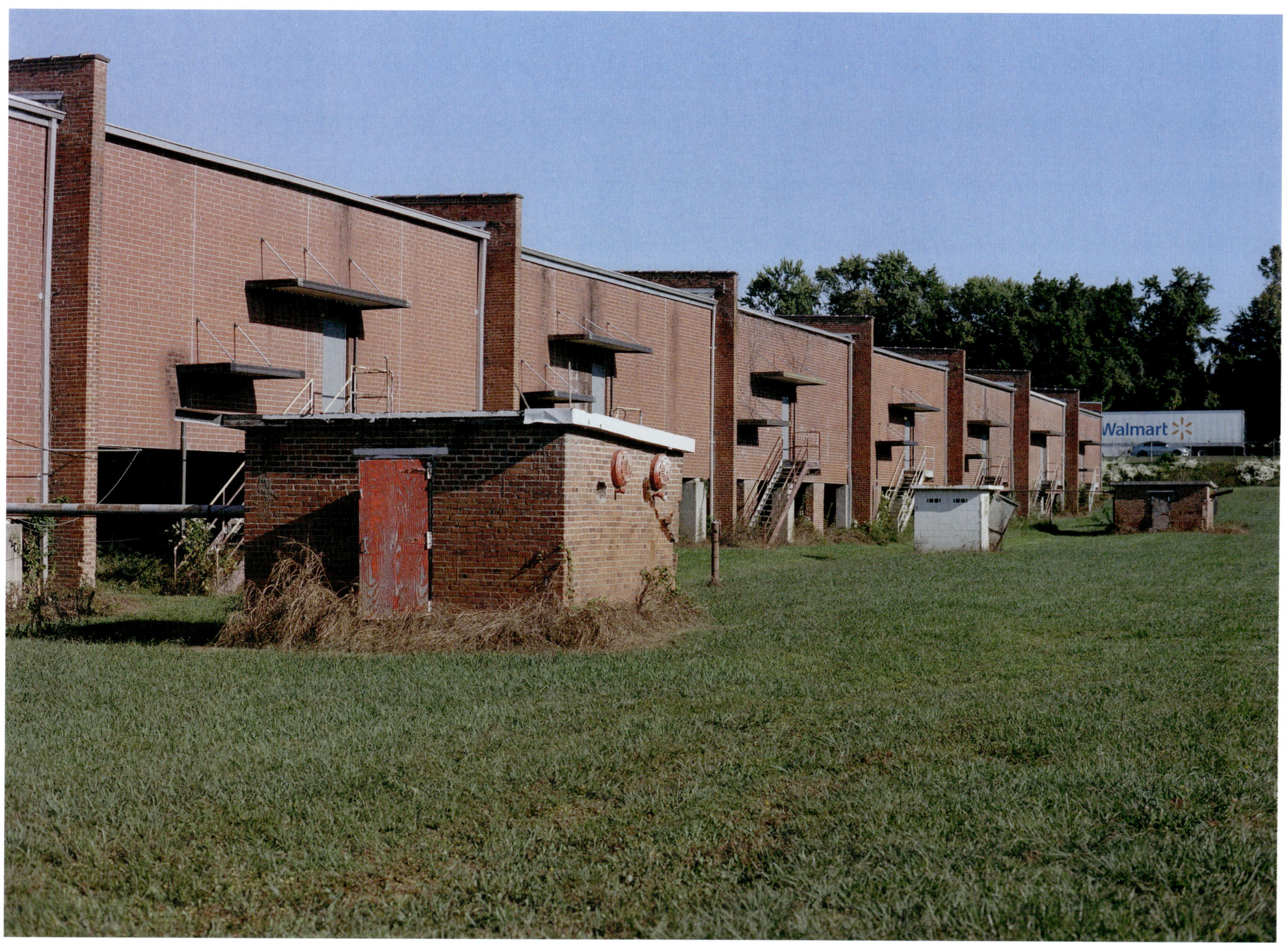

Rear view of the Cotton Warehouses

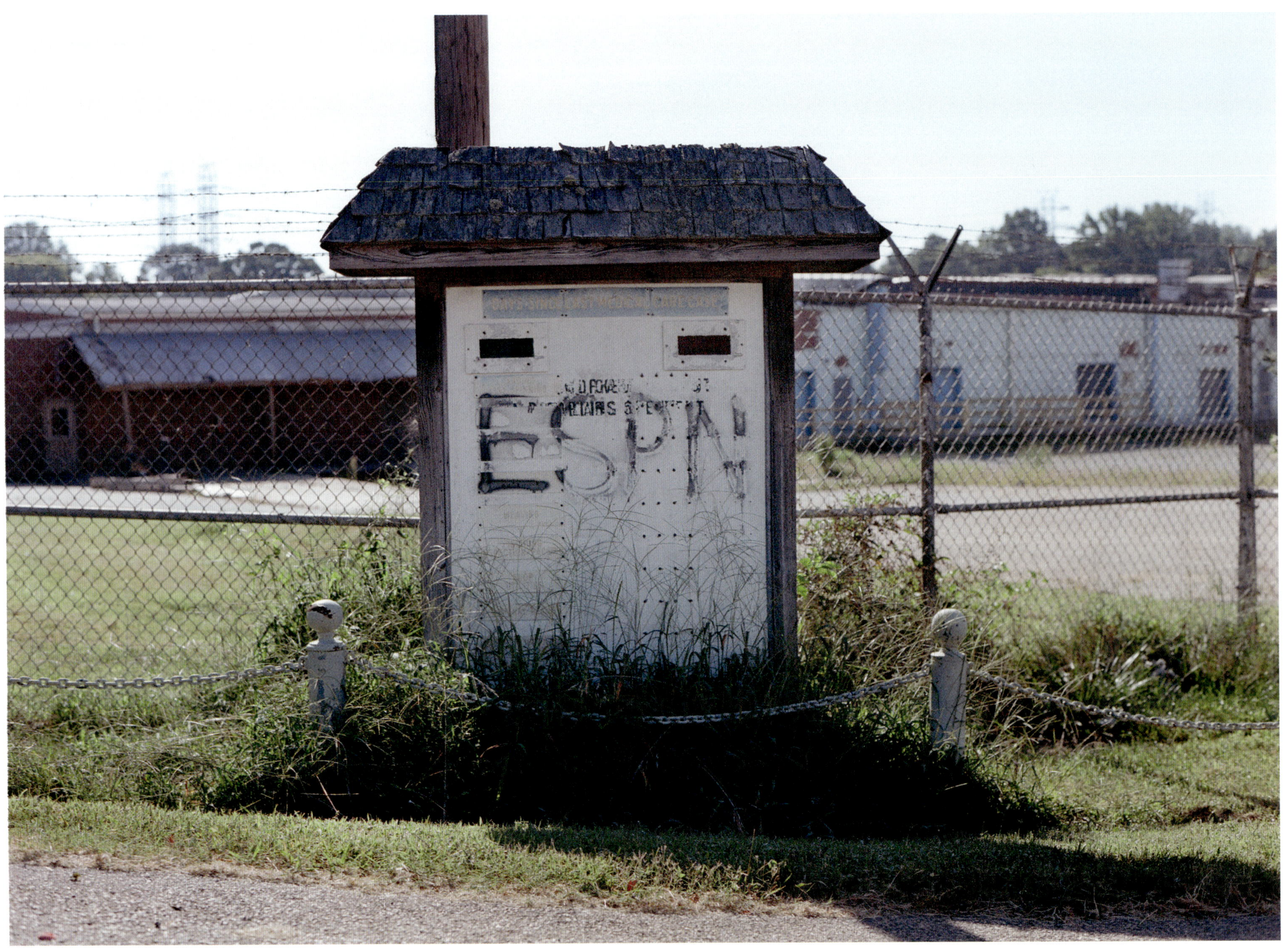

Bulletin board at the 16th Street trucking entrance

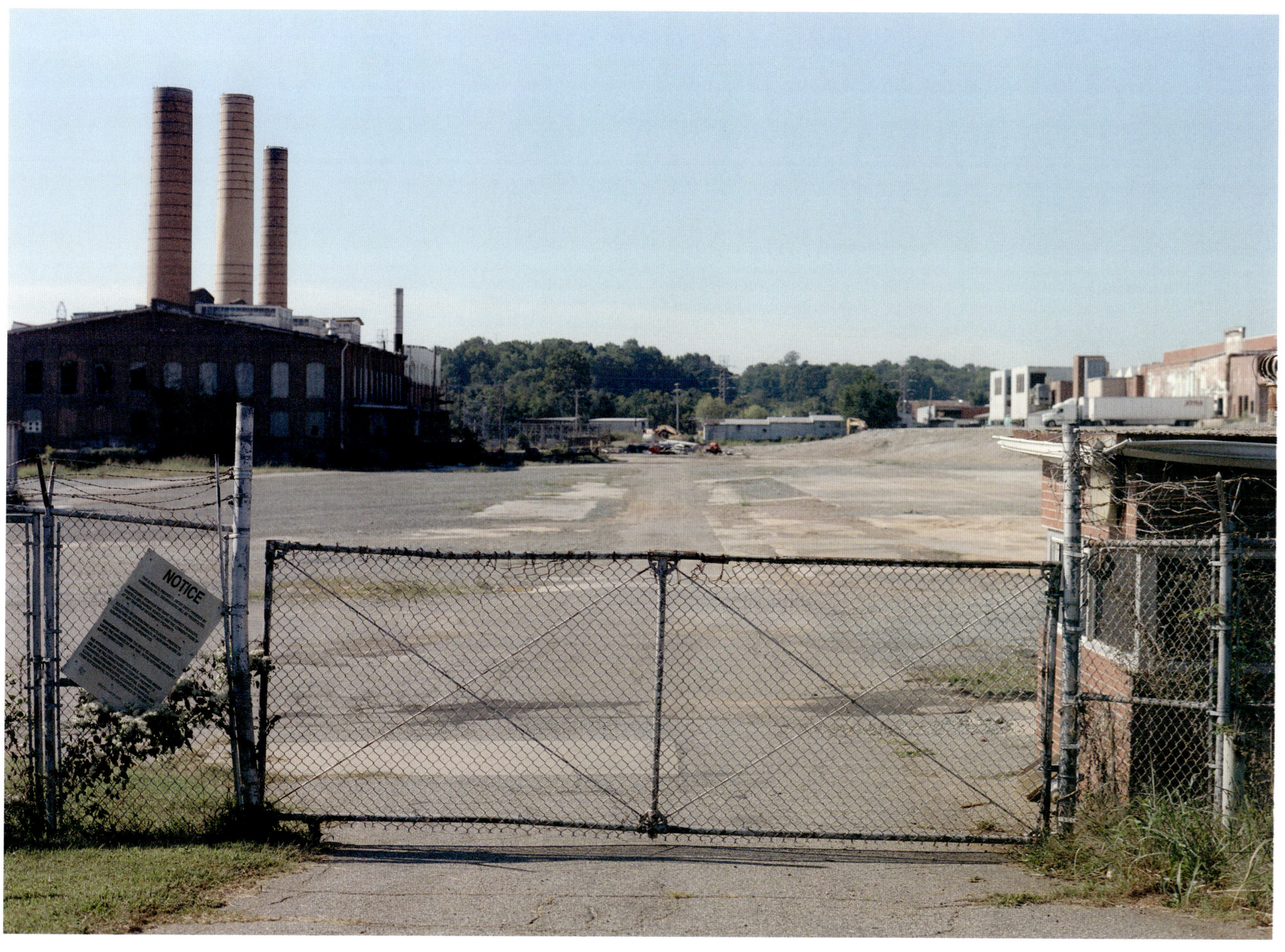

Powerhouse in view from the 16th Street entrance. Flattened area is the missing and recently demolished Dye House, Slashing Department, Spinning Room, and Electrical Room

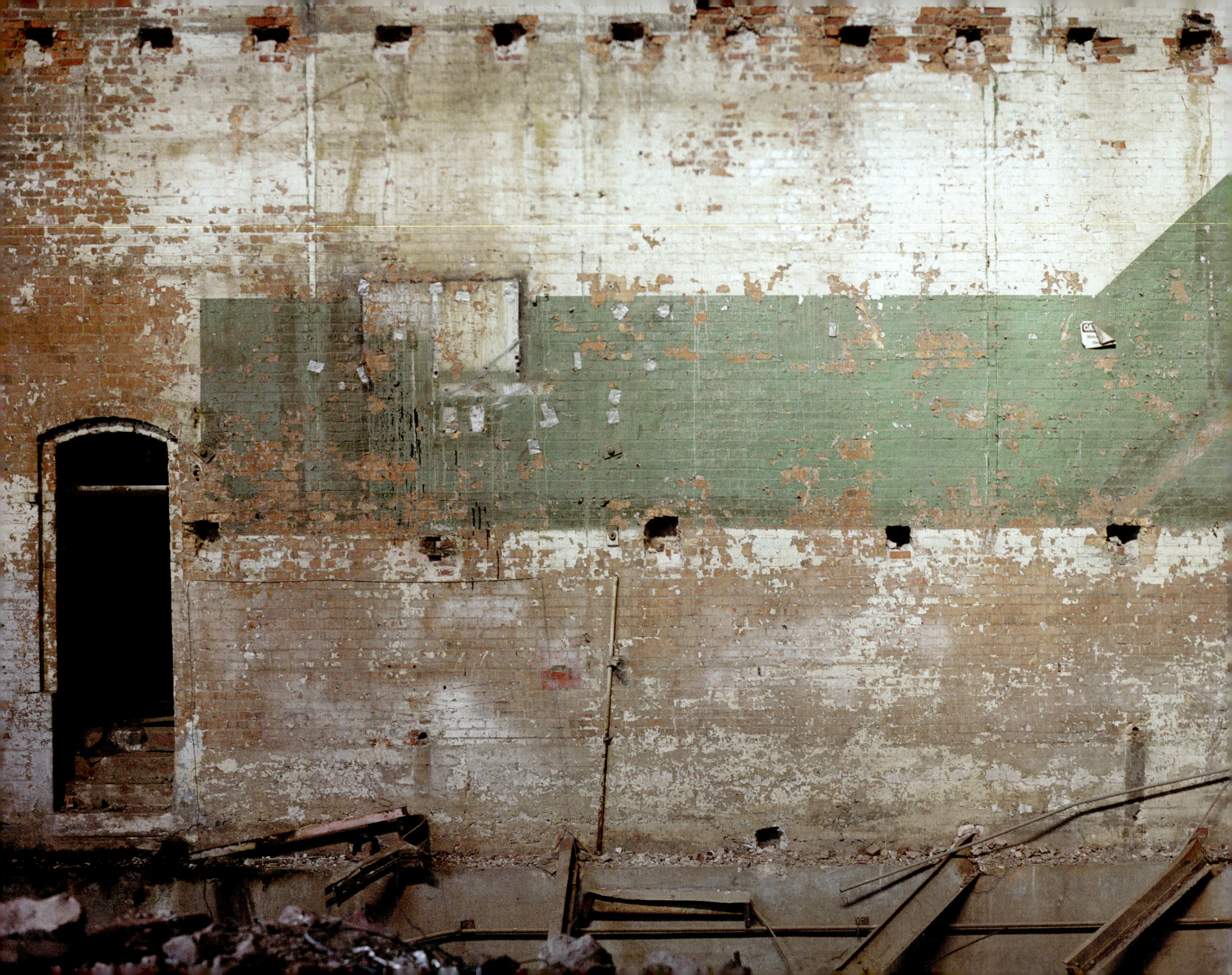

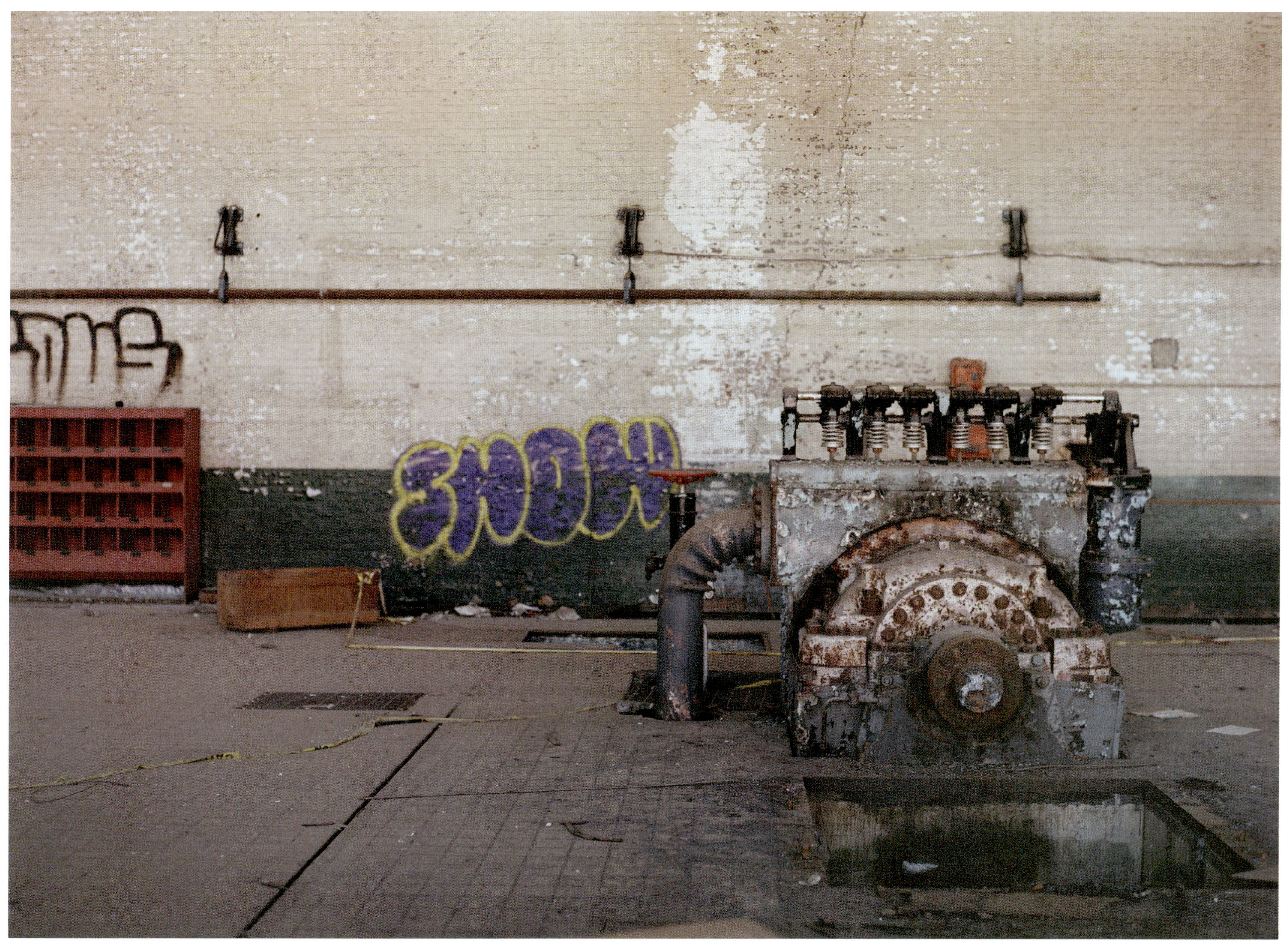

Boiler room with missing floors in the Powerhouse // Electrical-conversion generator in the Powerhouse

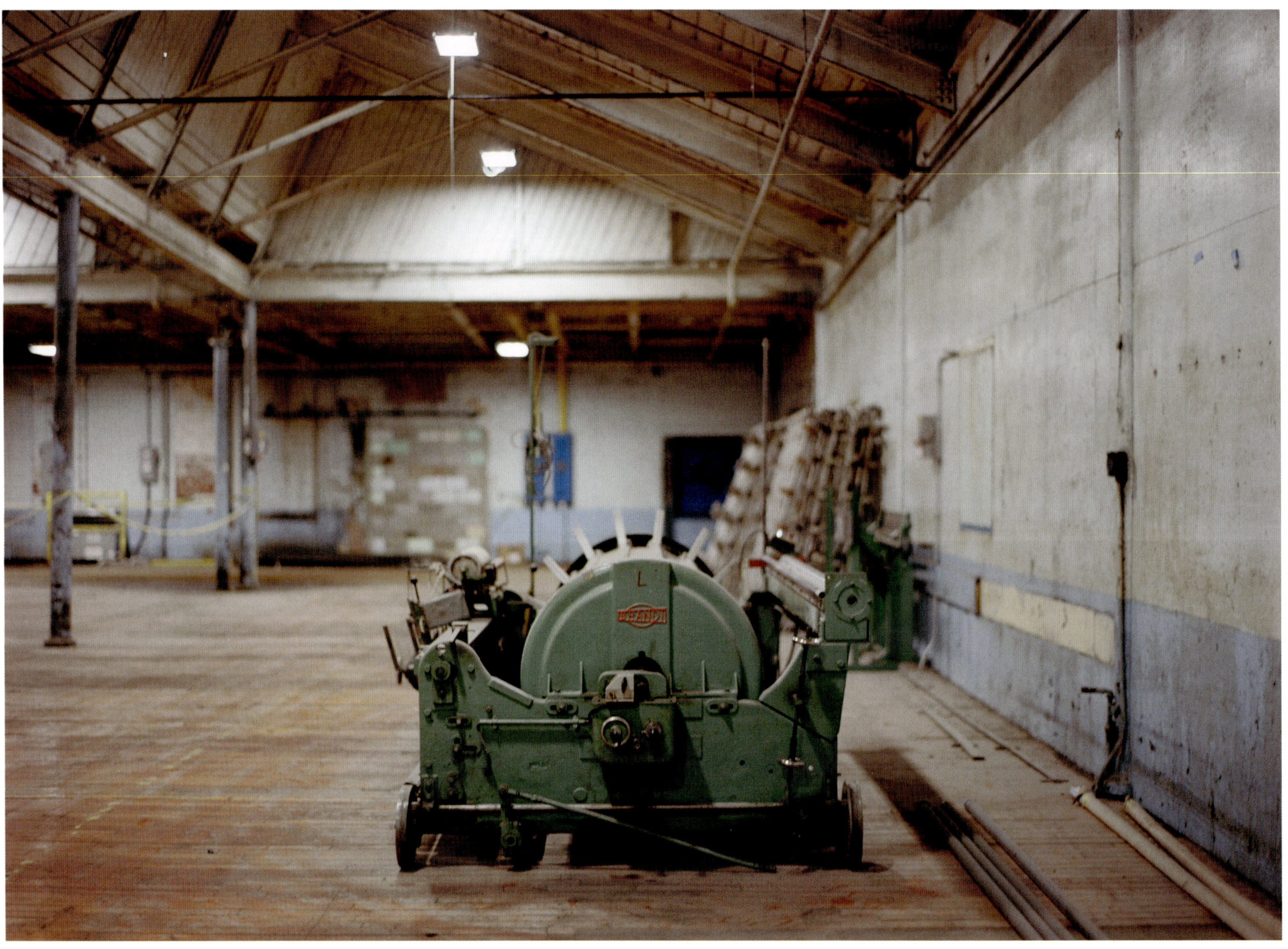

Brandt sectional warper and empty creel in the Weave Room that previously housed
the Draper X-3 looms

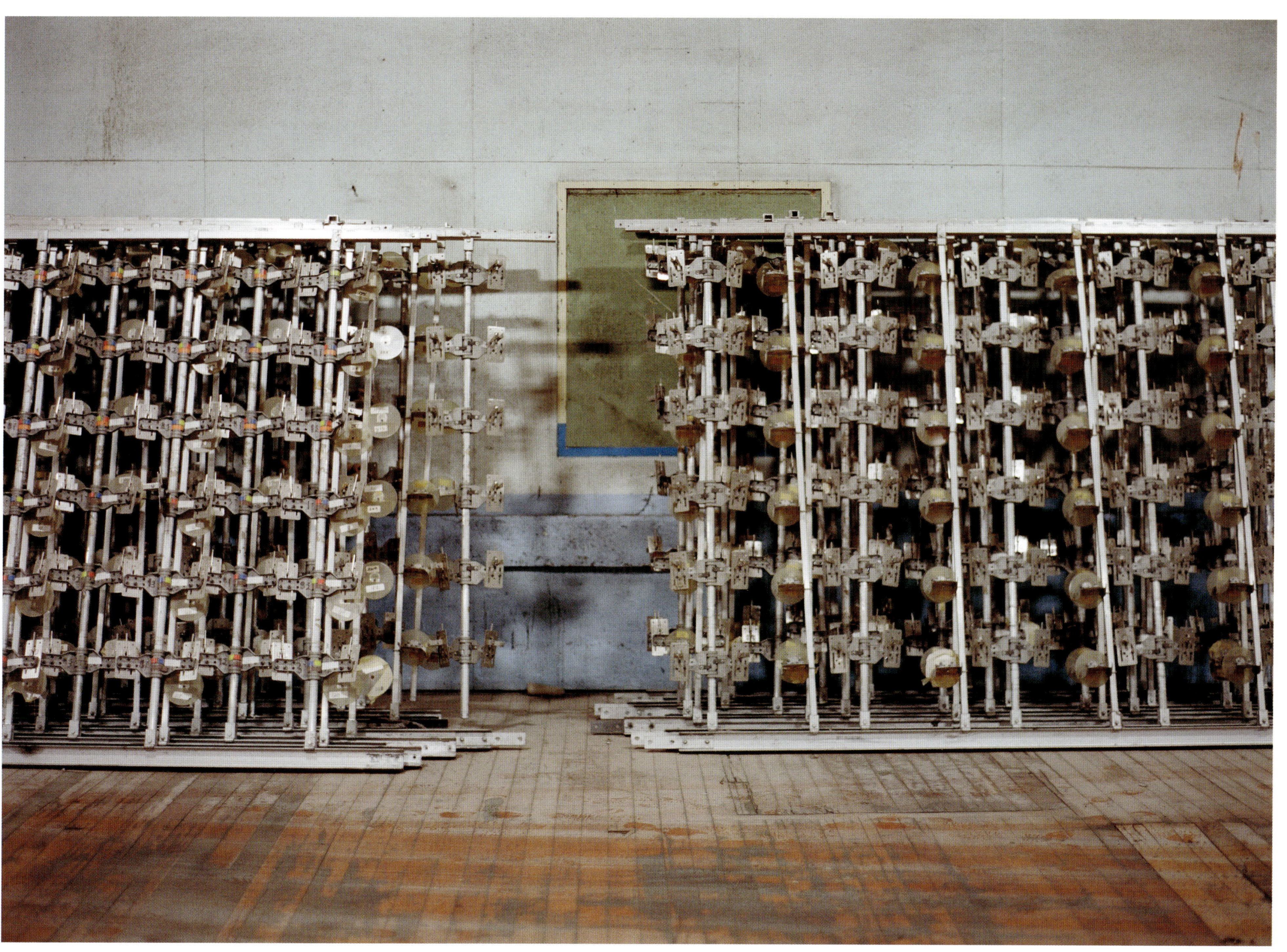

Yarn creel against the wall in the Weave Room

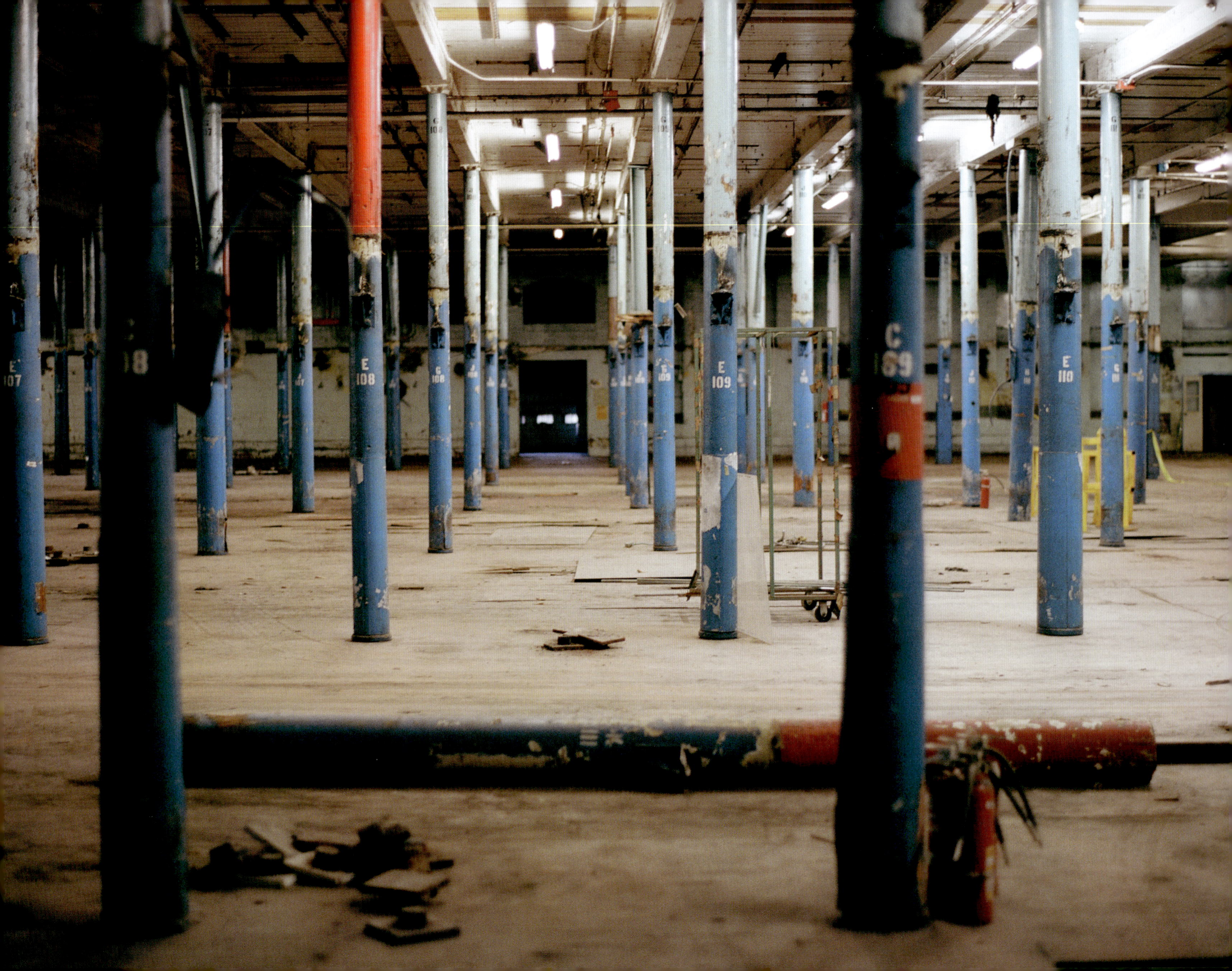

E 107
E 108
G 108
E 109
G 109
E 109
C 89
E 110

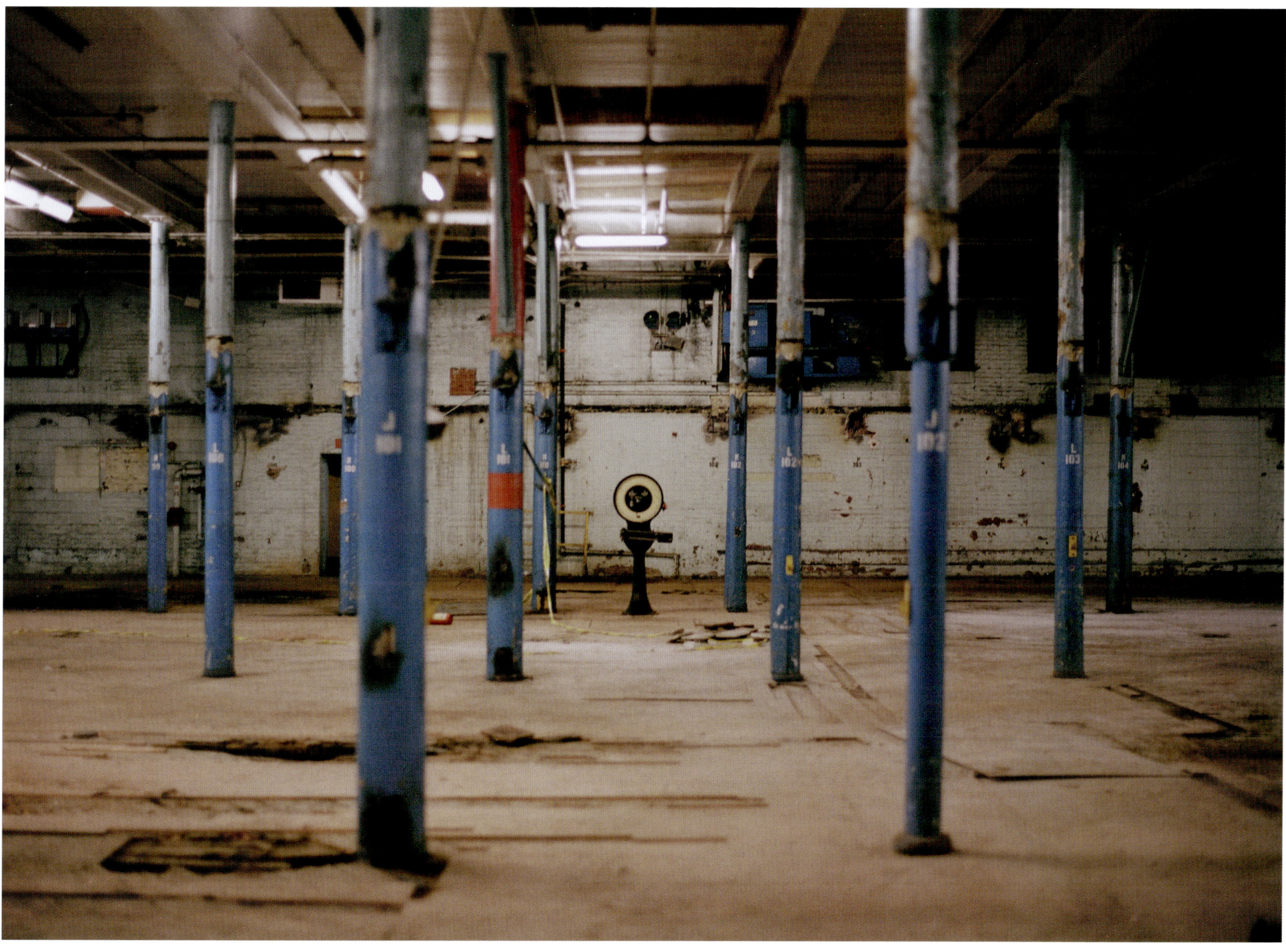

Finishing Department, with the mezzanine floor cut out in the Weaving Building   //   Fairbanks scale in the Finishing Department of the Weaving Building

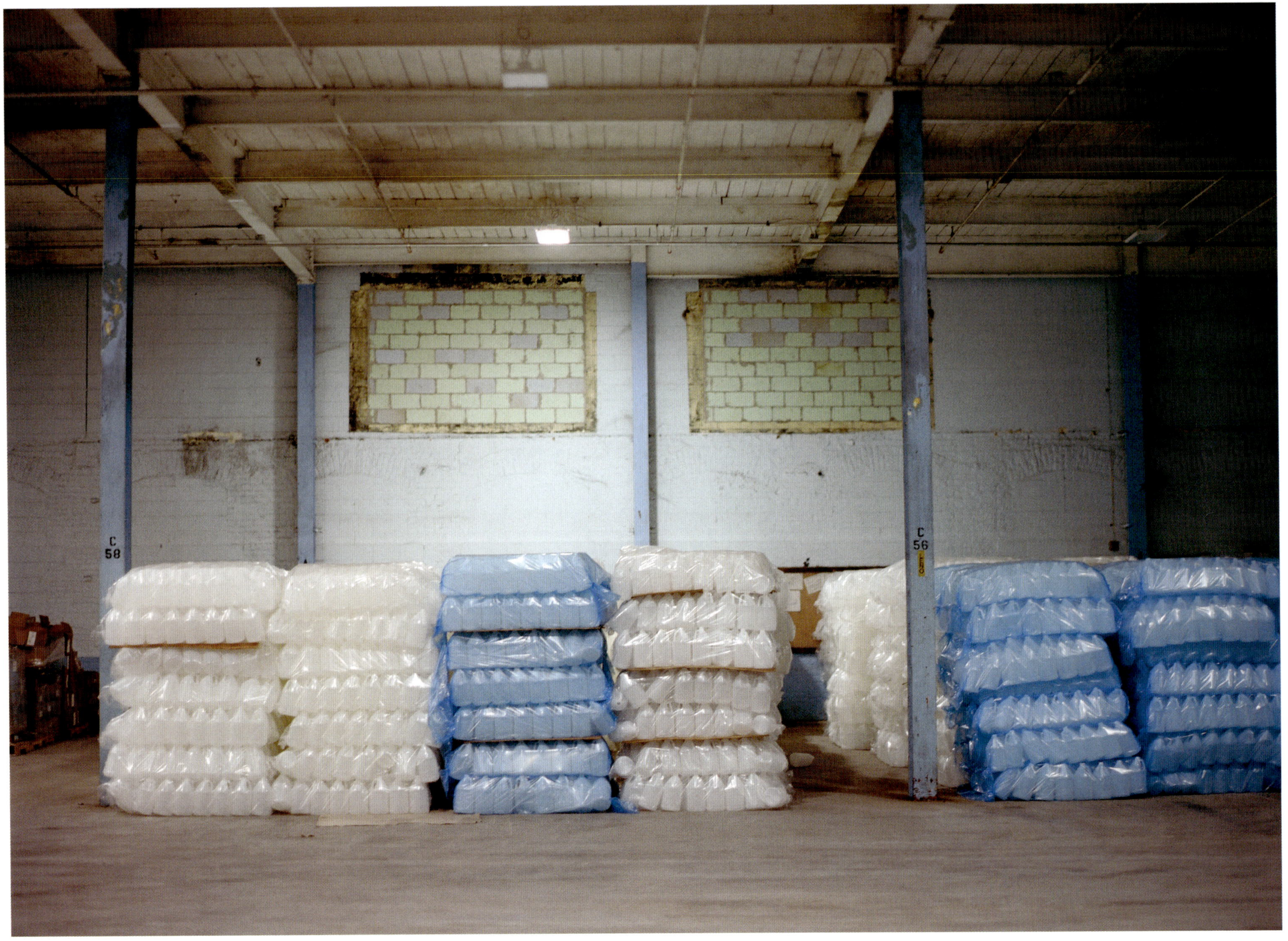

Cases of empty plastic bottles in the former Weave Room. Now used as a regional
distribution center for an automotive-solutions company.

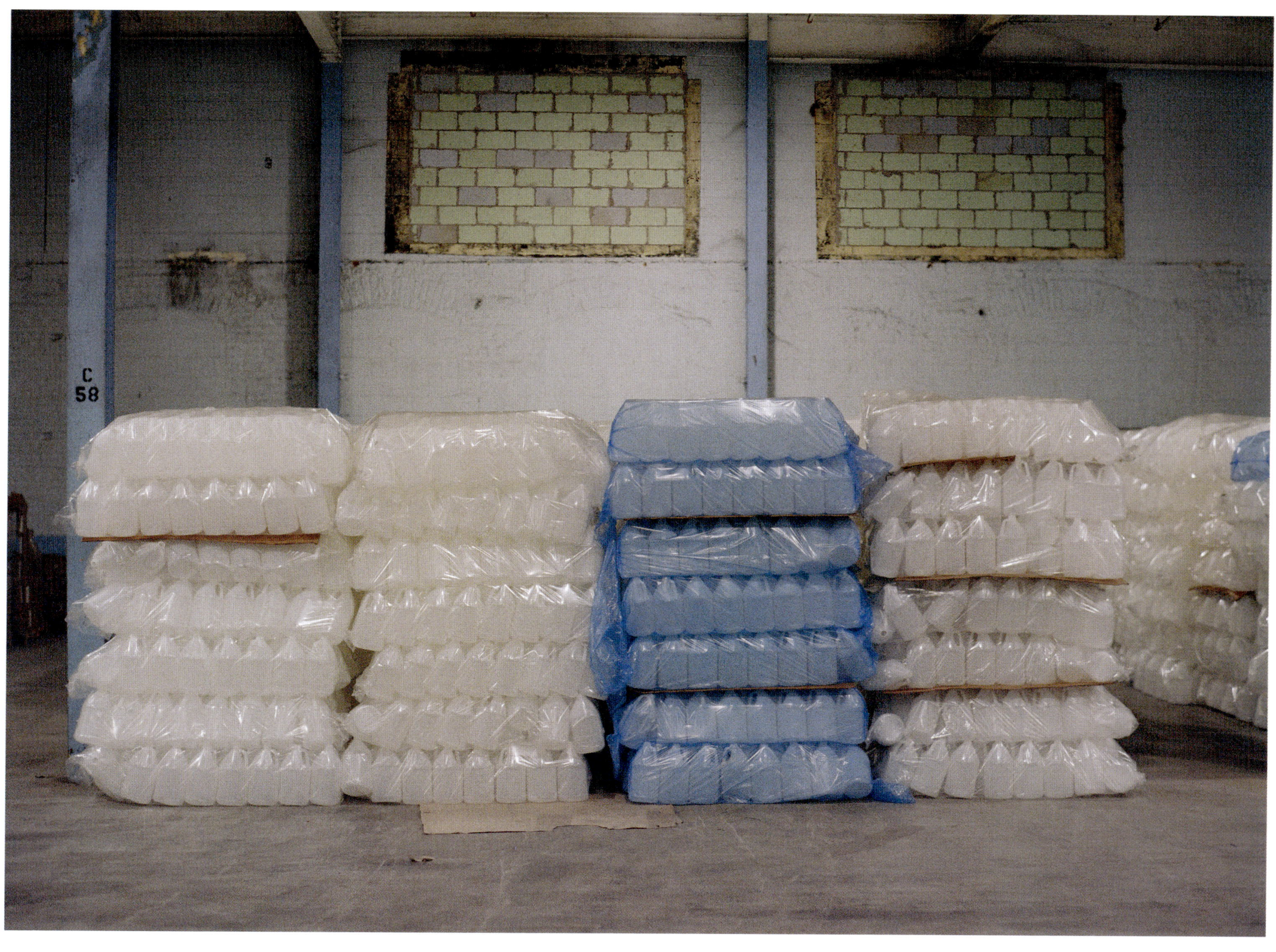

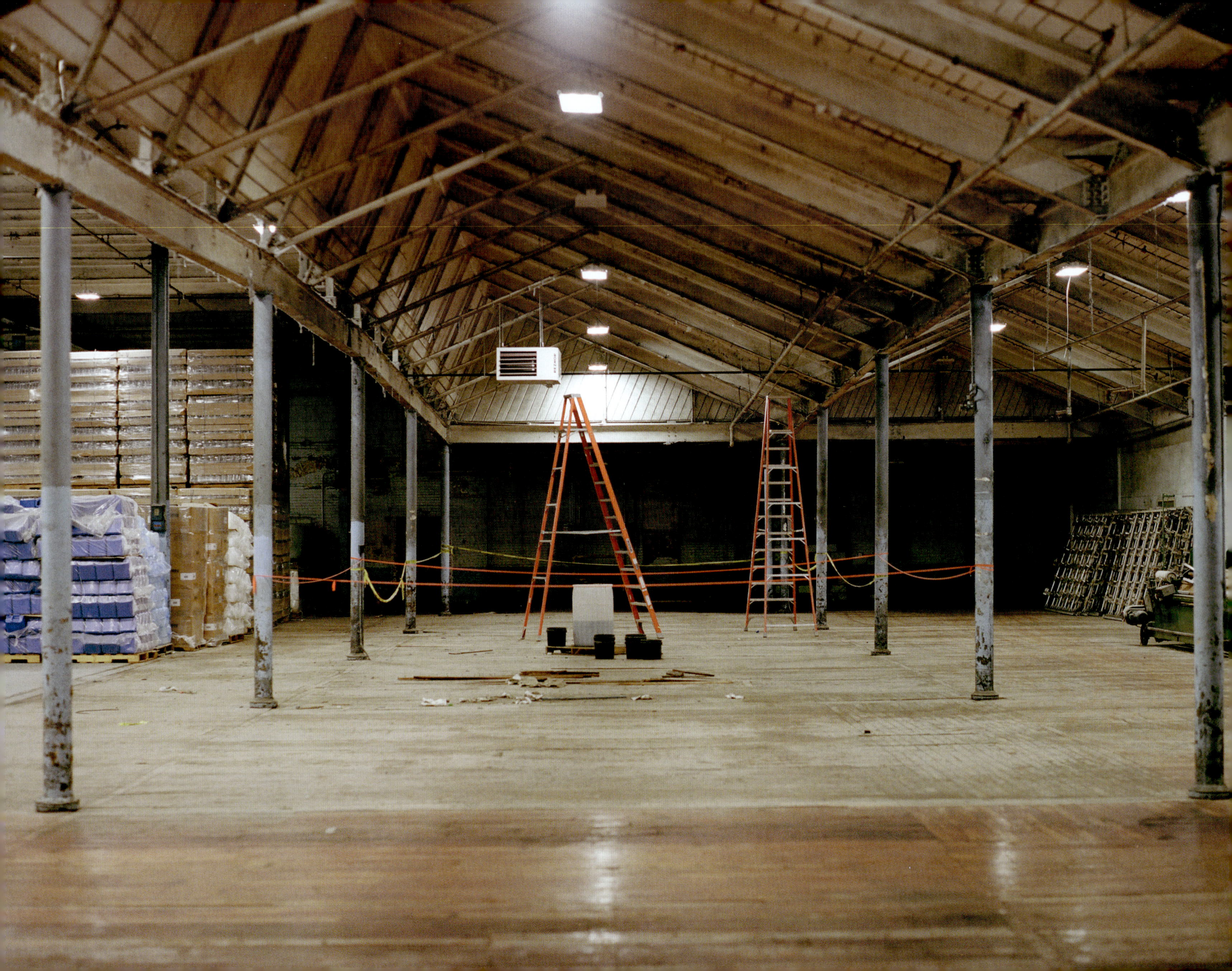

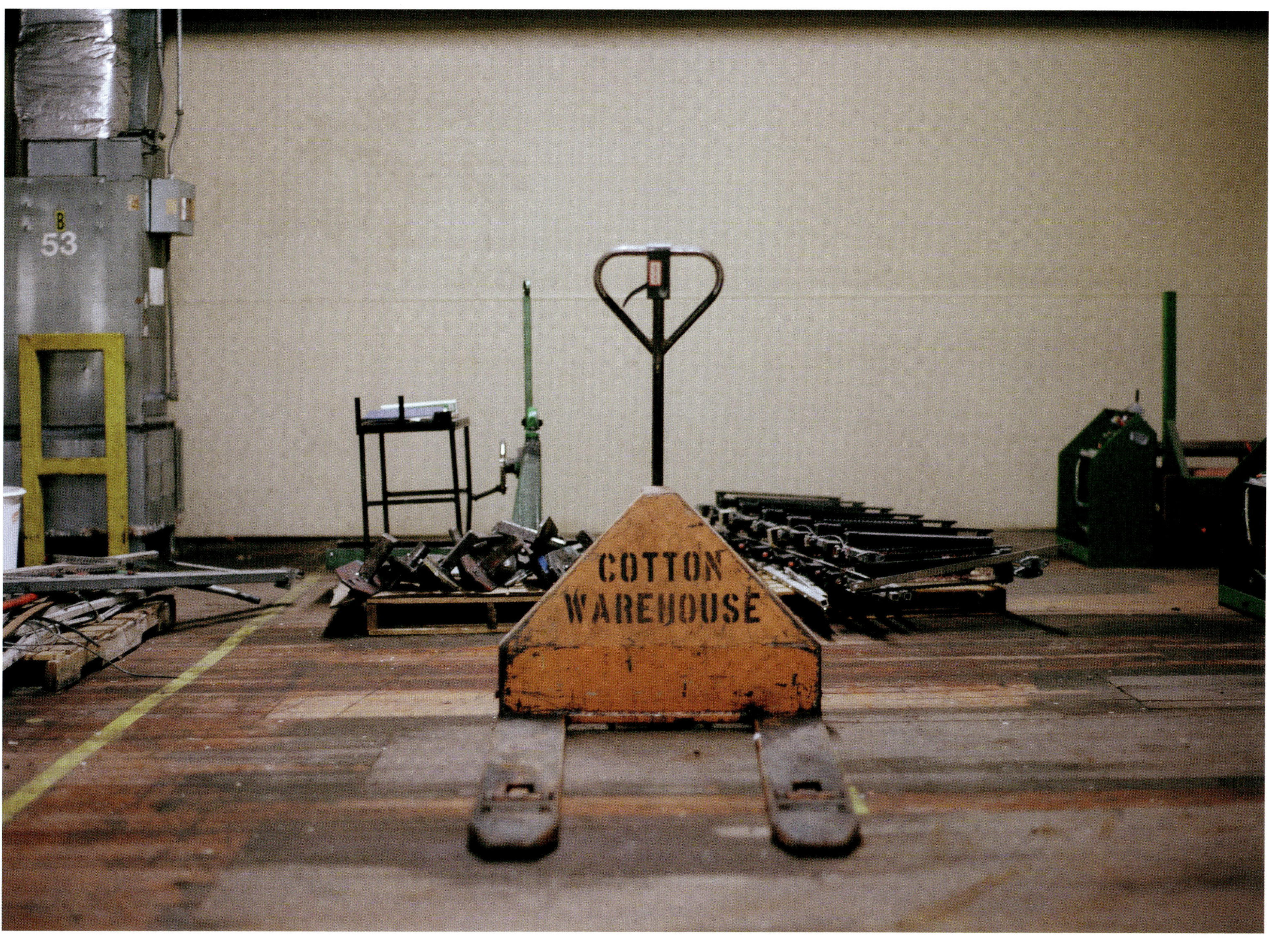

Former Draper X-3 weave floor in the Weave Room   //   Off-frame take-up and creel parts with pallet jack in the Weave Room

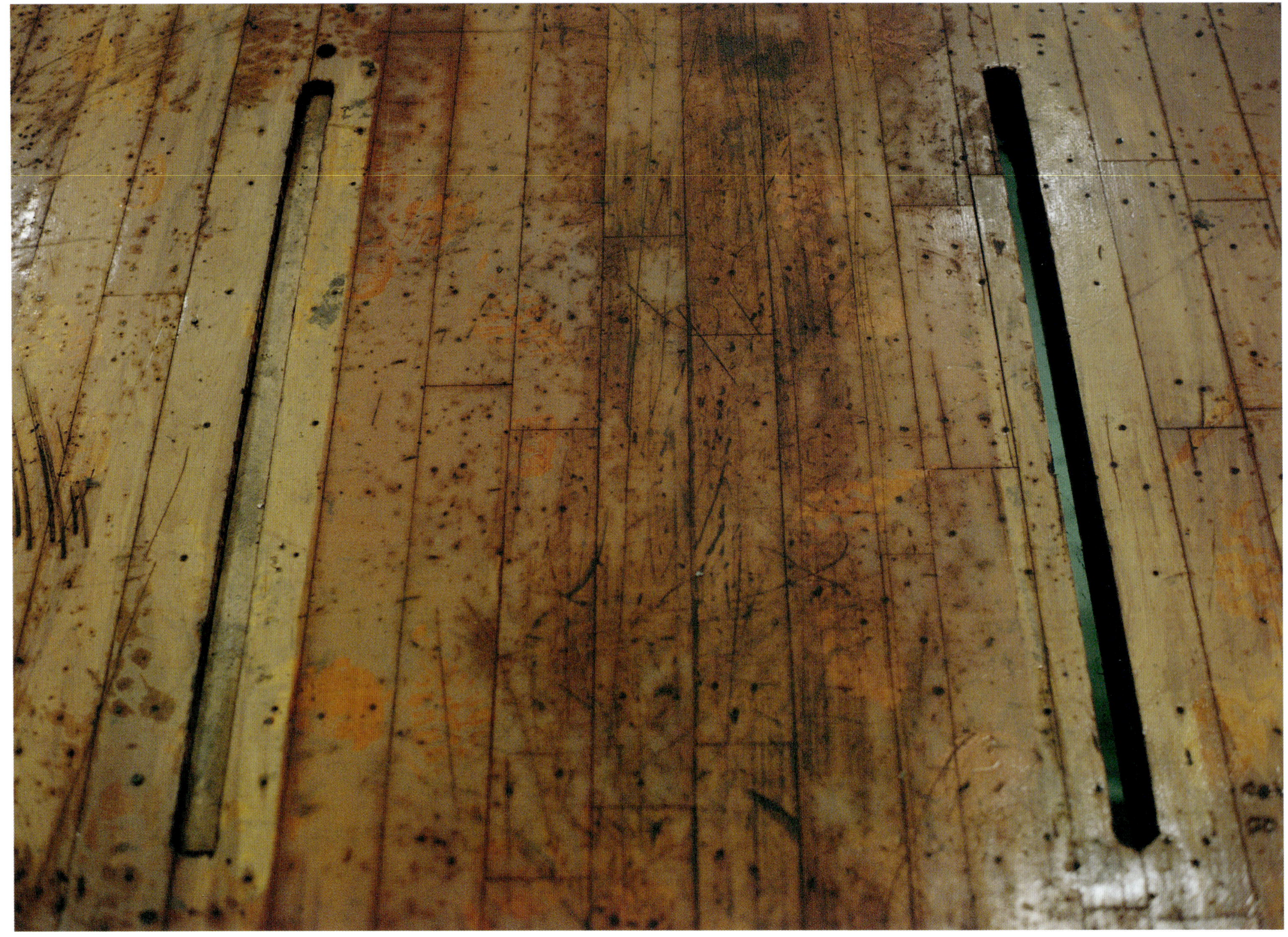

200

Slots leading from the X-3 weave floor, leading to inspection tables in the Quality Control Department

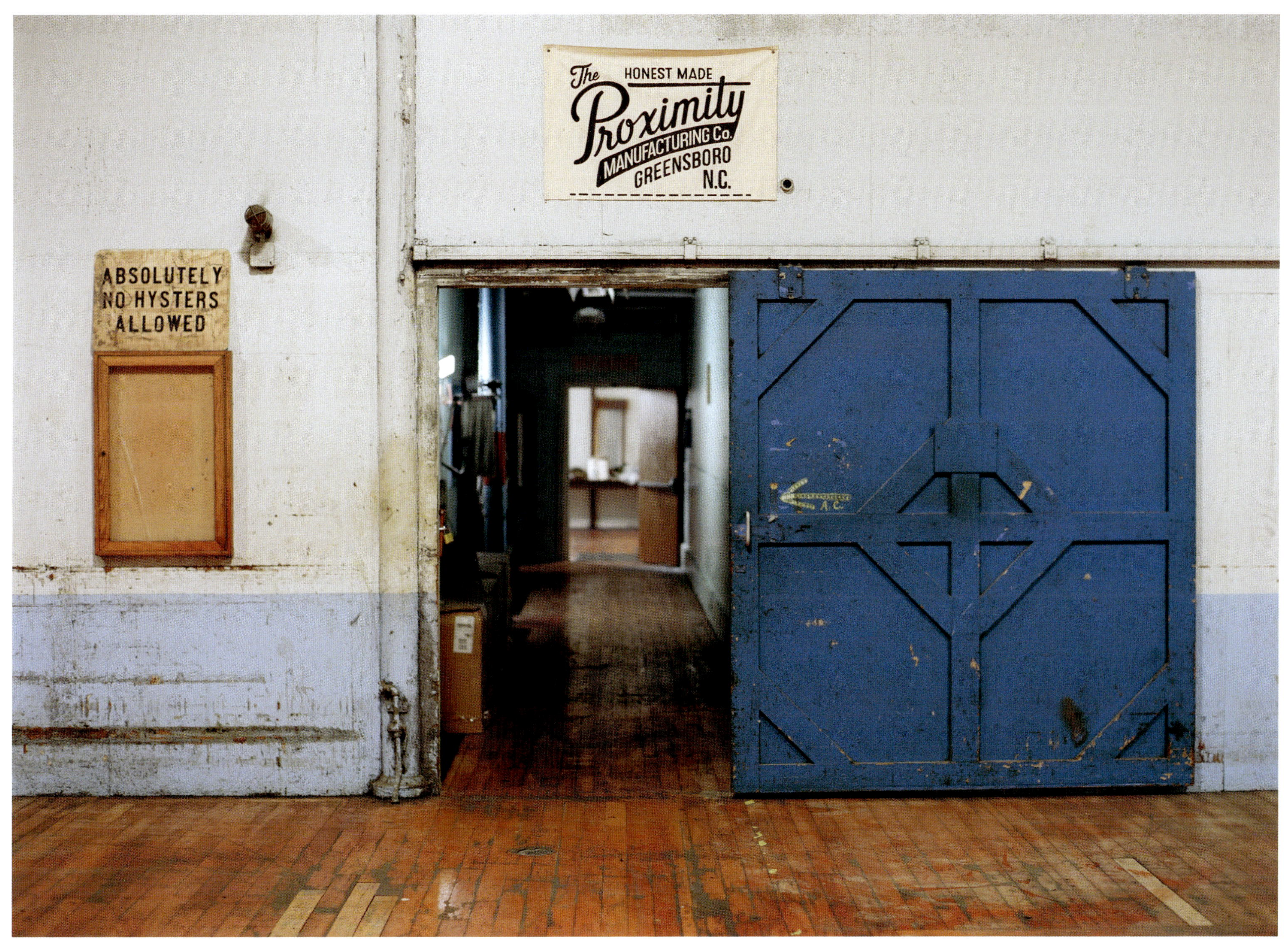

Entrance to operations of Proximity Manufacturing Company's shuttle-weaving room.
Adjacent to the former Draper X-3 Weave Room.

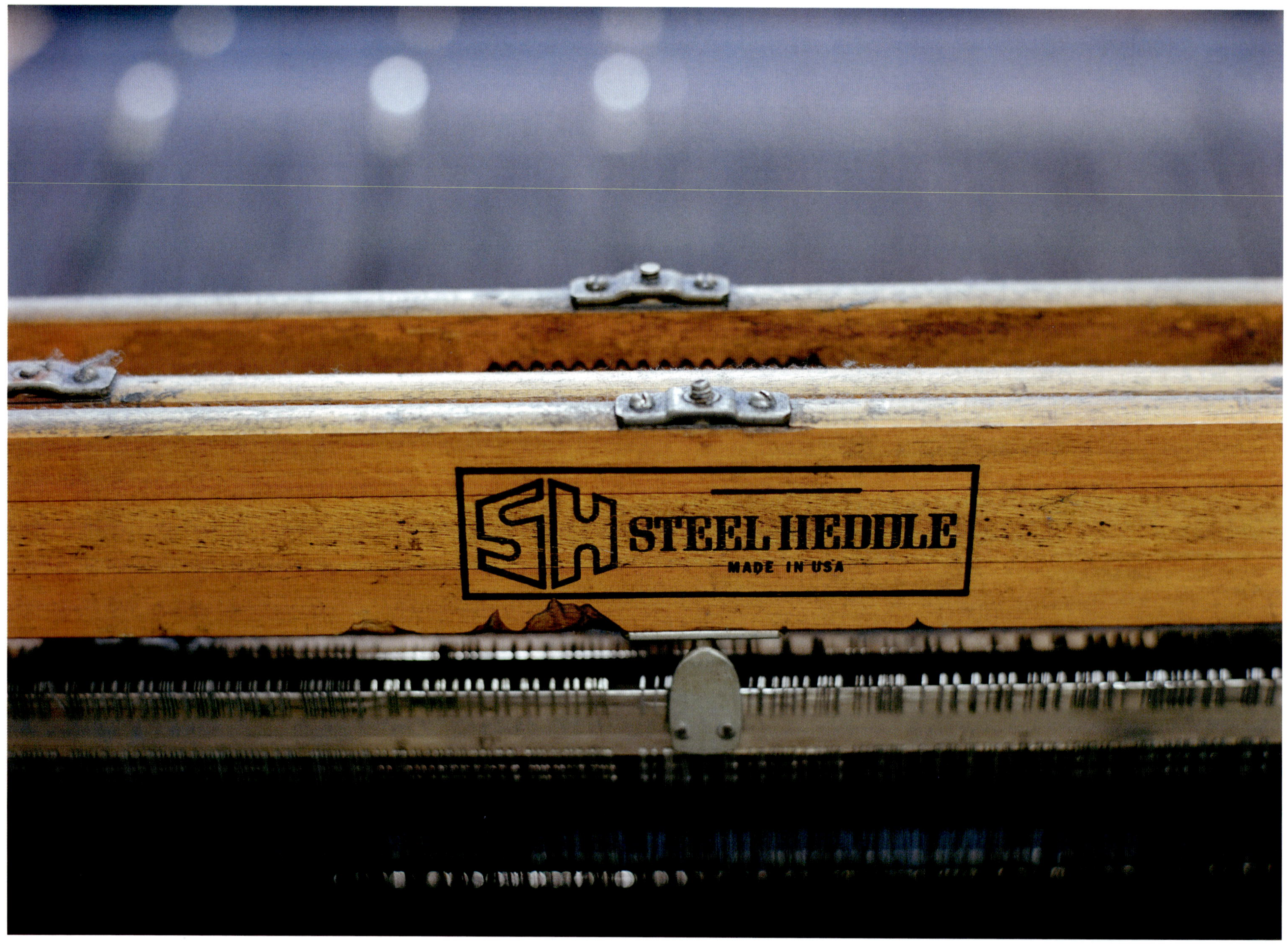

X-3 loom harnesses // Drop wires on X-3 loom

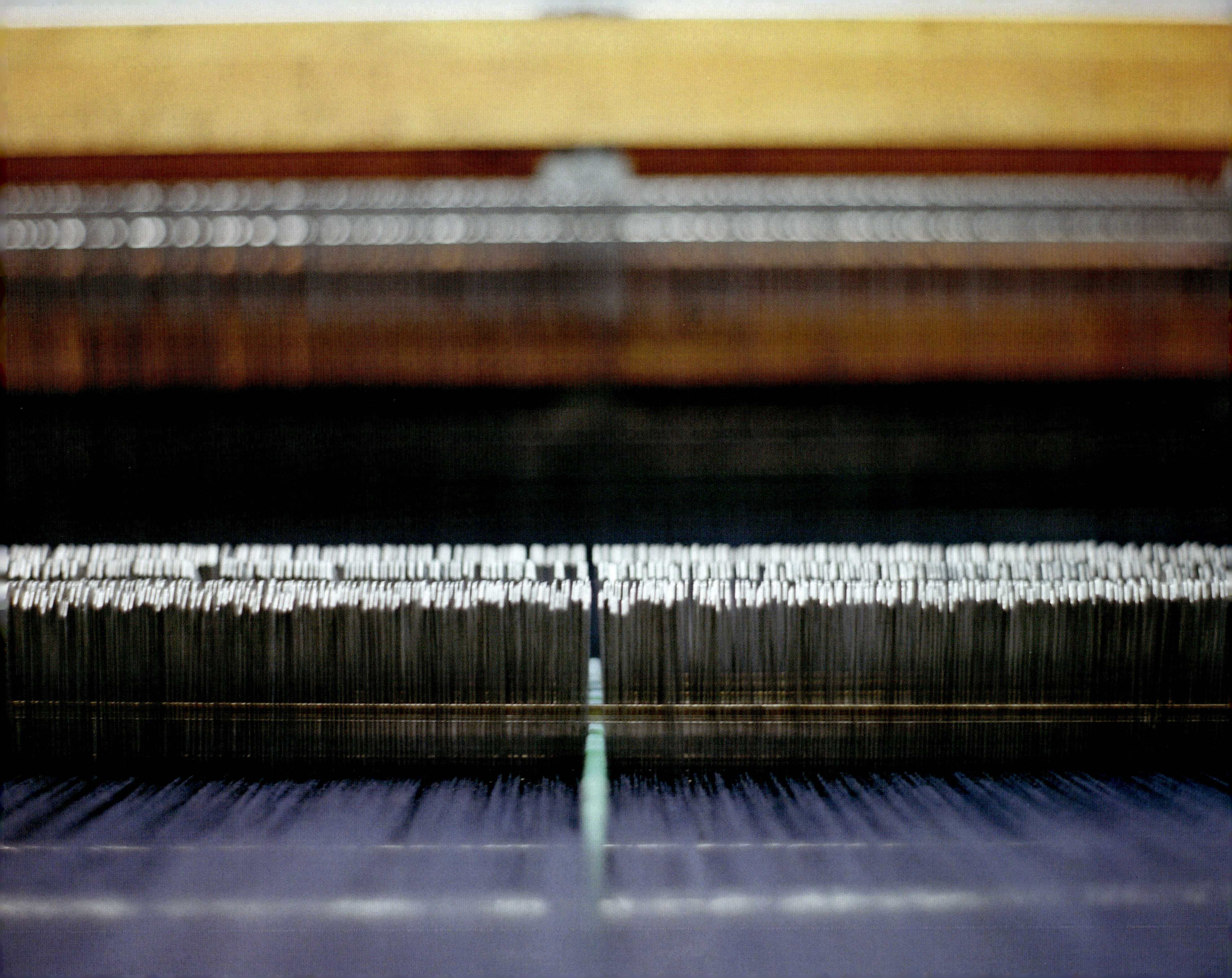

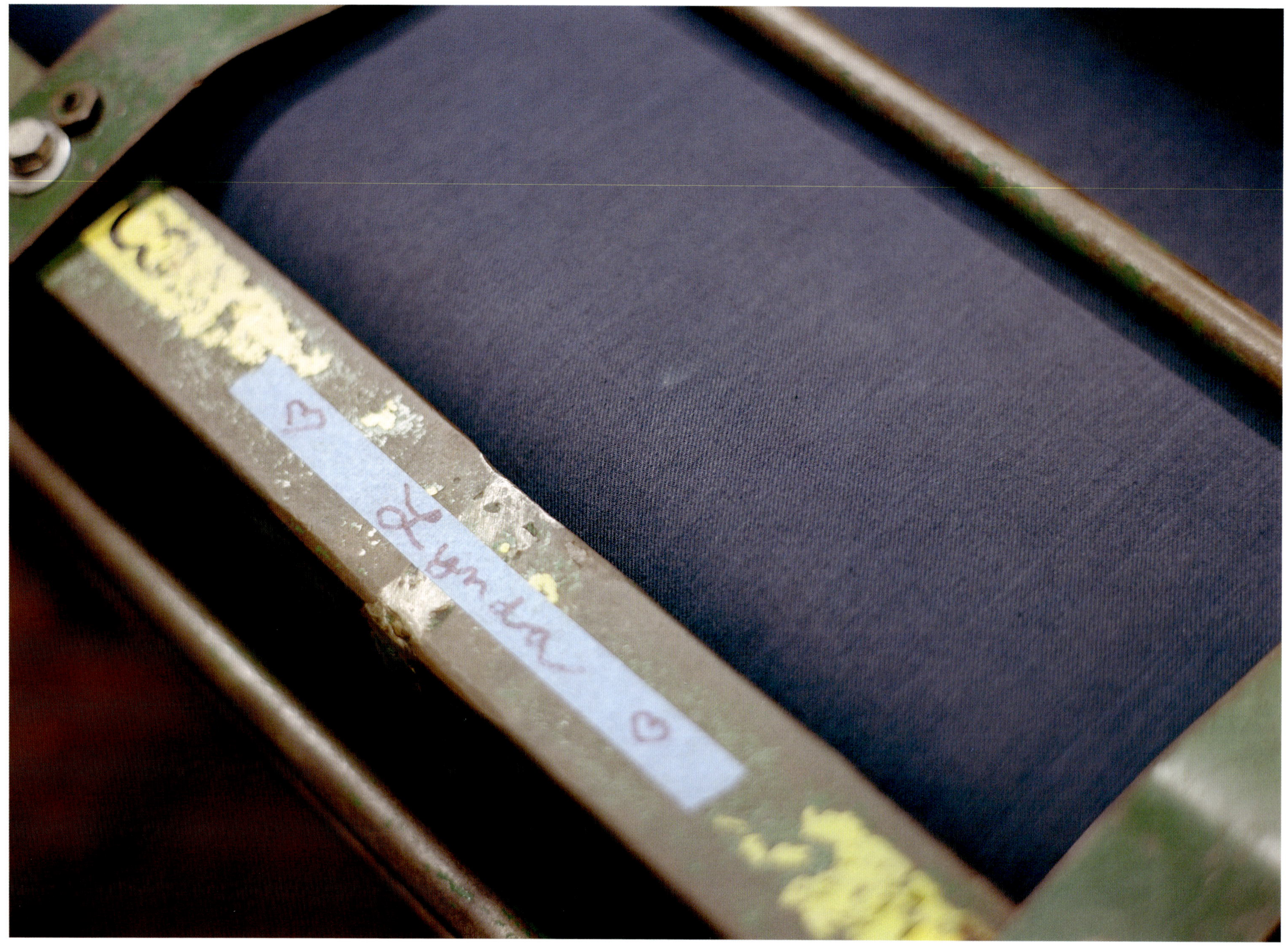

3 x 1 denim on X-3 loom named after former Head of the Garment Lab, Lynda Layton

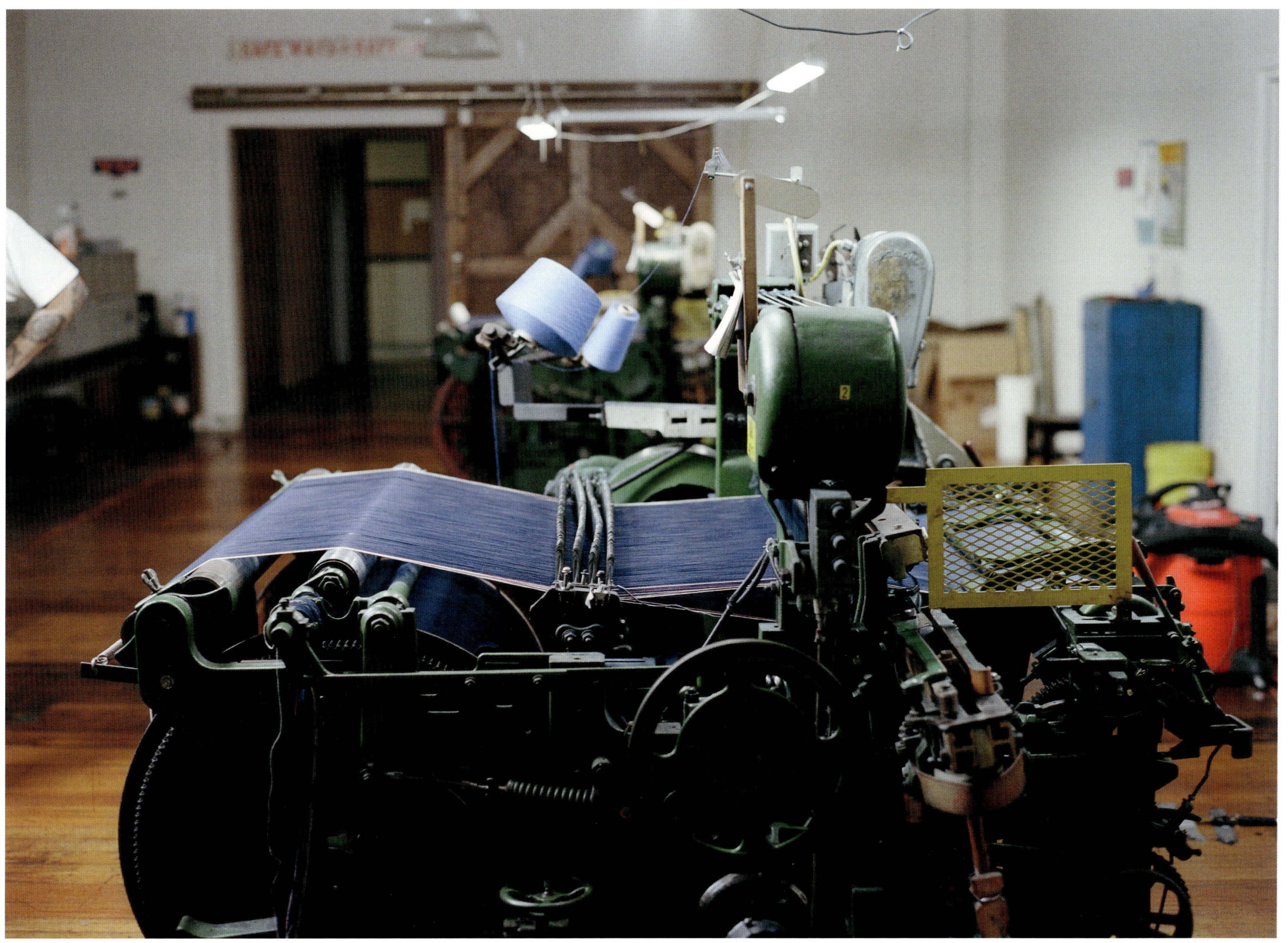

Proximity Manufacturing Weave Room. Formerly the White Oak Museum archive room.

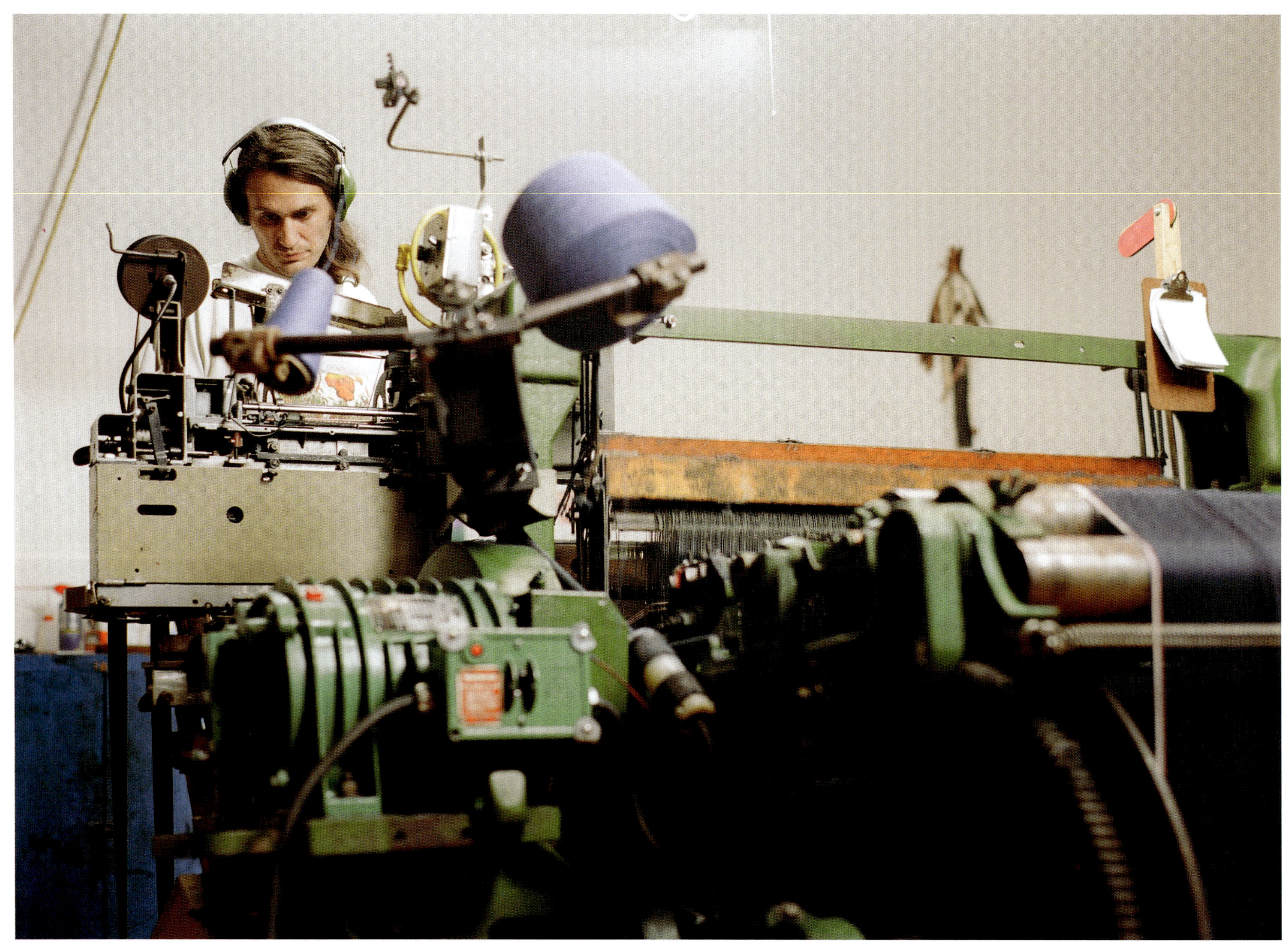

Greg Redelico operating a Draper X-3 loom and Unifil quill winder

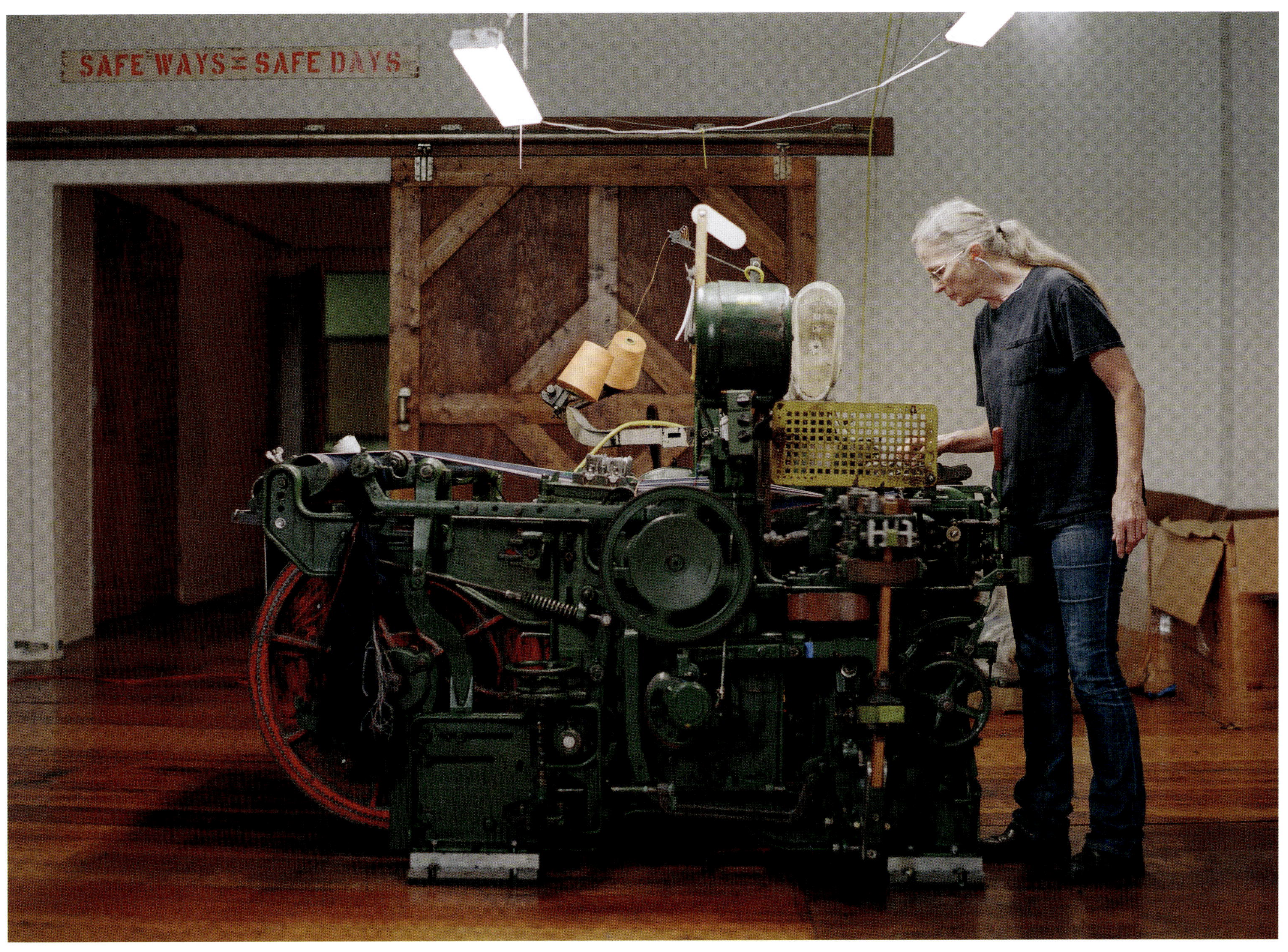

Debbie Lindsey inspecting the quality of a Draper X-3 shuttle loom in operation

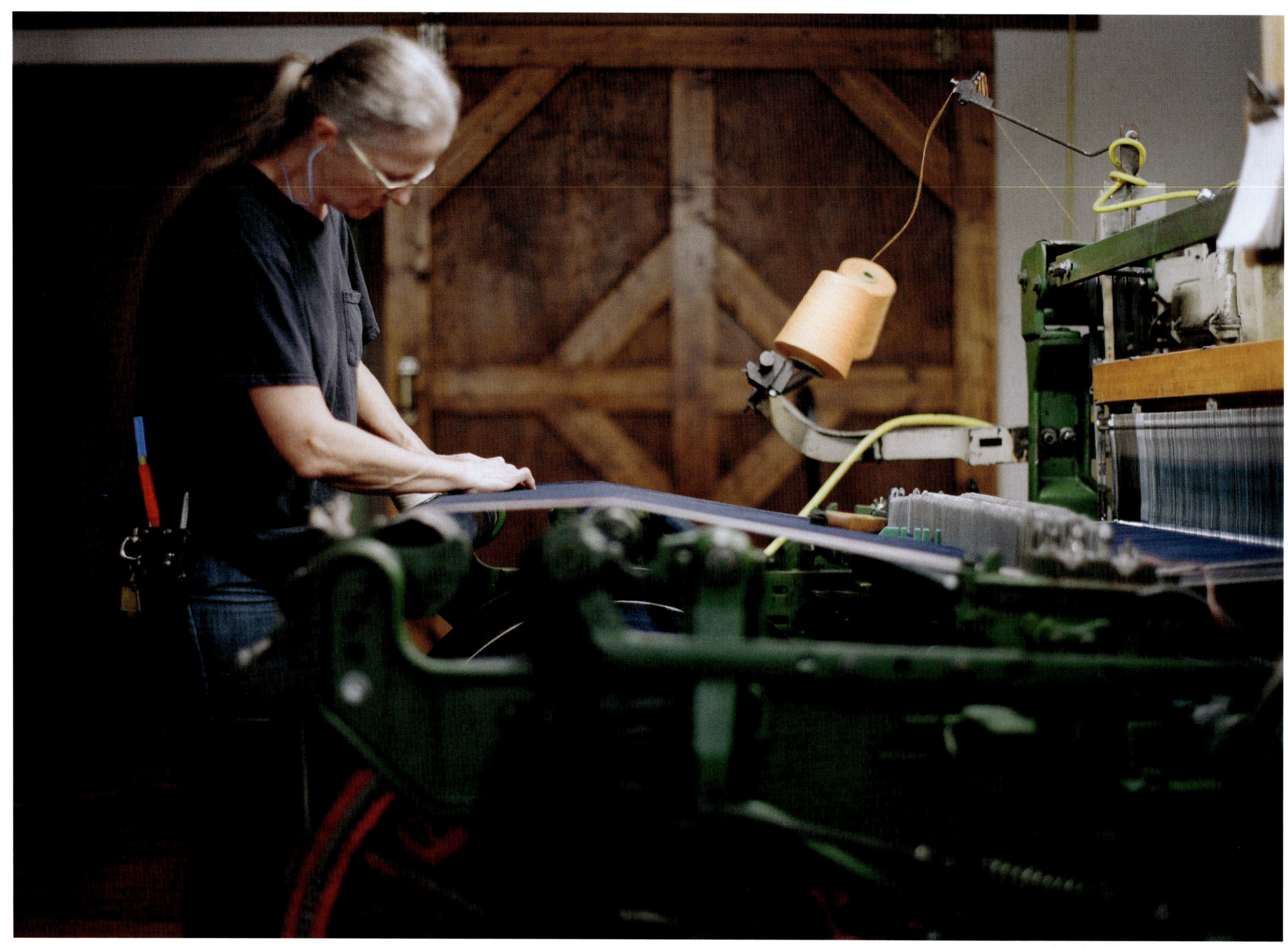

Debbie Lindsey, third-generation Weaver at this property, straightening yarns on an X-3 shuttle loom

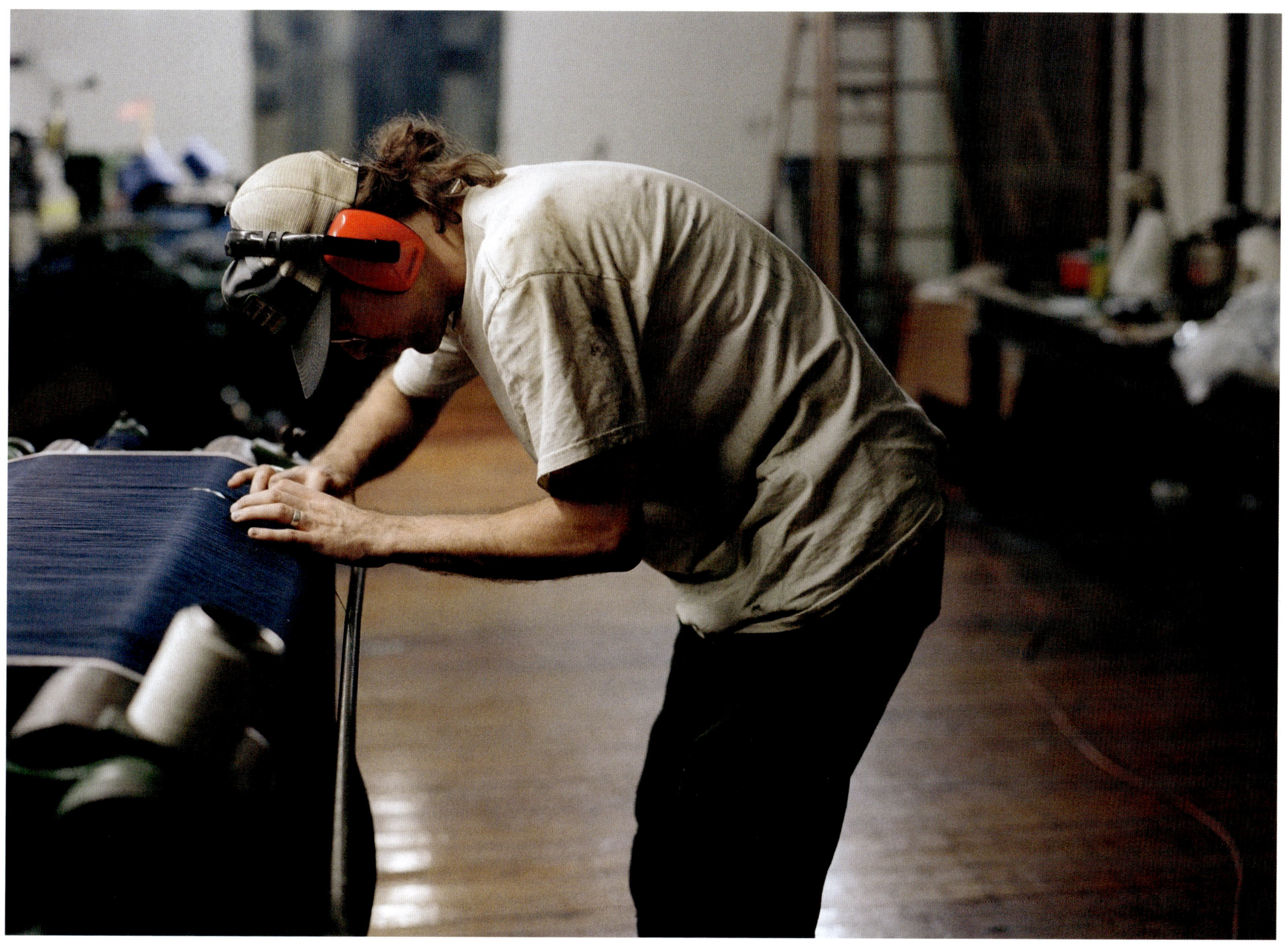

209

Evan Morrison inspecting yarns on a shuttle loom

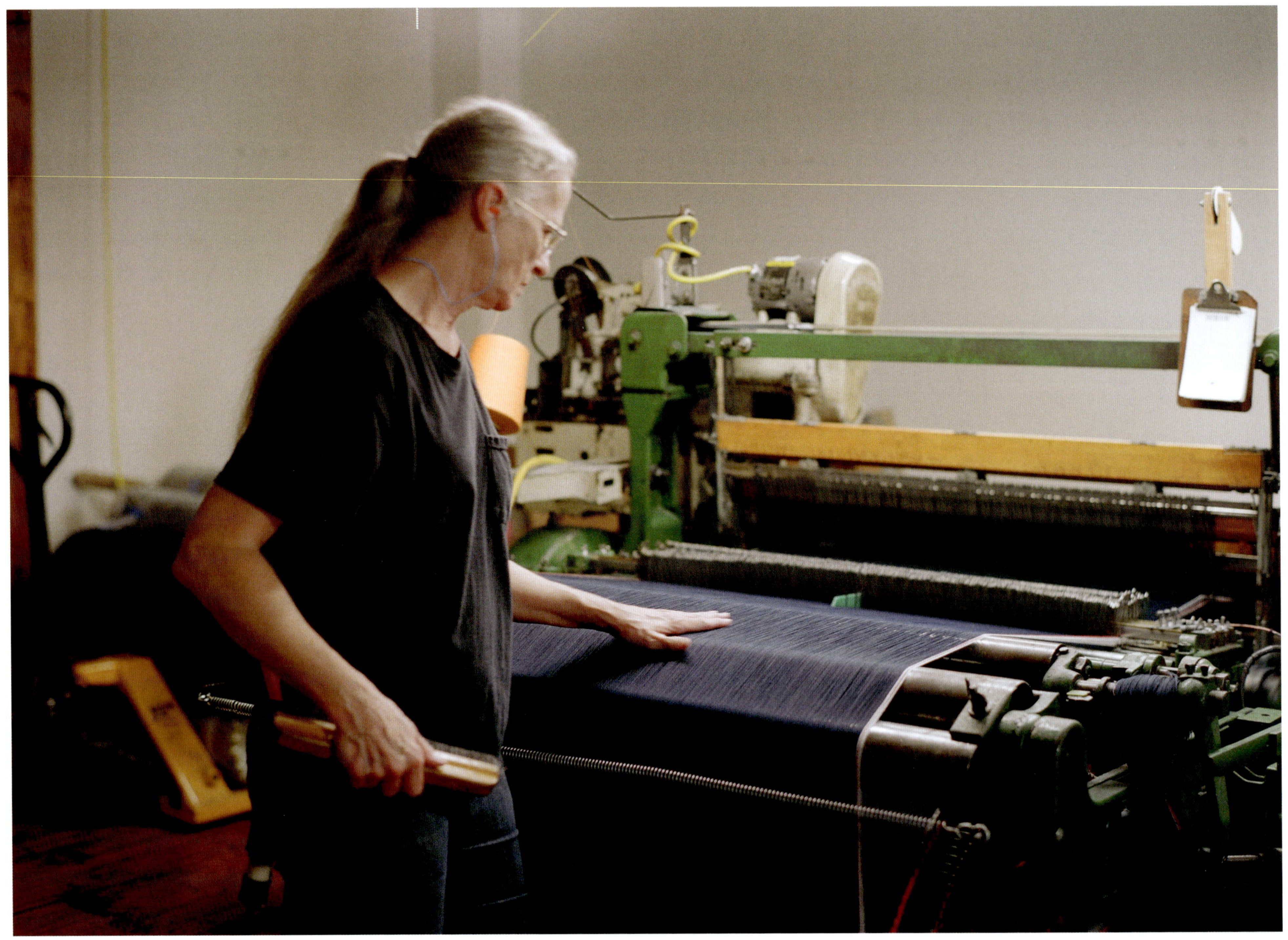

Debbie brushing warp to straighten yarns on an X-3 shuttle loom

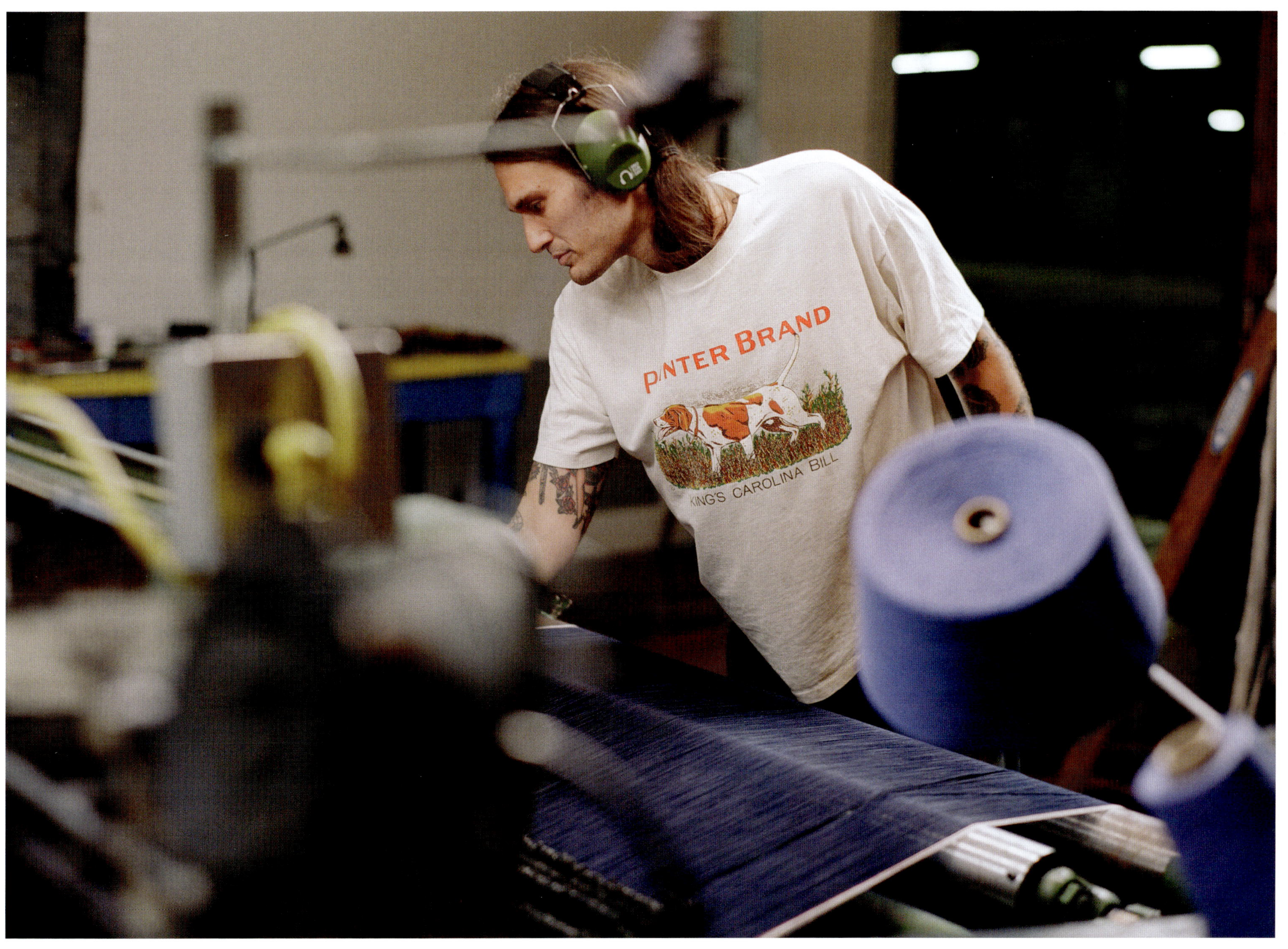

Greg inspecting yarns on a shuttle loom

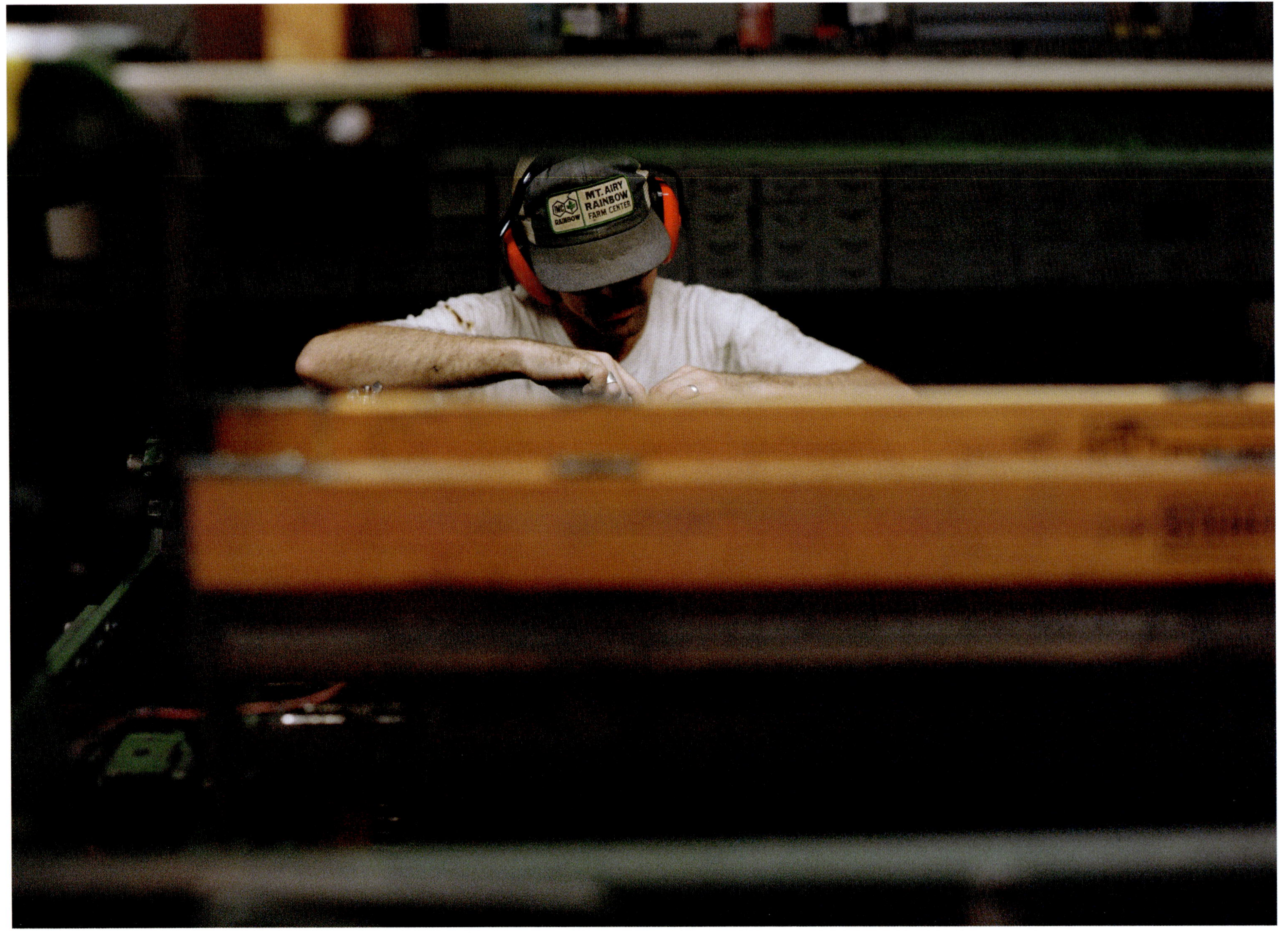

Evan straightening yarns on shuttle loom // Evan alleviating a yarn matte-up at drop wires on a shuttle loom

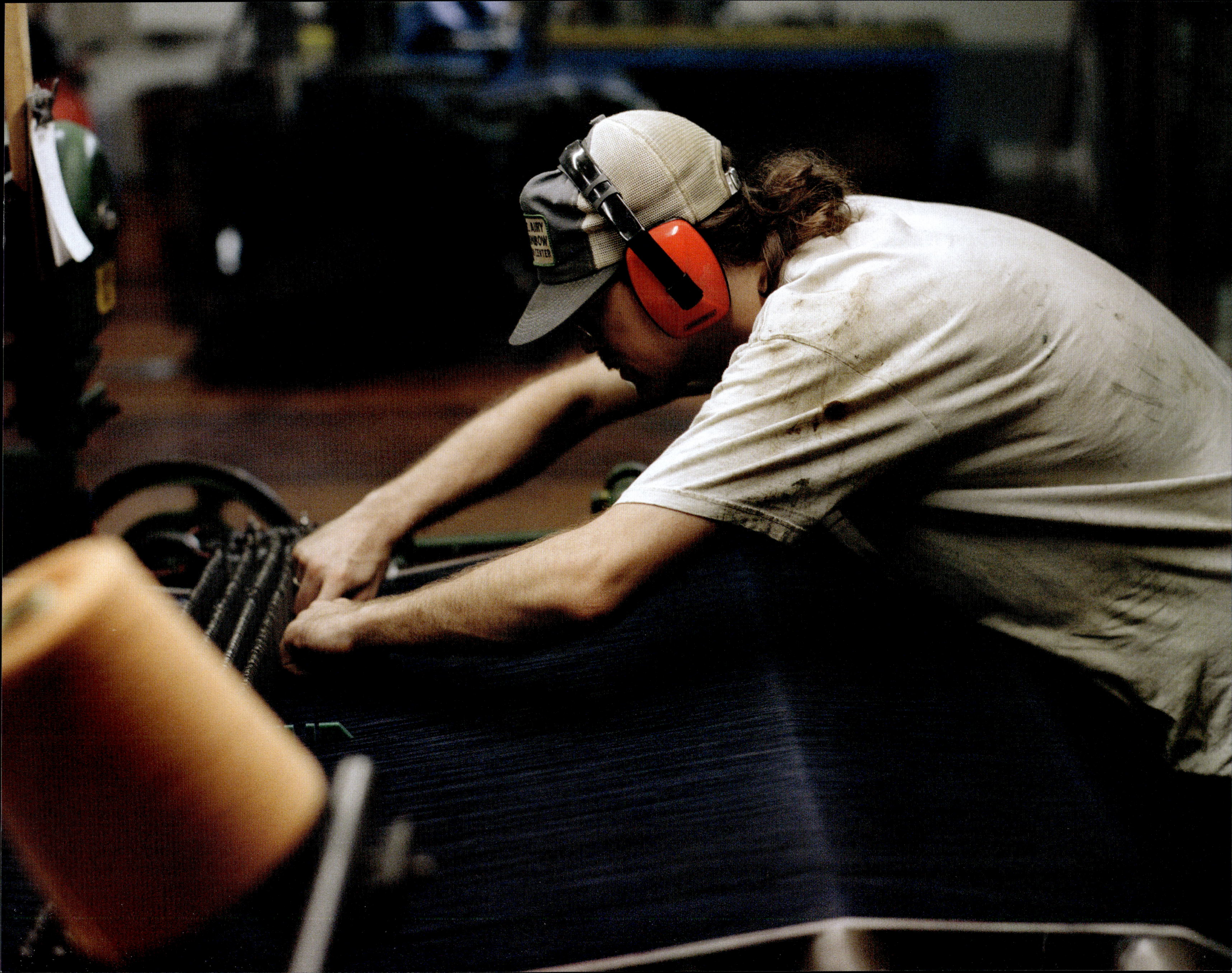

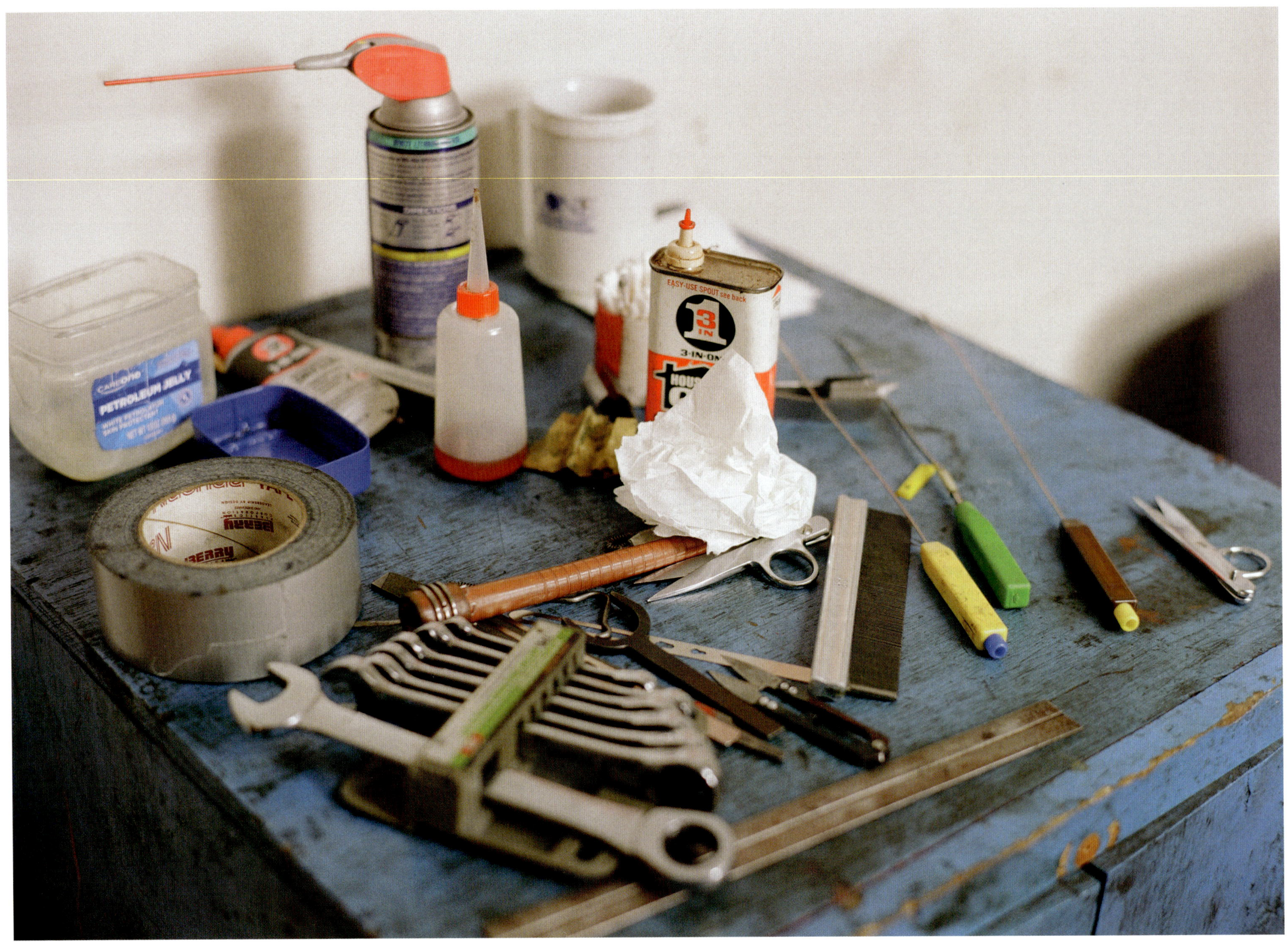

Weaving and loom-fixing tools

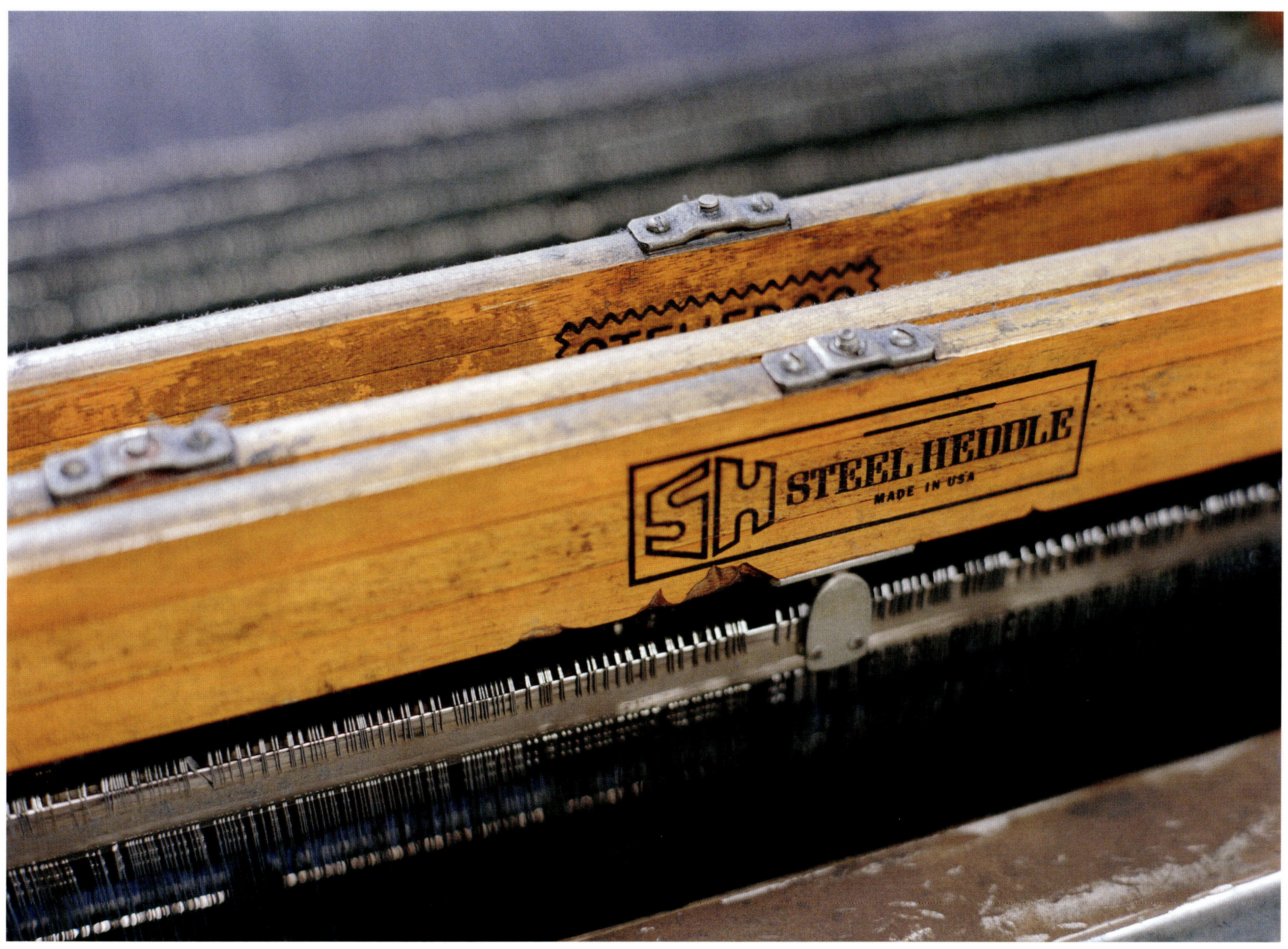

Draper X-3 loom harnesses

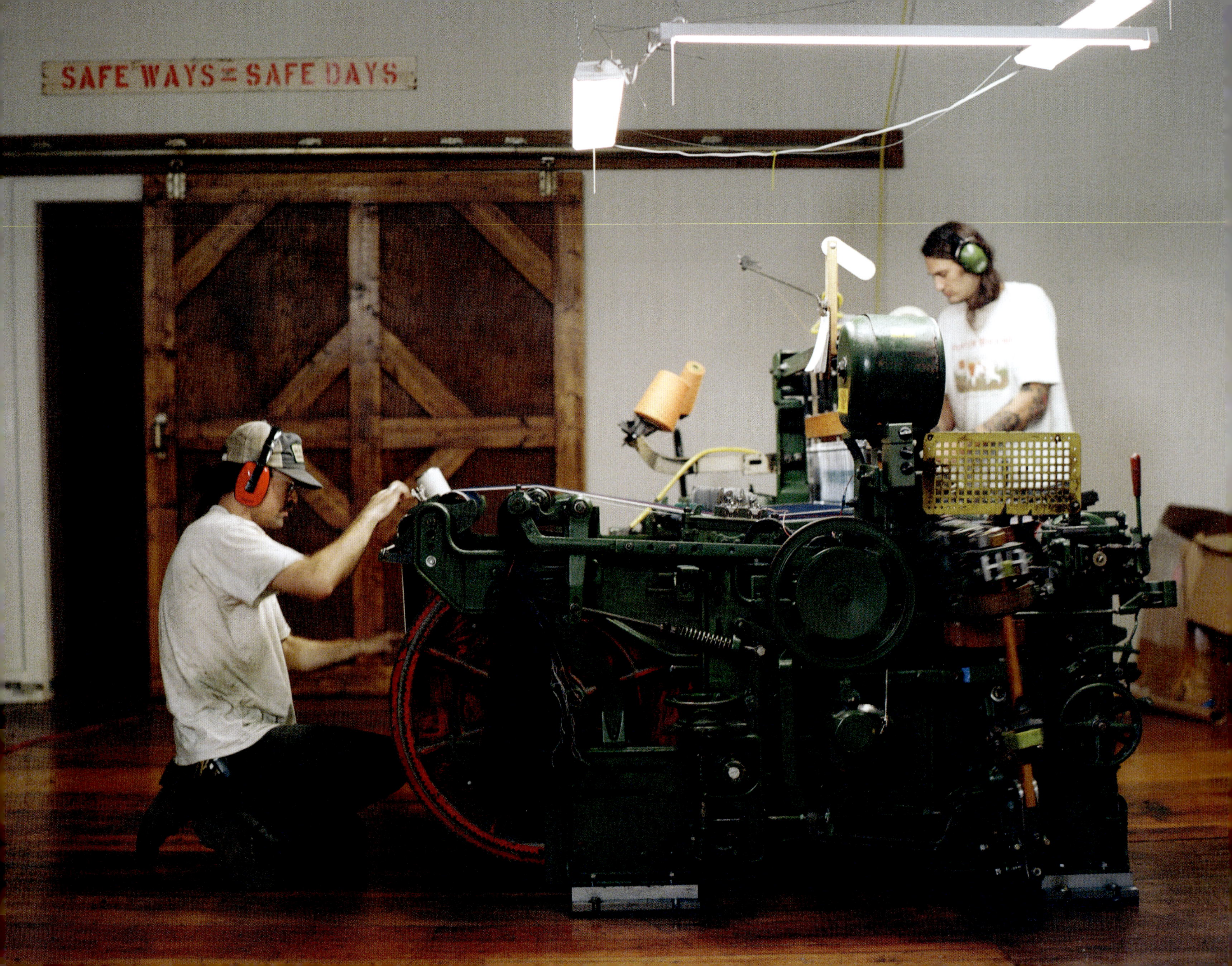
SAFE WAYS = SAFE DAYS

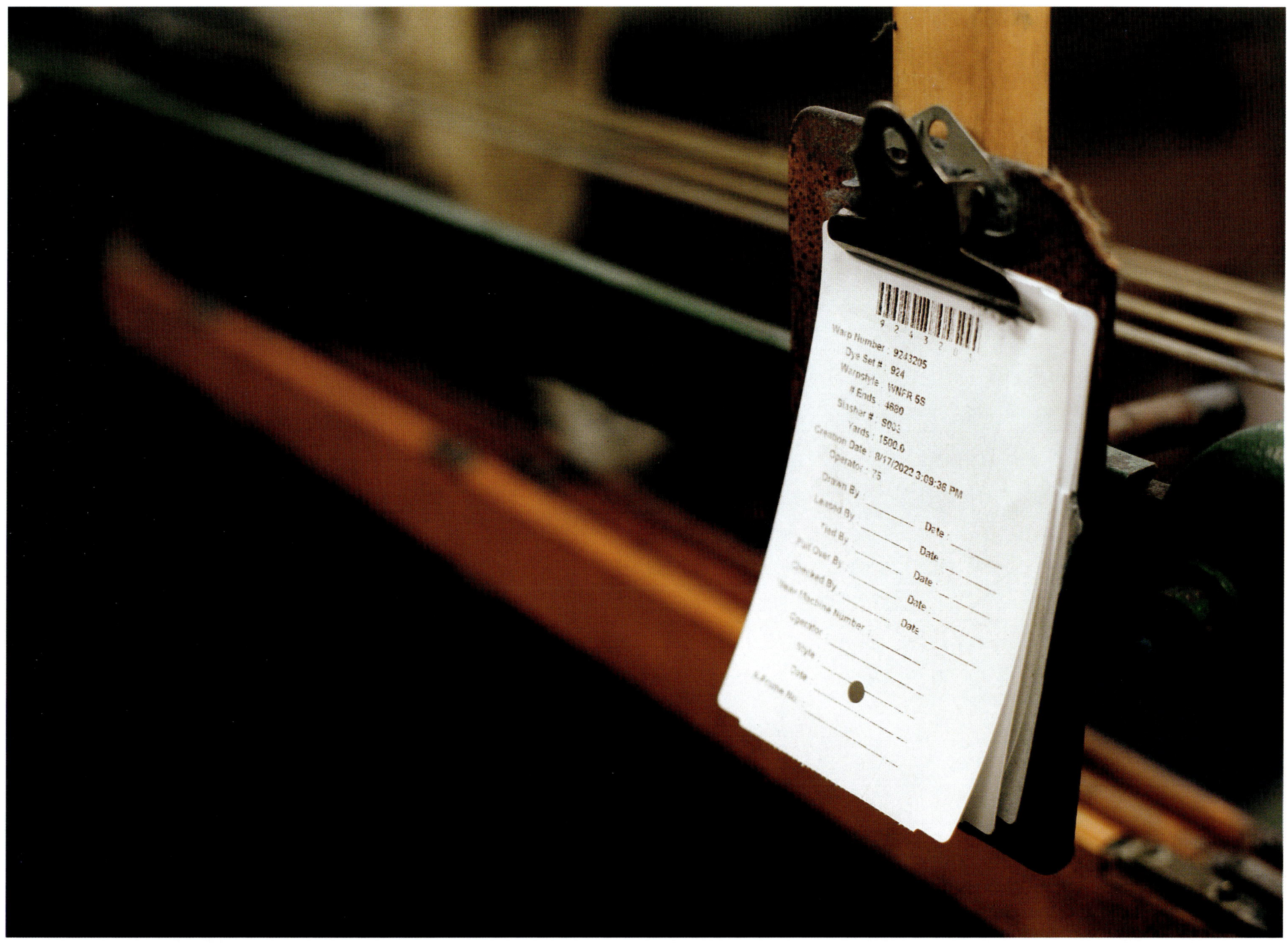

Evan and Greg operating and inspecting Draper X-3 shuttle loom   //   Warp ticket attached to a shuttle loom, denoting current production

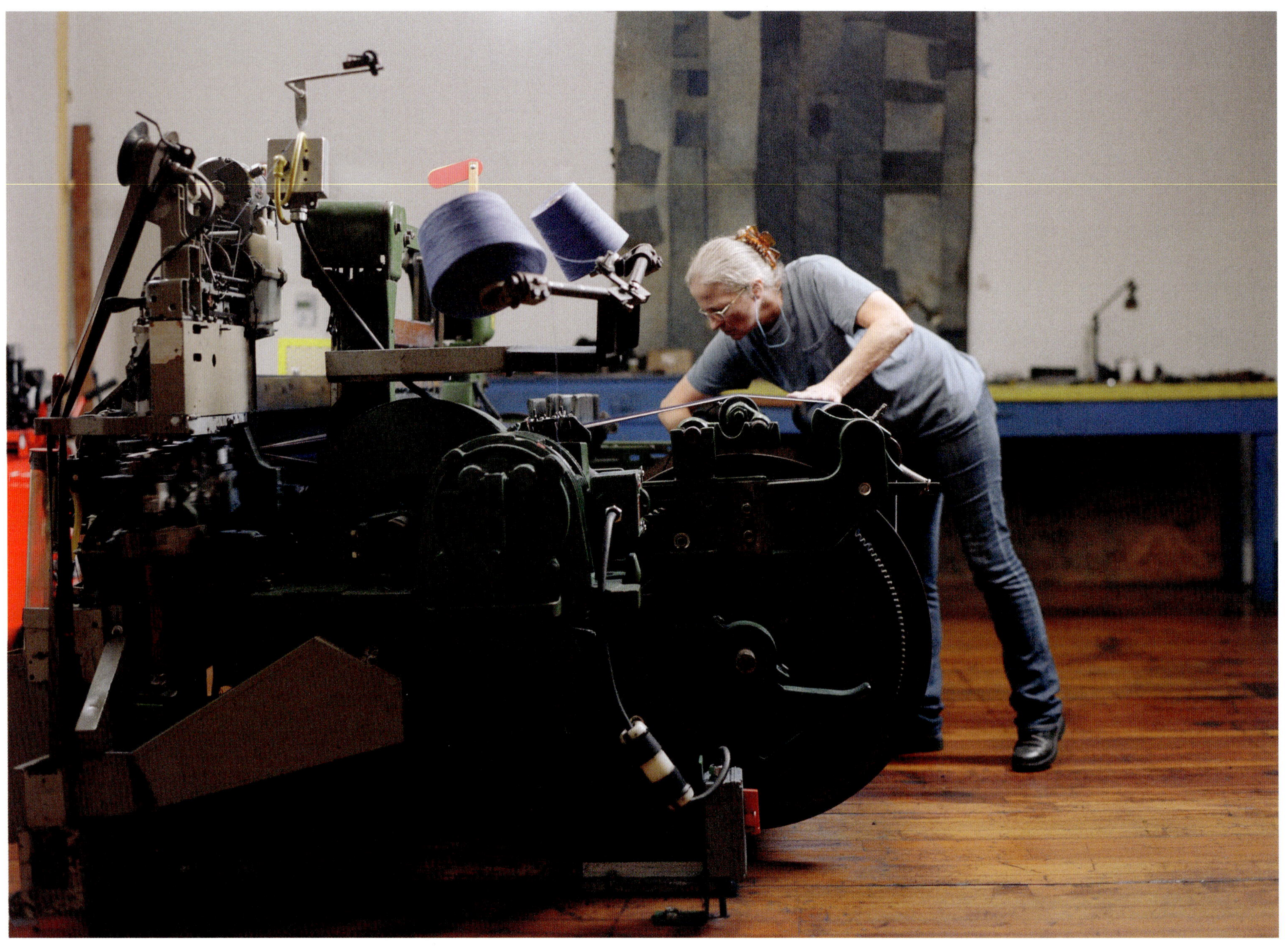

Debbie lapping extra warp yarns on a Draper X-3 shuttle loom // Straightening warp yarns

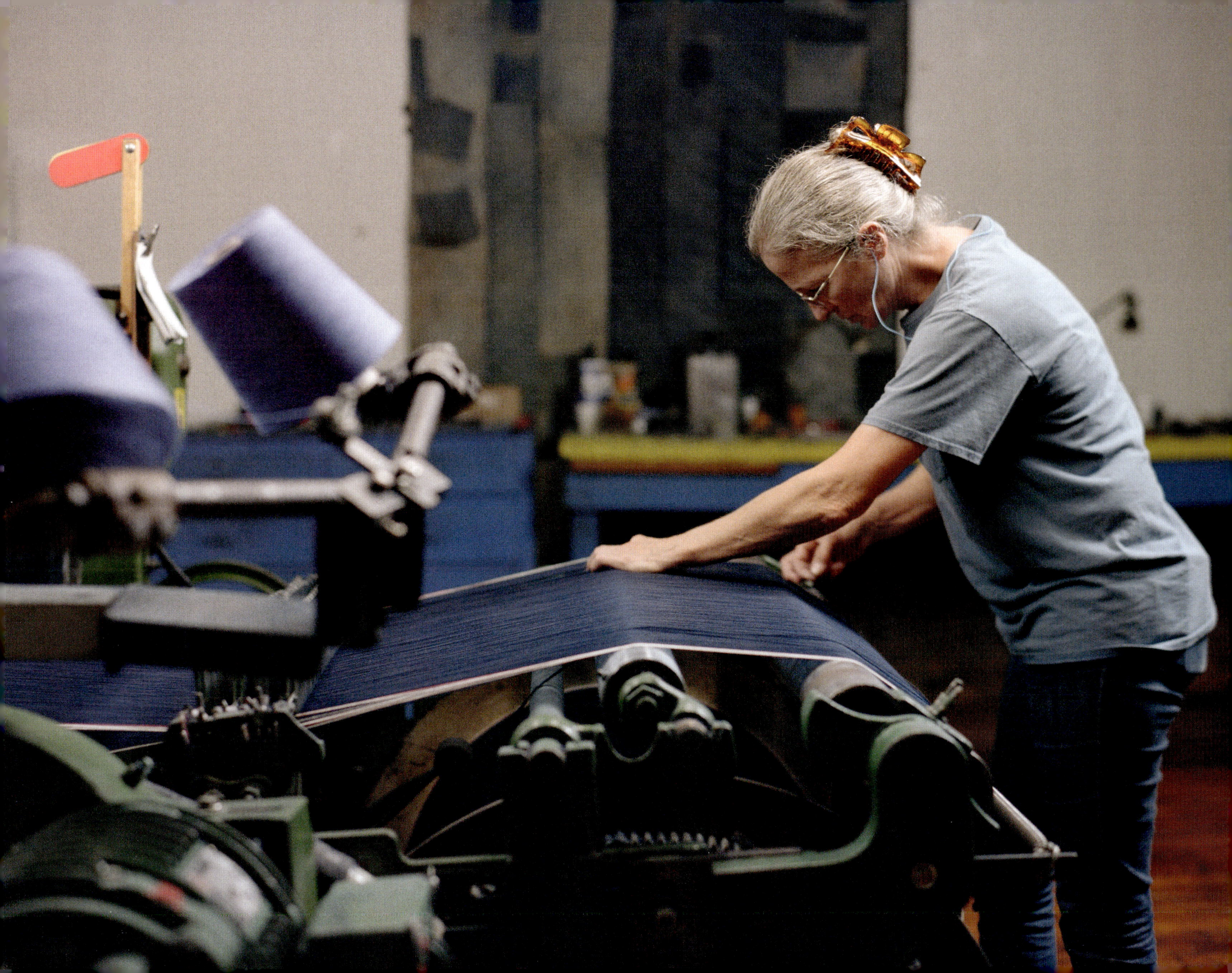

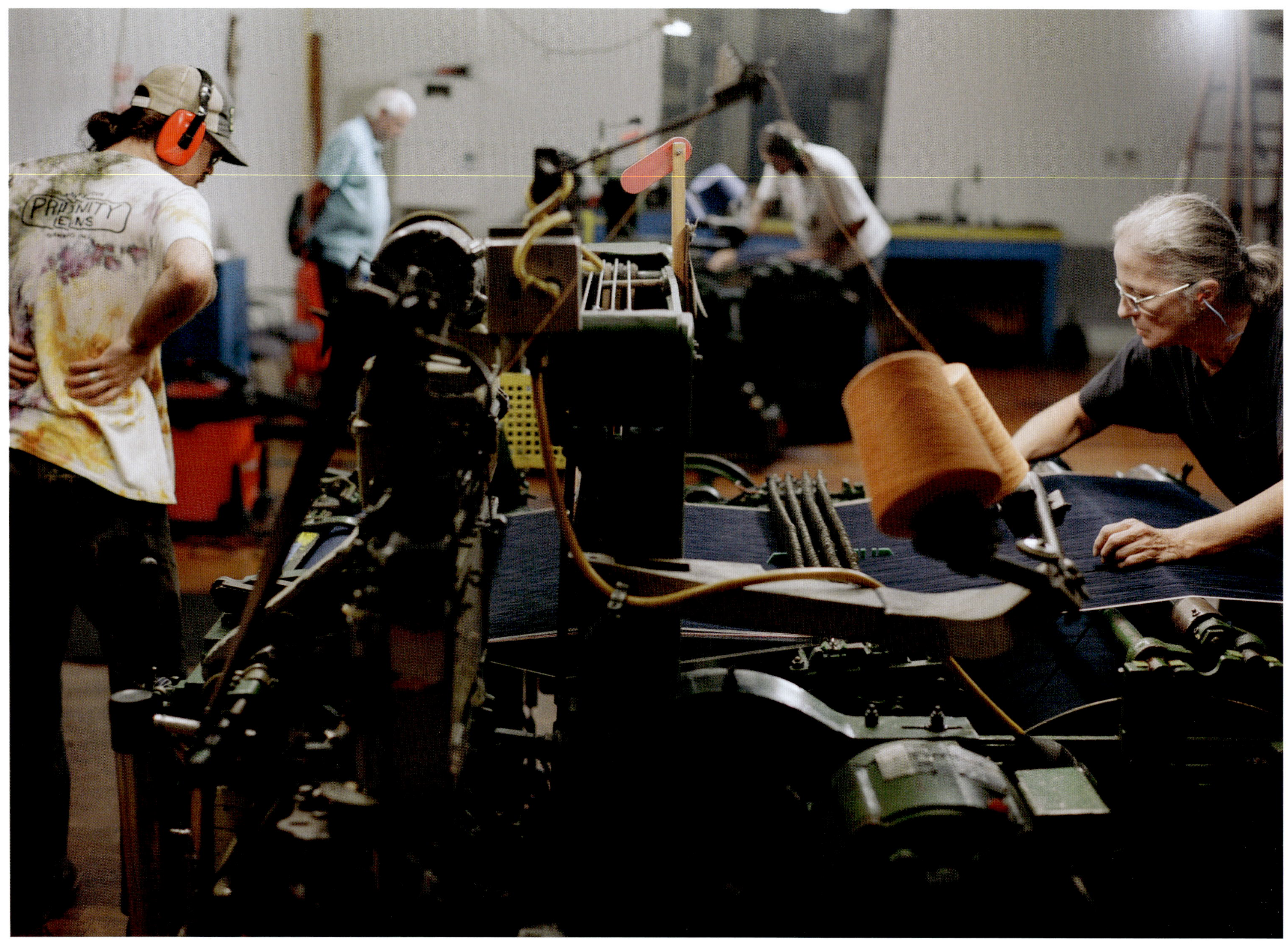

All hands on deck at Proximity Manufacturing. Evan Morrison, Ralph Tharpe, Greg Redelico, and Debbie Lindsey weaving and inspecting production.

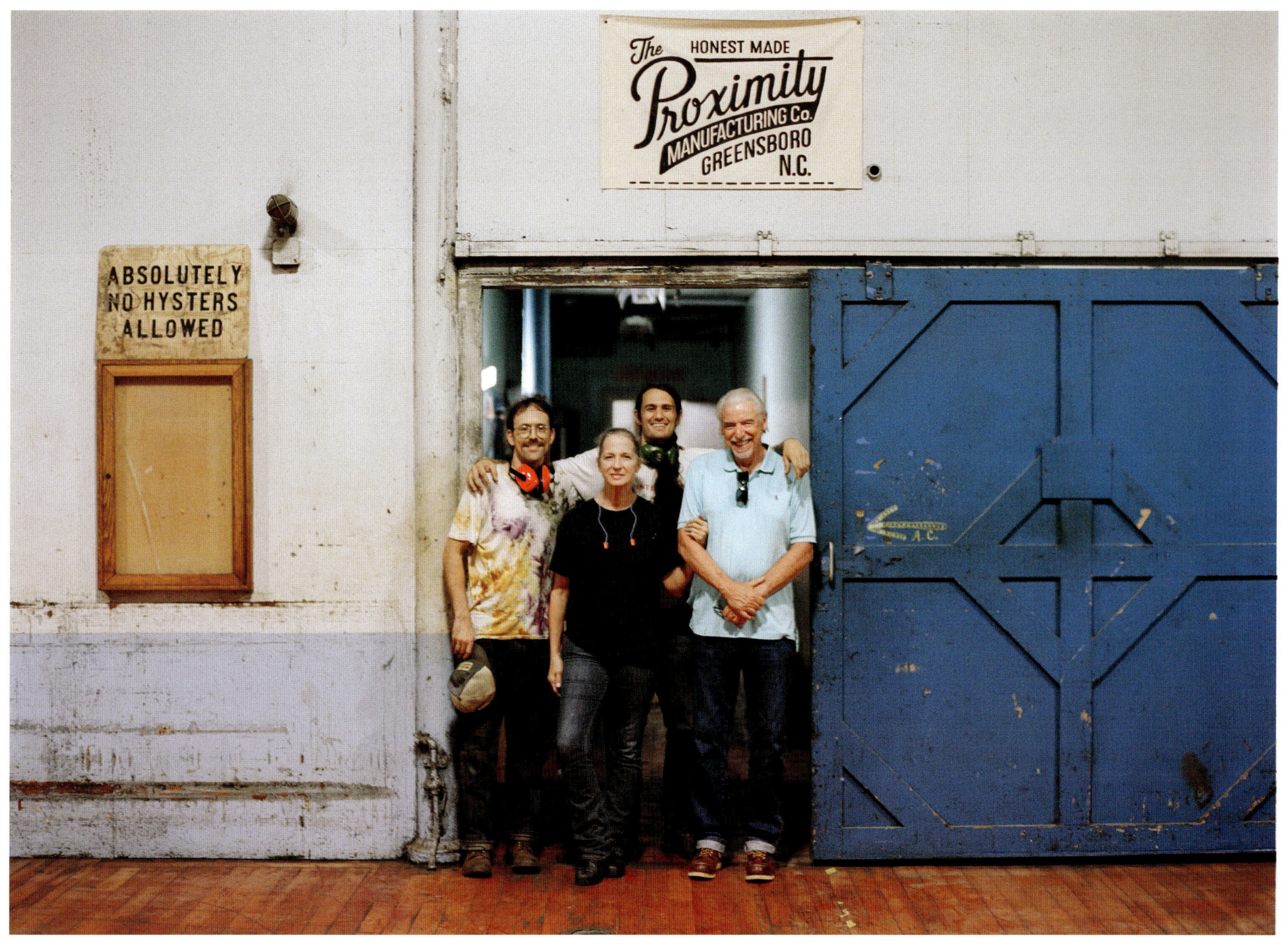

The entire Proximity Manufacturing Co. team as of September 2022 at the entrance to the Weave Room:
Evan Morrison, Debbie Lindsey, Greg Redelico, and Ralph Tharpe (*left to right*)

ACKNOWLEDGMENTS

To my mother and her father for their love of adventure and for putting cameras in my hands at a very young age. To my own father and his appreciation for all things well made. To my brother, for bestowing me the best tools of the craft and inspiring me through his own captures and masterful prints. To my wife, Dalila, for inspiring me through her artistic process and for helping to realize the dream of owning a clothing store that ultimately gifted us our community. To Quinten and Aiden, who went weeks without me in your formative years as I documented the things that I find fascinating, hoping to inspire you to chase your own curiosities. To Tony Patella and Pete Searson of Tellason for first exposing me to the greater history of White Oak. To Delores Sides for inviting me to capture that final week of production in 2017. To Noelle Gaberman for the borrowed camera that allowed me to capture chapters 1 and 2. To the team at Photoworks in San Francisco for being such great lab partners throughout this project. To Evan Morrison for his steadfast friendship, his vision and his unwavering commitment to preserving American textiles. To Will Dellinger for giving me the opportunity to explore the property on my own and for enabling denim production at that site to be born again. To Ralph Tharpe, Debbie Lindsey, and Greg Redelico of Proximity Manufacturing Co., and all at the White Oak Legacy Foundation for preserving the legacy of textile manufacturing in Greensboro. To Pete Schiffer, Jesse Marth, Molly Shields, Elizabeth Peters, and all at Schiffer Publications for believing in the importance of this story being told and for sharing it with the world. And lastly, to the generations of mill workers at White Oak who made the uniquely American fabric that kept the world clothed for over a century. This book would not have been possible nor had a purpose without you. All photographs in this book were captured exclusively with Kodak film.

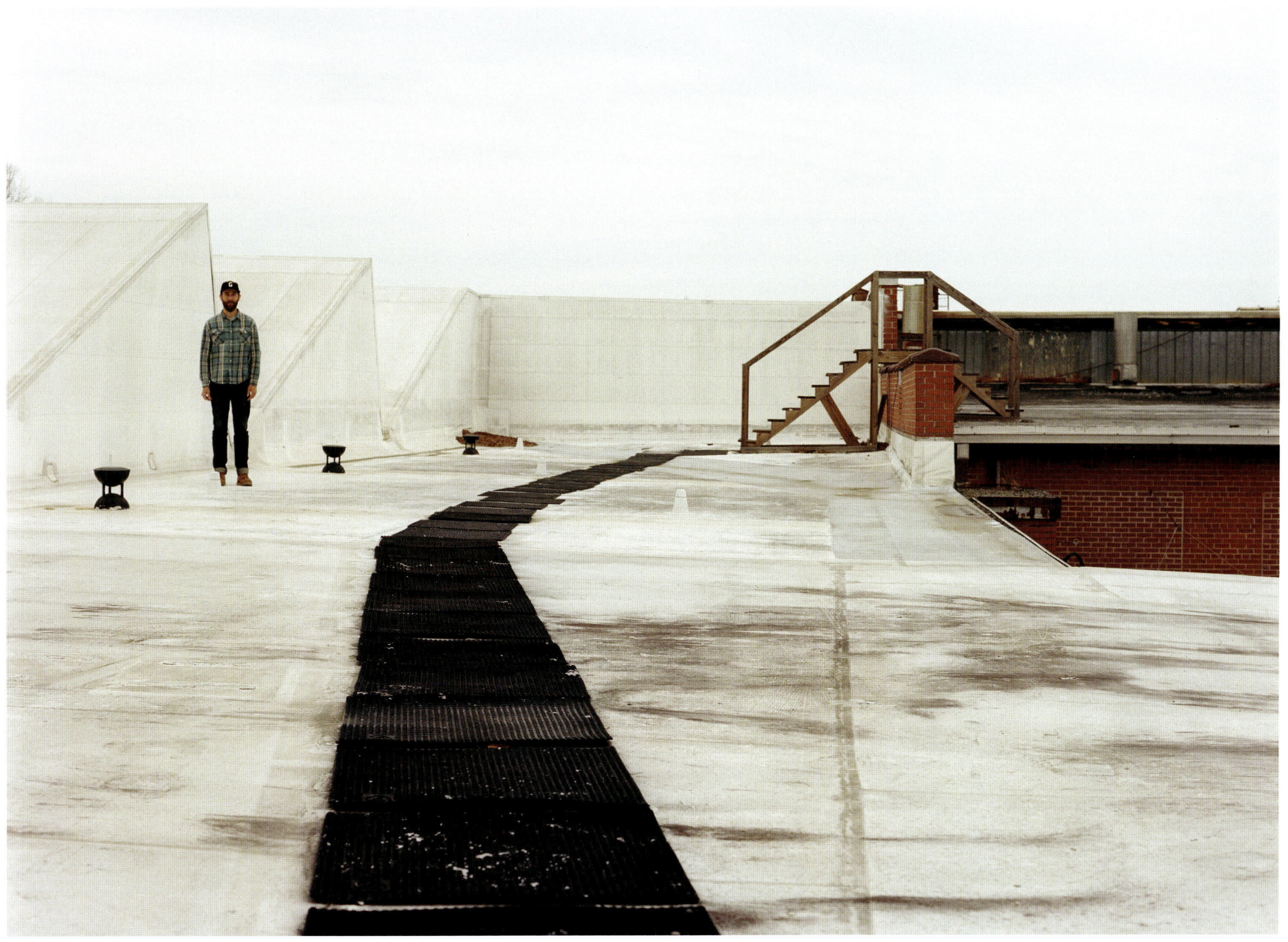

Matt Sharkey photographed on the roof of the Weave Room by Evan Morrison

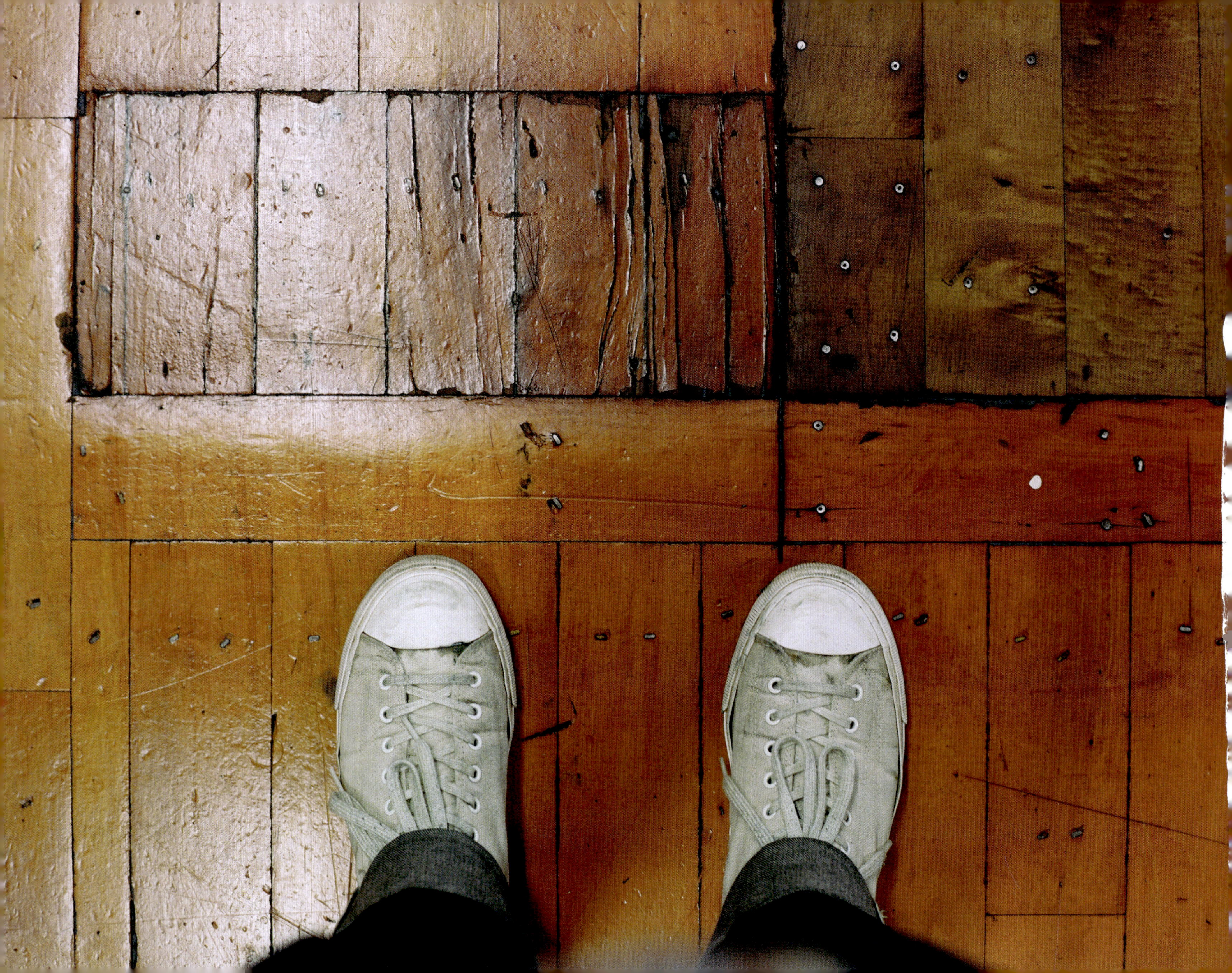

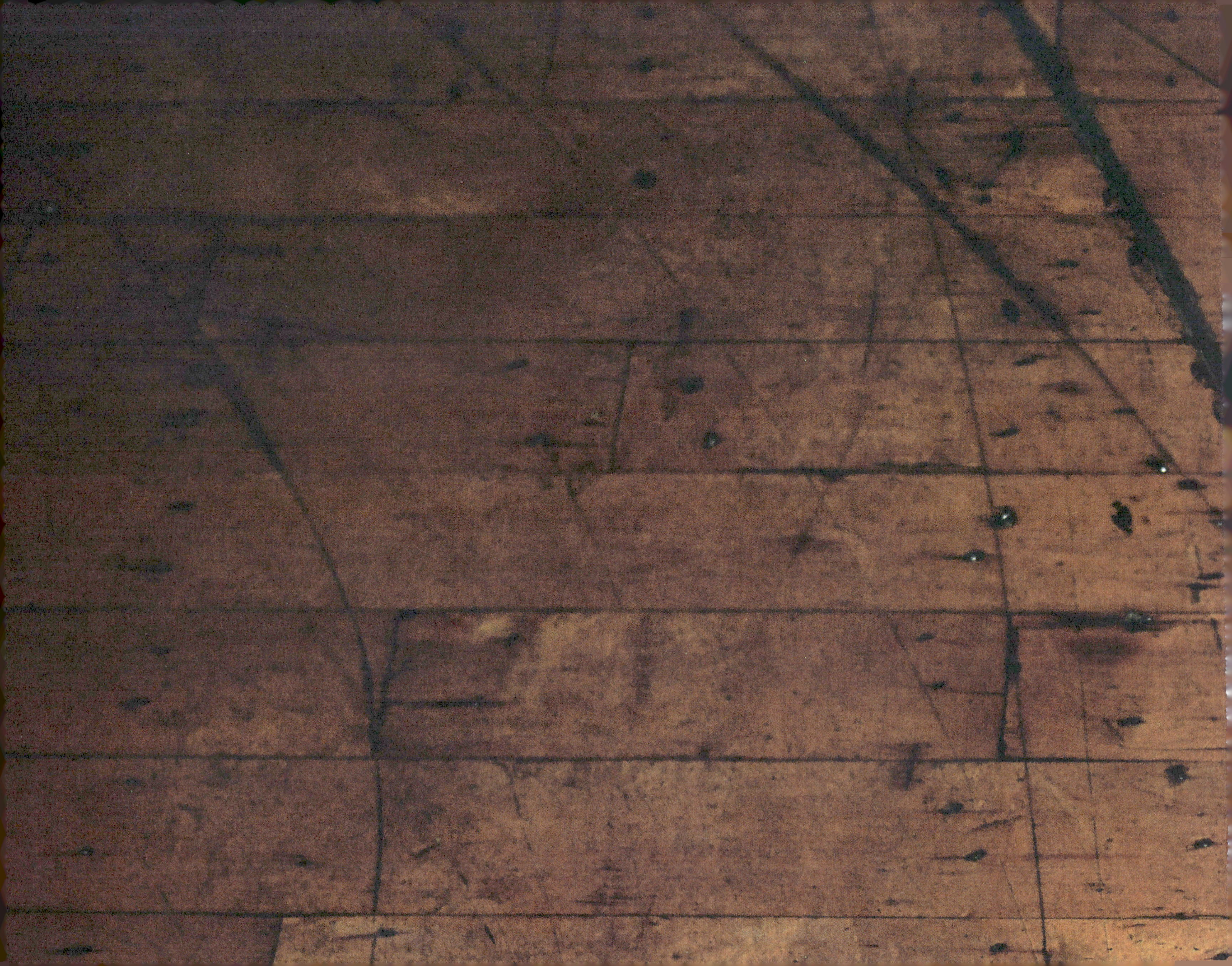